Artificial Intelligence based Solutions for Industrial Applications

Artificial Intelligence based Solutions for Industrial Applications examines the utilization of artificial intelligence (AI) technologies to tackle difficult industrial issues and offers readers a thorough understanding of how these technologies are employed to address intricate industrial challenges and to stimulate innovation. This book explores the fundamental principles of AI and its practical use in industrial environments. This book improves understanding of core concepts, the present state of the art and real-time implementation of AI in many industrial applications.

This book describes the detailed implementation of AI in the industrial sector as well as related case studies for in-depth understanding. Basic concepts, related work reviews, illustrations, empirical results, and tables are integrated within each chapter to give the readers the opportunity to gain maximum knowledge and to easily understand the methodology and results presented.

This book introduces a variety of smart algorithms to filter important information and to solve problems in the application domains. Application of machine learning and deep learning in the industry demonstrates the capabilities by which it may be used to solve practical problems in the 'Fourth Industrial Revolution', and it equips readers with the necessary knowledge and tools to design solutions by themselves with the help of theory and practical examples. The fourth industrial revolution and its consequences on society and organizations are discussed in this book.

Features

- Detailed understanding of the industrial application of AI.
- Discussion of core concepts of different machine learning and deep learning techniques such as artificial neural networks, support vector machines, K-nearest neighbour, decision tree, logistic regression, and many more.
- Detailed study on various industrial applications of machine learning and deep learning in healthcare, education, entertainment, share market, manufacturing, and many more.
- Case studies on industrial application of AI
- Summation of the fourth industrial revolution and its consequences on society and organizations.

This book is primarily written for graduate students, engineers, academic researchers, industrial practitioners, and anyone who wants to optimize production processes, explore AI technology, or stay ahead in the industrial field. It covers the complexities of AI in industrial contexts from core basic understanding to complex implementation.

Computational Methods for Industrial Applications

Series Editor:
Bharat Bhushan

In today's world, IoT platforms and processes in conjunction with disruptive blockchain technology and path-breaking AI algorithms lay out a sparking and stimulating foundation for sustaining smarter systems. Further computational intelligence (CI) has gained enormous interest from various quarters in order to solve numerous real-world problems and enable intelligent behaviour in a changing and complex environment. This book series focuses on varied computational methods incorporated within the system using artificial intelligence, learning methods, analytical reasoning, and sense making in big data. Aimed at graduate students, academic researchers, and professionals, the proposed series will cover the most efficient and innovative technological solutions for industrial applications and sustainable smart societies in order to alter green power management, effect of carbon emissions, air quality metrics, industrial pollution levels, biodiversity, and ecology.

Blockchain for Industry 4.0: Emergence, Challenges, and Opportunities
Anoop V.S, Asharaf S, Justin Goldston and Samson Williams

Intelligent Systems and Machine Learning for Industry: Advancements, Challenges and Practices
P.R Anisha, C. Kishor Kumar Reddy, Nguyen Gia Nhu, Megha Bhushan, Ashok Kumar, Marlia Mohd Hanafiah

Sustainability in Industry 5.0: Theory and Applications
Edited by C Kishor Kumar Reddy, P R Anisha, Samiya Khan, Marlia Mohd Hanafiah, Lavanya Pamulaparty and R Madana Mohana

For more information about this series please visit: **https://www.routledge.com/Computational-Methods-for-Industrial-Applications/book-series/CRCCMIA**

Artificial Intelligence based Solutions for Industrial Applications

Edited by
Pooja Jha, Shalini Mahato, Prasanta K. Jana,
Sudhanshu Maurya, and Inès Chihi

CRC Press
Taylor & Francis Group
Boca Raton London New York

CRC Press is an imprint of the
Taylor & Francis Group, an **informa** business

Front cover image: LeoWolfert/Shutterstock

First edition published 2025
by CRC Press
2385 NW Executive Center Drive, Suite 320, Boca Raton FL 33431

and by CRC Press
4 Park Square, Milton Park, Abingdon, Oxon, OX14 4RN

CRC Press is an imprint of Taylor & Francis Group, LLC

ISBN: 978-1-032-54801-2 (hbk)
ISBN: 978-1-032-87840-9 (pbk)
ISBN: 978-1-003-53476-1 (ebk)

DOI: 10.1201/9781003534761

Typeset in Sabon
by codeMantra

Contents

Preface

The Fourth Industrial Revolution, referred to as Industry 4.0, has currently found a place almost in every industry or organization. It has a primary effect on different industrial sectors, which include automation, healthcare, agriculture, education, and so on. The aim of this book *Artificial Intelligence based Solutions for Industrial Applications* emerges from a profound conviction in the profound potential of AI to revolutionize traditional industrial processes, enhance efficiency, and unlock new opportunities for growth and competitiveness. This book demonstrates how artificial intelligence (AI)-based approaches like machine learning and deep learning methods can be applied to real-world industrial problems allowing Industry 4.0 and providing the essential knowledge and tools to empower readers to develop their own solutions based upon a compact theoretical and practical groundwork. AI comprises several methods enabling renovation in such a way that the resulting software systems can offer actionable perceptions towards best possible decisions. These decisions are evident in different industrial sectors in challenges such as diagnosis, predictive maintenance, conditioning monitoring, assets health management, and many more.

This book is organized into 15 chapters. The aim of this book is to serve as a complete resource for engineers, researchers, industrial practitioners, and students who are attempting to traverse the challenges of deploying AI solutions in industrial contexts.

Organization of this Book:

Chapter 1 discusses about industrial applications of deep learning under Industry 4.0. Deep learning allows machines to learn from data and improve their performance over time. These technologies have the potential to revolutionize industrial processes by providing better insights, optimizing production, and cut costs. Industry 4.0 is the fourth industrial revolution characterized by the integration of advanced technologies such as the Internet of Things (IoT), big data analytics, AI, and cloud computing. The ultimate goal of Industry 4.0 is to create a smart factory, where machines, processes, and people can communicate and work together seamlessly. Deep learning is an essential component of Industry 4.0 as it enables machines to learn from the vast amounts of data generated by the smart factory and make autonomous decisions. The industrial applications of deep learning under Industry 4.0 are vast and varied. From predictive maintenance to quality control, supply chain management and safety, machine learning, and deep learning have the potential to revolutionize industrial processes. As Industry 4.0 continues to evolve, it is likely that the use deep learning will become even more widespread, leading to increased efficiency, productivity, and profitability for manufacturers around the world. This chapter aims to provide a comprehensive overview of the fundamentals of deep learning for industrial applications. It covers the basic principles and concepts of deep learning, their applications in various industries, and the challenges and limitations of using these techniques in industrial settings.

Chapter 2 discusses machine learning and AI that aids to improve the performance of computer systems by learning from data without being explicitly programmed. AI has a lot of applications in healthcare wherein machine learning can be used to analyse huge volumes of complex data, such as electronic health records, medical imaging, and genomics which aids in improving patient outcomes, reducing costs, and enhancing healthcare delivery. In this chapter, some of the most common applications of machine learning and deep learning in healthcare are discussed. Machine learning algorithms are most used nowadays to analyse medical images generated from X-rays, CT scans, and MRI scans. It helps in the early diagnosis of chronic diseases like cancer, Alzheimer's, and heart disease. Additionally, several other applications of machine learning algorithms are discussed including extracting meaningful insights from large EHR datasets and analysing patients for disease prediction and progression. Based on these treatments, personalized treatment plans for diseased patients can be developed. However, some latest applications of machine learning and deep learning in healthcare include the identification of new drugs, development of medical robots, and virtual assistants to handle medical queries. Machine learning and deep learning in healthcare are expected to grow as technology advances and more healthcare providers adopt these tools to improve patient care.

Chapter 3 includes the need for improved efficiency in the agricultural sector in the 21st century is indispensable to address the numerous challenges, which are faced by farmers in the sector ranging from climate change to pest resistance. Shortage of labour and rising demand for food production is adding further strain to conventional agricultural practice. The advent of new technologies like Internet of Things (IoT) and Machine Learning (ML) are providing numerous benefits in agriculture to offset the challenges via advanced crop and soil monitoring, automation in agriculture processes, and smart water management systems. For instance, UAVs equipped with sensors are being considered to collect data in real-time via remote sensing technology to monitor crop health assessment on a large scale. In this chapter, smart data collection systems for parameters such as nitrogen, phosphorus, and potassium contents of the soil; average soil temperature; moisture; soil pH, and rainfall are described. Additionally, existing open-source datasets with the parameters and crop labels are assessed. Furthermore, suitable ML models for the selected datasets would be described, implemented programmatically, trained, and tested. The corresponding results will then be analysed and discussed in further detail in forthcoming sections in this chapter. This study will provide the reader with a global overview of some existing Io-based technologies for data collection in the agricultural context as well as some of the corresponding machine learning techniques that can be used in this case. The reader in this chapter will also be taken through the core aspects of the program implementation of the proposed ML model, as well as through its training and testing phases including the analysis of corresponding results.

Chapter 4 discusses how AI increasingly influences educational settings, it presents substantial opportunities and challenges for educators and learners alike. This study emphasizes the importance of adopting a human-centric approach, which considers human values, emotions, and ethics in the design of AI applications to better serve the learning process. Additionally, it delves into key aspects of deep learning techniques, such as neural networks, attention mechanisms, and transformer models, elucidating their roles in enhancing natural language processing, personalized learning, and collaborative learning environments. To ensure that AI applications align with the needs and values of learners and educators, the integration of human-centric frameworks—including user-centred design, participatory design, and value-sensitive design—is advocated.

Chapter 5 shows AI applications used to enable machines to think autonomously, without the need for human intervention. To accomplish this, machine learning, a subset of AI, offers statistical tools for data exploration. Deep learning, another subfield of machine learning, utilizes multi-neural architecture to achieve this goal. In recent times, machine learning has found extensive applications across various fields, including education, where it has played a crucial role. It analyses the learning capabilities of numerous students, aiding in the enhancement of their learning efficiency. A hybrid model is needed that combines the strengths of artificially enabled systems and teachers. When students interact with teachers, the focus should be on fostering creativity, exploration, and problem-solving. Simultaneously, the artificially enabled system can provide an individualized learning experience. This chapter explores several use cases of machine learning and deep learning in the education sector.

Chapter 6 discusses how social media reigns supreme in our daily lives, the social networking sphere, must confront a growing insidious problem that threatens the vitality of digital communities: trolling. Several attempts have been made to define what constitutes trolling but there remains an element of subjectivity in our best definitions. Trolls are social media users who engage in deceptive, disruptive and/or deviant behaviour, often linked to antisocial and sadistic tendencies. Troll interactions can lead to serious harm at various levels. It is no surprise that apart from users who troll for self-appeasement, there exist sponsored trolls who are backed by institutions. Given how far reaching this issue is, it is the need of the hour to develop intelligent troll detection systems using AI. Features quantifying insights about the characteristics of trolls combined with the predictive ability of approaches based on Machine Learning, NLP, or Deep Learning have proved to be immensely useful. With the aim of providing a thorough overview, this chapter lays out the definition of a troll, the psychology of a troll, characteristics of trolls, some approaches that have been developed for troll identification, and lastly, some challenges and recommendations.

Chapter 7 discusses AI for Smart Manufacturing Using Digital Twin Technology. With the rise of digitalization and the increasing utilization of AI, the theory of digital twin has immense potential to completely transform and reshape various industries. By leveraging AI technology, digital twin (DT) allows for real-time monitoring and prediction of an object's current status and future behaviour. This concept extends beyond physical entities to include processes, enabling optimization of entire procedures. DTs are integrated with AI, IoT, Extended Reality, and cloud computing. The integration of digital twin and AI has significant applications in smart manufacturing, unmanned vehicles, and smart city transportation. By replicating processes using DT technology, data can be gathered and future functionalities can be forecasted. By digitally twinning manufacturing procedures, malfunctions can be prevented, minimizing financial losses. DT technology enables simulation of manufacturing processes, reducing the need for costly prototyping. AI-driven DTs optimize resource allocation, requiring fewer manpower and financial resources. DTs also have challenges like other complex technologies. Scaling beyond proof of concept, talent acquisition, granularity of coverage, infrastructure requirements, implementation costs, and cybersecurity are to name a few. In conclusion, establishing a virtual domain for manufacturing and process management alongside the real world offers timely fault detection and extends equipment lifetimes. DTs identify resource bottlenecks, optimize processes, and provide efficiency gains and cost savings. The deployment of DTs and AI in monitoring manufacturing operations ensures comprehensive insights, leading to improved efficiency and sustained efforts.

Chapter 8 includes Machine learning and deep learning in cybersecurity challenges in recent technology by understanding the different thrust areas for evaluating the cybercrime activities and detection in an accurate manner. In recent days, humans are replaced by all the areas for automating and transmitting the applications such as AI, Robotics, image processing, and machine vision. During the time of data transmission, there may be a high chance for security attacks, Phishing, Network intrusion, spams, and malware functioning. To overcome the limitation, some new highlighted advancements are being used in cybersecurity along with machine learning and deep learning applications. These algorithms are widely preventing security attacks and minimize the risks of malware detection for solving and analysing complex data. The data in cybersecurity systems analyse the data from security incidents for learning and applying for preventing the new attack in it. Promoting flexibility in analysing data breaches related to malicious network activities, the system helps to proactively prevent intruders from accessing the systems and promptly notifies the user of any potential threats. This chapter mainly focuses on the application aspects using ML/DL which is more appropriate to people's daily lives and illustrates a real-time system for giving a solution for threats in public security. The insightful discussions on the state-of-the-art are expected to show that the development of machine learning and deep learning influence the field of cybersecurity and it is going to highlight the strengths in upcoming concepts.

Chapter 9 discusses AI and media as one of the most sought-after fields in the technology industry, as they have the potential to revolutionize the way we consume and interact with media content. From personalized recommendations to automated content creation, AI is changing the game for media companies and consumers alike. AI is revolutionizing the media industry by enabling personalized content recommendations, improving search results, automating content creation, improving the accuracy of data analysis, automating, analysing audience behaviour, and even creating original content. As AI technology continues to advance, it is expected to play an even bigger role in shaping the future of media. With AI, media companies can deliver more engaging and relevant content to their audiences while also streamlining their operations. As AI technology continues to advance, it is likely that we will see even more innovative applications of AI in the media industry. AI can also help media companies personalize their content based on the preferences and behaviours of individual users, leading to higher engagement and retention rates. Additionally, AI-powered tools can assist in identifying and combating fake news, ensuring that audiences receive accurate and trustworthy information.

Chapter 10 examines various machine learning feature selection methods for anatomic brain MRI in the context of dementia, encompassing CNN, Random Forest, and SVM. The study aimed to evaluate data-driven feature selection and classification strategies for whole-brain voxel-based classification of MRI scans.

Chapter 11 of this book provides the enhancement of heat transmission by including gyrotactic microorganisms in nanofluids using ANN is a popular issue in the twenty-first century due to industrial and biomedical applications. This chapter provides a computational analysis of the bio-nanofluid flow over a vertically inclined surface using a CI heuristic based on backpropagation Levenberg-Marquardt neural networks (LMNN).

Chapter 12 of this book addresses a novel intelligent fault diagnosis method that can extract useful information from the noisy vibration signals and help to detect the incipient bearing fault at an early stage. This chapter discussed a machine learning-based fault diagnosis method that will detect and diagnose bearing faults using the time domain statistical features with classification models.

Chapter 13 covers stock price prediction using various techniques, including technical analysis, fundamental analysis, time series analysis, statistical analysis, etc., but none of these techniques has been shown to be a reliable predictor of price over time. A popular method for finding complex and hidden patterns in data that are appropriate for share market prediction is the artificial neural network (ANN). Therefore, ANN is the best method for predicting stock market prices.

Chapter 14 discusses the emergence of machine learning and deep learning has revolutionized the industry by providing powerful tools for content creation, marketing, and audience engagement. This chapter provides an in-depth analysis of how machine learning and deep learning are used in the entertainment and media industry to improve various processes such as content creation, recommendation systems, marketing strategies, and audience engagement. It discusses some of the key applications such as the usage of software in various video-audio presentations of web series, anime, and movies. Software usage for such purposes is part of machine learning and deep learning in the industry.

Chapter 15 discusses the application of machine learning in the prediction of different heart diseases. Heart is one of the most vital organs in our body and crucial for proper bodily functioning. An unfit heart can seriously affect fitness and lifestyle and severely decrease the expected lifetime of an individual making a healthy heart necessary for survival. Recent trends show that heart diseases are on a steady incline which is a major health concern. Heart disease diagnosis is a highly complex and expensive process necessitating greater precision and accuracy requiring tools like predictive algorithms. Machine learning (ML) is an AI technique that offers assistance in predicting any type of heart disease using day-to-day data as training data. In this study, we investigate various types of heart diseases as well as various classification techniques used for diagnosis of cardiovascular diseases (CVD). Study includes state-of-the-art literature review used for the prediction of heart diseases using ML techniques.

Through this book, the reader would go through a journey from the basics of AI to the implementation of AI in industry. The topics covered would help the reader get an insight into how AI techniques have reshaped the traditional industry paradigms into modern industry.

Dr. Pooja Jha

Dr. Shalini Mahato

Dr. Sudhanshu Maurya

Prof. Prasanta K. Jana

Prof. Ines CHIHI

Scope of this book

AI has an immense impact in almost every industry. This book covers all the necessary content for the reader for understanding and applying AI to solve problems This book covers the basics of AI including basics of machine learning (ML) and deep learning (DL) to the industrial application of AI including real life application for problem solving and case studies. This book discusses the impact of the fourth industrial revolution and its consequences on industries which include healthcare, education, media and entertainment, agriculture, and many more. Application of AI in industry demonstrates the capabilities by which ML/DL may be used to solve practical problems in the fourth industrial revolution, and it equips readers with the necessary knowledge and tools to design solutions by themselves with the help of theory and practical examples discussed in this book.

For better insight for the readers, basic concepts, illustrations, real life application, case study, simulation, empirical results, and tables are integrated within each chapter to give the readers the opportunity to gain maximum knowledge. This chapter describes solving challenging issues using AI in different domains.

This book introduces a variety of smart algorithms to help in filtering important information and solve problems in the application domains discussed. The motivation of this book is to enhance the understanding of students, researchers, and industrial practitioners regarding basic concepts and theories of AI and apply them to real life problems to come up with solutions.

Introduction

The integration of artificial intelligence (AI) technology has emerged as a crucial element in fostering innovation, enhancing efficiency, and driving competitiveness in the rapidly evolving industrial environment of the present era. AI is transforming conventional industrial processes, opening up new opportunities, and changing how we address industrial challenges in a variety of industries, including manufacturing, healthcare, education, agriculture and many more. The advent of the Fourth Industrial Revolution, also known as Industry 4.0, has impacted almost each and every sector immensely.

The purpose of this book *Artificial Intelligence based Solutions for Industrial Applications* is to explore how AI technologies are being used to address challenging industrial problems and provide readers with a comprehensive understanding of how AI technologies are being leveraged to address complex industrial challenges and drive innovation. This book covers the core concept of AI and its applications in industrial settings. This book contains the fundamental principles of machine learning, deep learning, and examines how these techniques are being applied to solve real-world industrial problems.

AI has become increasingly important in almost all the industries. It has grabbed the attention of almost all the industries, and everyone has a desire to make best use of it. This book would help in better understanding of core concepts, present state-of-the-art and real-time implementation of AI in numerous Industrial Applications.

The chapters explore the various areas in which AI has been significantly contributing and bringing drastic change in the industry. This book mainly focusses on the influence of AI in the areas of healthcare, education, agriculture, share market, cyber security, entertainment, smart manufacturing, and many more. This book is planned to provide the detailed implementation of AI in the industrial sector as well as related case study for in-depth understanding.

This book aims to provide a comprehensive guide for engineers, researchers, and industrial practitioners who want to optimize production processes, explore AI technology, or stay ahead in the industrial field. It covers the complexities of AI in industrial contexts.

The goal is to provide readers a greater understanding of how AI may revolutionize traditional industries and lead us towards a more intelligent, efficient, and sustainable future.

About the editors

Pooja Jha received her Ph.D. in Information Technology from BIT Mesra, Ranchi, India in 2019 and has completed her Master's degree from BIT Mesra, Ranchi, India. She is currently working as Assistant Professor in the Department of CS-IT, Amity University Jharkhand, India since 2017. She has more than 17 years of teaching experience. She has contributed more than 11 research papers. She has published papers in many reputed journals such as *International Journal of System Assurance Engineering and Management*, *International Journal of Computer Applications in Technology*, *Journal of Theoretical and Applied Information Technology*, etc. She has been a reviewer of *International Journal of Software Innovation* (*IJSI*) and an active member of International Association of Engineers (IAENG). She has handled many university students with their scholarly research papers. Recently, she has been assigned as a reviewer of IEEE Madras Section International Conference, MASCON 2021. Her research interests mainly focus on Software metrics, Artificial Intelligence, Machine Learning and Soft Computing. She has been the Session Chairperson of various IEEE conferences. She is currently the Head of Department, IT at Amity University Jharkhand.

Shalini Mahato is currently working as Assistant Professor in Department of Electronics and Computer Engineering, National Institute of Advanced Manufacturing Technology, Ranchi (NIAMT), (formerly NIFFT) Ranchi, Jharkhand, India. She has done her Ph.D. in Computer Science and Engineering from Birla Institute of Technology, Mesra, Jharkhand, India in 2021. She has completed her M. Tech. (Computer Systems and Technology) degree in 2013 and B. Tech (Computer Science and Engineering) degree in 2008 from Maulana Abul Kalam Azad University (erstwhile WBUT), West Bengal, India. She has worked as a developer in Tata Consultancy Services (TCS). She has contributed to more than 20 research papers out of which 7 are SCI indexed and 6 are Scopus indexed. She has a project selected in MSME Idea Hackathon 2022 sponsored by Ministry of Micro, Small & Medium Enterprises (MSME) Innovative Scheme, Government of India.

She is the winner MSME Idea Hackathon 2022. She has three Indian patents filed under her name. She organized an international conference as a Co-convenor, International Conference on "Global Mental Health and Public Health Challenges and Innovation (GMHPHCI'22)" organized by the Indian Institute of Information Technology (IIIT), Ranchi, Jharkhand, India. She has been a resource person in many FDP, Seminar and Conferences. She has been the reviewer of *IEEE Access*, *Journal of Medical Systems*, *Journal of Intelligent & Fuzzy Systems*, and many others. She has been the organizer of many conferences and FDP. Her area of interest includes machine learning, soft computing, and biomedical signal processing.

Prasanta K. Jana received M. Tech. in Computer Science from the University of Calcutta in 1988 and Ph.D. in 2000 under the joint supervision of Indian Statistical Institute and Jadavpur University, Kolkata. Currently, he is a Professor (HAG) in the Department of Computer Science and Engineering at Indian Institute of Technology (Indian School of Mines) in Dhanbad, India. He has contributed 216 research publications in his credit and co-authored five books, four book chapters, and three edited volumes. He has also produced 20 Ph. Ds. As a recognition of his outstanding research contributions, he has been awarded *Senior Member of IEEE* in 2009. He is also the recipient of *Canara Bank Research Publication Award* in 2015 and 2017. He is also among the World ranking of top 2% (all fields) Indian scientists in Artificial Intelligence surveyed by Stanford University, USA in four consecutive years, i.e., 2020–2024. His current research interest includes Cloud Computing, Fog/edge Computing, and Machine Learning with a special interest on Federated Learning. He is on the editorial board of two international journals. He has served as a referee in many reputed international journals. Prof. Jana has acted as the General Chair of the International Conferences ICADCML-2022, RAIT-2018, and RAIT-2012, Co-chair of the national conference RAIT-2009 and Convener of the workshop WPDC-2008. He has also served as a member of Advisory Committee and Programme Committee of several international conferences. He has made academic visit to several foreign universities including the University of Aizu, Japan, Imperial College of London, UK, University of Macau, Macau and Curtin University, Malaysia.

Sudhanshu Maurya is a visiting post-doctoral researcher at the University of Luxembourg. He has completed his post-doctoral research at the School of Computer & Communication Engineering, Universiti Malaysia Perlis (UniMAP), Kangar, Perlis, Ministry of Education, Malaysia. He is currently working as Associate Professor CSE & Research Head at Symbiosis Institute of Technology, Nagpur Campus, Symbiosis International (Deemed University), India. He is also associated with the Center of Artificial Intelligence & Robotics, Indian Institute of Technology (IIT) Mandi, for AI-based Quantum

Cryptographic Techniques research. Dr. Maurya has completed his Ph.D. in Computer Science, with a particular focus on the "Security and Privacy aspect of Mobile Cloud Computing." He acquired M. Tech CSE (Gold Medalist) from Hemvati Nandan Bahuguna Garhwal Central University, Uttarakhand. His area of research is dedicated to Artificial Intelligence, the Internet of Everything, and Cloud Computing. He has two international Patents (Granted), 13 Indian patents, two edited books, and five-course books, and authored/co-authored more than a hundred research papers indexed in Web of Science, SCOPUS, or UGC-approved journals. He has attended several FDPs, short-term courses, conferences, seminars, and workshops in various renowned Institutions like IIT Kharagpur, IIT Guwahati, IIT Patna, NIT Surat, JNU New Delhi, Central University of Rajasthan, and many others. As far as his Professional Affiliations are concerned, he is a senior member of IEEE, IEEE Branch Counsellor at SIT Nagpur under IEEE Bombay Section, IEEE Young Professionals, a Professional Member of ACM, a Member of CSI, Computer Society Technical Committee on Cloud Computing, an Associate Member Technologist of the Institute of Engineers (India), and Computer Science Teachers Association (csta.acm.org).

Ines Chihi was born in Tunisia, Inès Chihi obtained her PhD degree in electrical engineering in 2013 and her "Habilitation Universitaire" in 2019 from the National Engineering School of Tunis. Until the end of 2021, she was a professor in electrical engineering in Tunisia. She is the founder and president of the Association of Energy Efficiency and the Environment (AEEE) and a member of the first Tunisian Network for Energy Transition. Since January 2022, Inès has joined the Department of Engineering at the University of Luxembourg. Her research focuses on electrical measurement and sensor technologies, with a particular emphasis on smart sensors and the estimation and identification of complex systems with unpredictable behaviours. Her innovative approaches find application in a wide array of fields, including bioengineering, energy, and Industry 4.0. Her impactful contributions shape academia and foster a sustainable future.

Contributors

R. Badlishah Ahmad
Electronic Engineering and Technology
Centre of Excellence Advanced Computing
Universiti Malaysia Perlis
Perlis, Malaysia

R. S. Anand
Department of Electrical Engineering
Indian Institute of Technology
Roorkee, India

Rabab Anjum
Department of Biotechnology
IMS Engineering College
Ghaziabad, India

Arti Arya
Department of CSE
PES University
Bengaluru, India

Anupam Kumar Bairagi
Department of Computer Science and Engineering
Khulna University
Khulna, Bangladesh

Yogesh Beeharry
Department of Electrical and Electronic Engineering
University of Mauritius
R´eduit, Mauritius

Priyanka Chandra
Department of Mathematics
Vellore Institute of Technology (VIT)
Vellore, India

Sumedha Chaudhury
Amity School of Communication
Amity University Jharkhand
Ranchi, India

Ayontika Das
Department of Computer Science and Engineering
Adamas University
Kolkata, India

Biva Das
Department of Pharmacy
Khulna University
Khulna, Bangladesh

Moupriya Das
Department of Computer Science and Engineering
Adamas University
Kolkata, India

Raja Das
Department of Mathematics
Vellore Institute of Technology (VIT)
Vellore, India

S. Divya
Department of Biomedical Engineering
Dr. N.G.P Institute of Technology
Coimbatore, India

Ahmed A. Elngar
Department of Computers and Artificial Intelligence
Beni-Suef University
Beni-Suef City, Egypt

Sapan Kumar Gupta
Amity School of Communication
Amity University Jharkhand
Ranchi, India

Shahid Shadab Hassan
Department of Humanities, Social Sciences and Management
Indian Institute of Information Technology Ranchi
Ranchi, India

Jayashree J
Department of Database Systems
Vellore Institute of Technology (VIT)
Vellore, India

Vijayashree J
Department of Databases Systems
Vellore Institute of Technology (VIT)
Vellore, India

Nipun Jain
Department of Electrical Engineering
Indian Institute of Technology
Roorkee, India

M. R. Mano Jemila
Department of Biomedical Engineering
Dr. N.G.P Institute of Technology
Coimbatore, India

Susan Mathew K
Department of CSE
PES University
Bengaluru, India

Dr. Ravneet Kaur
Department of Commerce
Pt. Mohan Lal SD college for Women
Gurdaspur, India

Mohit Kumar
Department of Life Sciences
Sharda School of Basic Sciences and Research
Sharda University
Greater Noida, India

Rajeev Kumar
Department of Electrical Engineering
Indian Institute of Technology
Roorkee, India

Sudhir Kumar
Journalism and Mass Communication
Sarala Birla University
Ranchi, India

Jayanthi M
School of Science and Computer Studies
CMR University
Bengaluru, India

Lakshmi M
Department of Mathematics
SRM Institute of Science and Technology-Ramapuram Campus
Chennai, India

Shalini Mahato
Department of Electronics & Computer Engineering
National Institute of Advanced Manufacturing Technology
Ranchi, India

Balakrishnama Manohar
Department of Mathematics and Statistics
School of Applied Sciences and Humanities
Vignan's Foundation for Science, Technology and Research
Vadlamudi, Guntur, India

Anindya Nag
Department of Computer Science and Engineering
Khulna University
Khulna, Bangladesh

Kunal Nagarkoti
Department of Biotechnology
IMS Engineering College
Ghaziabad, India

Dev Narula
Department of Computer Science & Information Technology
Amity University Jharkhand
Ranchi, India

Jaykumar Ishvarbhai Prajapati
Department of Mathematics
Vellore Institute of Technology (VIT)
Vellore, India

Zeeshan Rafi
Integral University
Lucknow, India

B. Ragavi
Department of Electronics and Communication Engineering
Dr. N.G.P Institute of Technology
Coimbatore, India

Rishi Raghu
College of Engineering
Northeastern University
Boston, Massachusetts

Raviduth Ramful
Mechanical & Production Engineering Department
University of Mauritius
R′eduit, Mauritius

K. Madhusudhan Reddy
Math and Computer Science Unit
University of Technology and Applied Sciences
Shinas Sultanate of Oman

Sumit Kumar Rohit
Amity School of Communication
Amity University Jharkhand
Ranchi, India

Manju Sadasivan
School of Science and Computer Studies
CMR University
Bengaluru, India

Vinithra Sankaran
Department of Technology
City University of Seattle
Seattle, Washington

Richa Sharma
Department of CSE
PES University
Bengaluru, India

Surkhab Shelly
Department of Computer Science
Pt. Mohan Lal SD College for Women
Gurdaspur, India

Priya Shome
Amity School of Communication
Amity University Jharkhand
Ranchi, India

Md. Rashid Iqubal Siddique
Amity School of Communication
Amity University Jharkhand
Ranchi, India

Bam Bahadur Sinha
Department of Computer Science & Engineering
National Institute of Technology Sikkim
Ravangla, India

B. Sivasakthi
Department of CSE
MVJ College of Engineering
Bengaluru, India

Radha Sridharan
Department of Computer Science
CMR University,
Bangalore, India
India

Nimish Srivastava
Department of Biotechnology
IMS Engineering College
Ghaziabad, India

N. S. Sukanya
Department of MCA
New Horizon College of Engineering
Bengaluru, India

M. T. Vasumathi
Department of MCA
New Horizon College of Engineering
Bengaluru, India

Abbreviations

ACO	Ant colony optimization
AD	Alzheimer's disease
ADNI	Alzheimer's disease neuroimaging initiative
AI	Artificial intelligence
ANN	Artificial neural network
AR	Augmented reality
BCI	Brain-computer interfaces
BERT	Bidirectional encoder representations from transformers
CCPA	California Consumer Privacy Act
CFAS	Cognitive function and ageing study
CNN	Convolutional neural network
COTS	Commercially available off-the-shelf
CPPS	Cyber-physical production systems
CPS	Cyber-physical systems
CRM	Customer relationship management
DLP	Data loss prevention
DT	Decision tree
DT	Digital twin
DW	Diffusion-weighted
EHR	Electronic health records
ERP	Enterprise resource planning
FFT	Fast Fourier transform
FN	False negatives
FP	False positives
GAN	Generative adversarial network
GDPR	General data protection regulation
HC	Healthy control
IID	Independent and identically distributed
IoT	Internet of things
ITS	Intelligent tutoring system
KPI	Key performance indicators
LIDAR	Light detection and ranging

LR	Logistic regression
MDM	Mobile device management
MFA	Multi-factor authentication
ML	Machine learning
MMSE	Mini-mental state examination
MRI	Magnetic resonance imaging
MSD	Manufacturing system design
NLP	Natural language processing
OASIS	Open access series of imaging studies
OPSEC	Operational security
PD	Participatory design
PDE	Partial differential equations
PINN	Physics-informed neural networks
PLM	Product lifecycle management
PTSD	Post-traumatic stress disorder
RBAC	Role-based access controls
RF	Random forest
RINI	Retrospective informant interviews
RL	Reinforcement learning
RMS	Root mean square
RNN	Recurrent neural networks
SaaS	Software-as-a-service
SL	Supervised learning
SMS	Smart manufacturing system
SSL	Semi-supervised learning
SVM	Support vector machine
TN	True negatives
TP	True positives
UCD	User-centred design
UL	Unsupervised learning
VD	Vascular dementia
VDCN	Very deep convolutional network
VPN	Virtual private networks
VR	Virtual reality
VSD	Value-sensitive design
VSM	Vibration signal monitoring
WIP	Work-in-progress

Chapter 1

Understanding the fundamentals of deep learning for industrial applications

Surkhab Shelly and Ravneet Kaur

1.1 INTRODUCTION

Data-driven solutions have undergone a revolution with the introduction of deep learning techniques, which has resulted in an increase in their use across numerous industries. These strategies are crucial for resolving complicated issues due to their capacity to recognize patterns and anticipate outcomes from massive datasets. However, the successful implementation of deep learning in industrial applications requires a deep understanding of their fundamental concepts, applications, and challenges. Artificial intelligence (AI) subfields, known as deep learning, use training algorithms to discover patterns and make predictions from big datasets. Deep learning and machine learning can be applied in industrial settings to solve difficult issues and increase productivity in a variety of industries, including manufacturing, healthcare, finance, and transportation. Machine learning algorithms use statistical techniques to analyze data and learn patterns that can be used to make predictions or classifications. These algorithms can be supervised, unsupervised, or semi-supervised, depending on whether they use labeled or unlabeled data to learn from neural networks, a subset of machine learning that mimics the structure and operation of the human brain and is used in deep learning. Layers of connected nodes that analyze information and recognize patterns make up neural networks. Deep learning algorithms can handle complex and unstructured data, such as images, speech, and natural language, making them useful for various industrial applications. In industrial settings, machine learning and deep learning can be used for predictive maintenance, quality control, anomaly detection, demand forecasting, and optimization. For instance, both deep learning and machine learning can be used in manufacturing to improve production processes and find product flaws. Machine learning and deep learning are used in the healthcare industry for the diagnosis of diseases, the discovery of new drugs, and personalized treatment. Deep learning can be used, for instance, to analyze medical images, spot irregularities, or forecast patient outcomes. In finance, machine learning and deep learning can be used for fraud detection, risk assessment, and portfolio optimization. For example,

DOI: 10.1201/9781003534761-1

machine learning can be used to detect anomalies in financial transactions, whereas deep learning can be used to predict stock prices. Overall, machine learning and deep learning have a vast potential for improving efficiency and solving complex problems in various industries, making them essential tools for industrial applications.

However, implementing machine learning and deep learning in industrial applications can be challenging due to several factors. One of the critical challenges is the availability and quality of data. Industrial applications often involve complex and heterogeneous data, which may require extensive preprocessing and cleaning before being used for machine learning or deep learning. Moreover, data privacy and security concerns can also limit access to data, making it challenging to train and evaluate machine learning and deep learning models. Another challenge is selecting the right machine learning or deep learning algorithms and architectures that are suitable for the problem at hand. The choice of algorithms and architectures can affect the accuracy and performance of the models and require extensive experimentation and tuning. Evaluation metrics and model interpretation are also critical factors to consider when implementing deep learning in industrial applications. The accuracy of the models should be evaluated using appropriate metrics and validated on different datasets. Furthermore, model interpretation is essential to understand how the models make predictions and gain insights into the underlying patterns in the data. Deep learning has revolutionized the field of data-driven solutions and has enormous potential for industrial applications. However, their successful implementation requires a deep understanding of their fundamental concepts, applications, and challenges. Industrial practitioners must carefully consider the essential steps, including data preparation, model selection, evaluation metrics, and hyper parameter tuning, when implementing machine learning and deep learning models to ensure their effectiveness in solving complex problems.

1.2 LITERATURE REVIEW

Recent literature on the fundamentals of deep learning for industrial applications has continued to explore the potential of these technologies in various sectors. In 2020, several studies focused on the applications of deep learning in industrial automation, quality control, and supply chain management. For example, a study by Wang et al. proposed a deep learning-based approach for fault diagnosis in rotating machinery, which could improve the accuracy and efficiency of equipment maintenance [1]. Another study by Zhang et al. (explored the potential of machine learning in optimizing the performance of supply chains by predicting demand and reducing inventory costs [2]. A study by Molnar et al. highlighted the ethical considerations in the development and deployment of machine learning algorithms in industrial settings, including issues related to fairness, transparency, and accountability [3]. In manufacturing, recent studies have explored the potential of machine learning and deep learning in predicting equipment failures, improving supply chain management, and enhancing product quality. For example, a study by Xu et al. proposed a deep learning-based approach to predict the remaining useful life of equipment in a manufacturing plant. The study demonstrated the potential of deep learning in improving predictive maintenance and reducing downtime [4].

In healthcare, recent studies have focused on the potential of deep learning in personalized medicine, drug discovery, and medical image analysis. For example, a study by Li et al. proposed a deep learning-based approach for the early diagnosis of Alzheimer's disease using MRI images [5]. The study demonstrated the potential of deep learning in improving the accuracy of disease diagnosis and providing early intervention for patients. In finance, recent studies have explored the potential of machine learning and deep learning in fraud detection, portfolio optimization, and risk assessment. For example, a study by Zheng et al. proposed a deep learning-based approach for portfolio optimization that could consider multiple objectives and constraints [6]. The study demonstrated the potential of deep learning in improving the efficiency of portfolio management and reducing financial risks. Moreover, recent research has also explored the latest developments and advancements in machine learning and deep learning algorithms and architectures for various industrial applications. For example, a study by Chen et al. [7] proposed a novel deep learning architecture for speech recognition in noisy environments. The study demonstrated the potential of deep learning in improving the accuracy of speech recognition and enabling better human–machine interaction in industrial settings.

In 2022, recent studies have focused on the latest developments and advancements in machine learning and deep learning for industrial applications, including the integration of these technologies with Internet of Things (IoT) devices and edge computing platforms. For example, a study by Tang et al. proposed a deep learning-based approach for anomaly detection in IoT-enabled manufacturing systems [8]. The study demonstrated the potential of deep learning in improving the efficiency and reliability of industrial automation systems. Another study by Xu et al. explored the potential of edge computing and machine learning in improving the accuracy and speed of predictive maintenance in manufacturing plants [9]. Overall, recent literature on the fundamentals of machine learning and deep learning for industrial applications has highlighted the potential of these technologies in improving various aspects of industrial operations. The studies have explored the latest developments and advancements in machine learning and deep learning algorithms and architectures and emphasized the need to address the challenges and implications of these technologies.

1.3 DEEP LEARNING AND INDUSTRY 4.0

1.3.1 Deep learning

The most crucial technique for the development of AI is undoubtedly deep learning as part of machine learning. It is pertinent for the adoption of Industry 4.0 in this position. It's a kind of AI and machine learning that mimics how people learn specific subjects. Data science, which also encompasses statistics and predictive modeling, contains deep learning as a key component [10]. Deep learning makes this process quicker and simpler, which is very advantageous to data scientists who are entrusted with gathering, analyzing, and interpreting massive amounts of data. An input layer and one or more hidden layers are connected one after the other in a fully connected deep neural network. Each neuron receives information from neurons in the input layer or neurons in the layer below. One neuron's output becomes the input for more neurons in the layer below it, and so on until the last layer of the network generates the network's output. The neural network's layers alter the input data in a variety of nonlinear ways, enabling the network to learn intricate representations of the data [11].

1.3.2 Industry 4.0

Industry 4.0, also known as the fourth industrial revolution, refers to the integration of advanced technologies and digitalization in manufacturing and industrial processes. It represents a paradigm shift in how industries operate, leveraging emerging technologies to create smarter, more efficient, and interconnected systems. The term "Industry 4.0" originated in Germany as part of a government initiative to promote the computerization of manufacturing. It builds upon the three previous industrial revolutions. The first revolution involved mechanization and the use of steam power, whereas the second revolution introduced mass production through electricity and assembly lines. The third revolution involved the adoption of computers and automation. The IoT, AI, big data analytics, cloud computing, and cyber-physical systems (CPS) are among the technologies that have come together to form Industry 4.0 [12]. These technologies enable communication, data analysis, and autonomous decision-making between machines, devices, and systems, enhancing productivity, efficiency, and flexibility in manufacturing processes [13].

1.3.3 Key components of Industry 4.0

The key components of Industry 4.0 are represented in Figure 1.1. The key components are as follows [14].

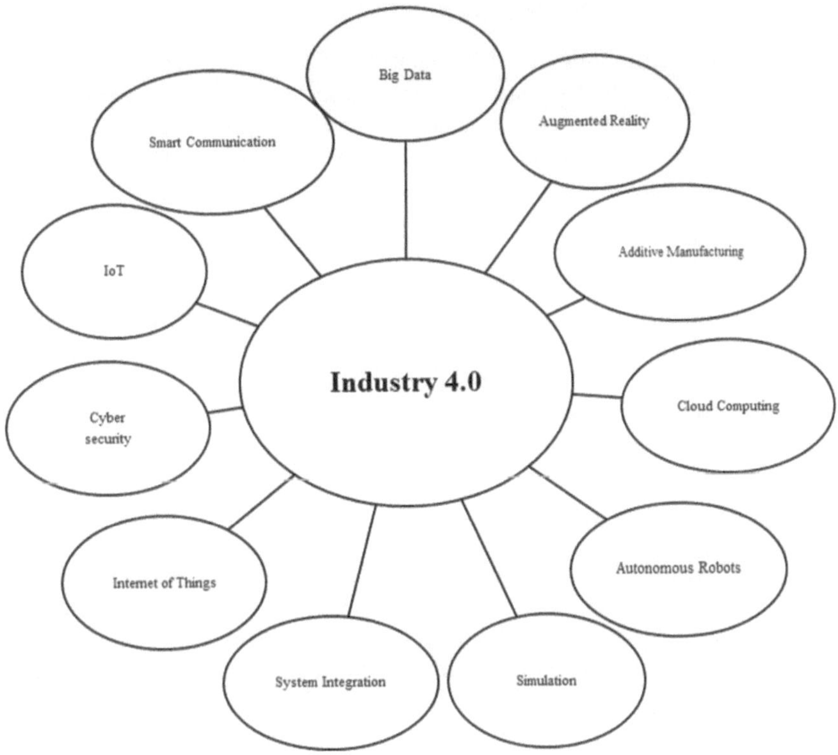

Figure 1.1 Industry 4.0.

1.3.3.1 Internet of Things

IoT refers to the network of physical devices embedded with sensors, actuators, and connectivity, enabling them to collect and exchange data. In Industry 4.0, IoT plays a crucial role in connecting machines, equipment, and products to create a seamless flow of information and enable real-time monitoring and control.

1.3.3.2 Cyber-physical systems

CPS combines physical elements with advanced computing, communication, and control capabilities. It involves the integration of physical systems with software and networked sensors, allowing for real-time data exchange and control of physical processes. CPS forms the backbone of Industry 4.0 by enabling the interaction between the physical and digital worlds.

1.3.3.3 AI and machine learning

AI and ML technologies enable machines and systems to analyze large amounts of data, identify patterns, and make intelligent decisions. These technologies can be used in predictive maintenance, quality control, demand forecasting, and optimization of manufacturing processes, leading to improved efficiency and reduced costs.

1.3.3.4 Big data analytics

Industry 4.0 generates vast amounts of data from various sources, including machines, sensors, and production processes. Big data analytics involves extracting valuable insights and patterns from this data to optimize operations, identify bottlenecks, and make data-driven decisions. It helps in improving efficiency, quality, and predictive capabilities.

1.3.3.5 Cloud computing

Cloud computing provides scalable and on-demand access to computing resources, storage, and services over the Internet. It enables the storage and processing of large amounts of data generated by Industry 4.0 applications. Cloud-based solutions offer flexibility, cost-effectiveness, and collaboration capabilities, allowing businesses to leverage advanced technologies without significant infrastructure investments.

1.3.3.6 Additive manufacturing (3D printing)

Additive manufacturing is a technology that builds objects layer by layer using digital models. It offers greater design freedom, customization, and rapid prototyping capabilities. 3D printing can reduce costs, lead times, and material waste, enabling more flexible and agile production processes.

1.3.3.7 Augmented reality (AR) and virtual reality (VR)

AR and VR technologies are used in Industry 4.0 for training, maintenance, and visualization purposes. AR can overlay digital information and instructions onto the physical environment, whereas VR creates immersive simulated environments. These technologies enhance worker productivity, facilitate remote collaboration, and support complex assembly and maintenance tasks.

1.3.3.8 Autonomous systems

Industry 4.0 emphasizes the use of autonomous systems that can operate and make decisions with minimal human intervention. These systems include autonomous robots and vehicles that can perform tasks such as material handling, assembly, and logistics. By automating repetitive and labor-intensive processes, businesses can achieve higher efficiency, accuracy, and cost savings.

1.3.3.9 Digital twins

Digital twins are virtual replicas of physical objects, processes, or systems. They are created by collecting real-time data from sensors embedded in physical assets and using it to build a digital model. Digital twins allow for the monitoring, simulation, and optimization of physical assets, enabling predictive maintenance, performance optimization, and rapid prototyping.

1.3.3.10 Smart factories

Industry 4.0 promotes the concept of smart factories, where interconnected systems and machines communicate and collaborate with each other. Smart factories leverage real-time data and advanced analytics to enable adaptive production processes. They can self-optimize, self-configure, and self-diagnose, leading to improved operational efficiency, reduced waste, and increased responsiveness to market demands.

1.3.3.11 Supply chain integration

Industry 4.0 facilitates seamless integration and collaboration across the entire supply chain, from suppliers to customers. With the help of IoT, real-time data sharing, and predictive analytics, companies can achieve greater visibility and transparency throughout the supply chain. This enables better demand forecasting, optimized inventory management, reduced lead times, and improved coordination between stakeholders.

1.3.3.12 Customization and personalization

Industry 4.0 enables mass customization and personalized production. With digital technologies and flexible manufacturing processes, companies can efficiently produce customized products to meet individual customer preferences. Through data-driven insights and advanced production techniques like 3D printing, businesses can offer a wide range of personalized products while maintaining cost efficiency.

1.3.3.13 Sustainability and energy efficiency

Industry 4.0 emphasizes sustainable and energy-efficient manufacturing practices. By leveraging real-time data monitoring, optimization algorithms, and intelligent energy management systems, companies can reduce energy consumption, optimize resource utilization, and minimize waste. This leads to a more sustainable approach to production, benefiting both the environment and the bottom line.

1.3.3.14 Human-machine collaboration

Industry 4.0 does not aim to replace human workers but rather focuses on enhancing their capabilities through collaboration with machines. Humans and machines can work together in symbiotic relationships, with machines handling repetitive tasks, data analysis, and physical labor, whereas humans provide creativity, problem-solving skills, and decision-making abilities. This collaboration improves productivity, safety, and overall job satisfaction.

1.3.3.15 Data security and privacy

With the increased connectivity and data exchange in Industry 4.0, ensuring data security and privacy becomes critical. Companies need robust cyber security measures to protect sensitive information and secure their systems from cyber threats. Additionally, regulations and ethical considerations regarding data privacy become crucial in the era of Industry 4.0.

Industry 4.0 represents a transformative journey for businesses, revolutionizing manufacturing processes and reshaping entire industries. It offers numerous opportunities for increased efficiency, productivity, and innovation. However, successful implementation requires strategic planning, investment in technology infrastructure, upskilling the workforce, and addressing the associated challenges to fully harness the potential of Industry 4.0.

1.4 APPLICABILITY OF DEEP LEARNING IN INDUSTRY 4.0

The relationship between humans and machines is becoming increasingly intimate, as machines now undertake tasks that were traditionally exclusive to humans. In the context of Industry 4.0, deep learning seamlessly blends various technologies, enabling software and machines to perceive, comprehend, execute, and acquire knowledge from human operations. This integration significantly enhances the efficiency of industrial production systems. Deep learning emerges as a pivotal evolving technology, driving the elevation of efficiency, product quality, and the reduction of operational expenses. In the industries like manufacturing, hospitality, digital assistants (IoT), automotive, etc., [15]. Deep learning is embryonic as one of

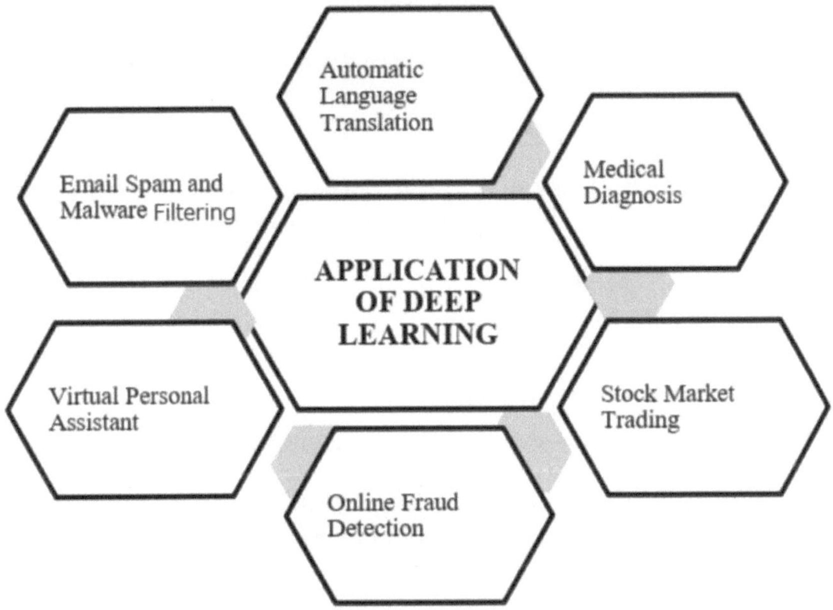

Figure 1.2 Application of deep learning.

the crucial practices. Industries are leveraging its applications to be part of Industry 4.0 because of the increased use of deep learning [16]. A few of the applications as represented in Figure 1.2 are discussed as follows.

1.4.1 Automobiles

Deep learning can play a significant role in the automobile industry by improving safety, efficiency, and performance. Here are some of the key applications of deep learning in the automobile industry.

1.4.1.1 Autonomous driving

Deep learning algorithms can be used to power autonomous driving systems, enabling vehicles to navigate complex environments and make decisions in real time. By analyzing data from sensors such as cameras and LIDAR, deep learning algorithms can help vehicles detect and avoid obstacles, interpret traffic signals, and make decisions about speed and direction.

1.4.1.2 Predictive maintenance

Deep learning algorithms can be used to predict equipment failures and schedule maintenance proactively. By analyzing data on vehicle performance and environmental conditions, deep learning algorithms can identify signs of impending equipment failure and alert maintenance teams before a breakdown occurs, reducing maintenance costs.

1.4.1.3 Quality control

Deep learning algorithms can be used to analyze images of vehicle components and detect defects. By comparing images of components to a database of known good parts, deep learning algorithms can identify any discrepancies and flag parts for inspection or replacement, improving quality control and reducing defects.

1.4.1.4 Personalized in-car experience

Deep learning algorithms can be used to personalize the in-car experience for each individual driver. By analyzing data on driving behavior, music preferences, and other factors, deep learning algorithms can make recommendations for music, climate control, and other features, improving the overall driving experience.

1.4.1.5 Supply chain optimization

Deep learning algorithms can be used to optimize supply chain operations by analyzing data on shipping times, delivery routes, and other factors. By identifying areas for improvement, deep learning algorithms can help automobile manufacturers reduce costs and improve delivery times, improving customer satisfaction.

Deep learning has the potential to revolutionize the automotive sector by enhancing performance, efficiency, and safety. As the technology continues to evolve and new applications are discovered, we can expect to see more automobile manufacturers adopting deep learning to improve their operations and better serve their customers.

1.4.2 Digital assistants: (voice search assistants)

Digital assistants, such as Amazon's Alexa, Apple's Siri, and Google Assistant, use deep learning to interpret natural language queries and provide users with relevant information or perform tasks. Here are some of the ways that deep learning is used in digital assistance.

1.4.2.1 Natural language processing

Deep learning algorithms can be used to interpret and respond to natural language queries. By analyzing the context of a user's query, deep learning algorithms can understand the intent behind the query and provide a relevant response.

1.4.2.2 Speech recognition

Deep learning algorithms can be used to recognize and transcribe speech accurately. By analyzing patterns in speech and noise, deep learning algorithms can distinguish between different speakers and filter out background noise, improving the accuracy of speech recognition.

1.4.2.3 Personalization

Deep learning algorithms can be used to personalize the digital assistant's responses based on the user's preferences and behavior. By analyzing data on previous queries, search history, and other factors, deep learning algorithms can tailor responses to each individual user, improving the overall user experience.

1.4.2.4 Task automation

Deep learning algorithms can be used to automate tasks such as scheduling appointments, setting reminders, or ordering groceries. By understanding the user's preferences and behavior, deep learning algorithms can perform these tasks without the need for explicit instructions, improving efficiency and saving time.

1.4.2.5 Contextual awareness

Deep learning algorithms can be used to provide contextual awareness to digital assistants, enabling them to understand the user's location, time of day, and other factors. By analyzing data on the user's surroundings, deep learning algorithms can provide more relevant and useful responses, improving the overall user experience.

It has played a notable role in advancing digital assistants, empowering them to offer responses that are not only more precise and tailored but also more valuable to users. With the ongoing progression of this technology, we can anticipate a wider array of deep learning applications in digital assistance, ultimately enhancing the holistic user experience and simplifying the process of retrieving information and accomplishing tasks.

1.4.3 Manufacturing

Deep learning can play a significant role in the manufacturing industry by improving efficiency, quality control, and predictive maintenance. Here are some of the key applications of deep learning in manufacturing [17–19].

1.4.3.1 Quality control

Deep learning algorithms can be used to analyze images of products and detect defects. By comparing images of products to a database of known good products, deep learning algorithms can identify any discrepancies and flag products for inspection or rejection, improving quality control and reducing defects.

1.4.3.2 Predictive maintenance

Deep learning algorithms can be used to predict equipment failures and schedule maintenance proactively. By analyzing data on equipment performance and environmental conditions, deep learning algorithms can identify signs of impending equipment failure and alert maintenance teams before a breakdown occurs, reducing downtime and maintenance costs.

1.4.3.3 Process optimization

Deep learning algorithms can be used to optimize manufacturing processes by analyzing data on equipment performance, production rates, and other factors. By identifying areas for improvement, deep learning algorithms can help manufacturers reduce costs and improve efficiency, improving overall productivity.

1.4.3.4 Inventory management

Deep learning algorithms can be used to predict demand for products and optimize inventory levels. By analyzing data on past sales and market trends, deep learning algorithms can predict future demand for products and ensure that inventory levels are optimized, reducing waste and improving profitability.

1.4.3.5 Supply chain optimization

Deep learning algorithms can be used to optimize supply chain operations by analyzing data on shipping times, delivery routes, and other factors. By identifying areas for improvement, deep learning algorithms can help manufacturers reduce costs and improve delivery times, improving customer satisfaction.

It has the potential to transform the manufacturing industry by improving efficiency, quality control, and predictive maintenance. As the technology continues to evolve and new applications are discovered, we can expect to see more manufacturers adopting deep learning to improve their operations and better serve their customers.

1.4.4 Banking industry

Deep learning has the potential to transform the banking industry by enabling intelligent automation, fraud detection, and personalized customer experiences. Here are some of the key applications of deep learning in banking.

1.4.4.1 Fraud detection

Deep learning algorithms can analyze large volumes of transaction data to detect fraudulent activity in real time. By identifying patterns and anomalies in transaction data, deep learning algorithms can help banks prevent financial losses due to fraud.

1.4.4.2 Credit risk analysis

Deep learning algorithms can help banks assess the creditworthiness of borrowers by analyzing large amounts of financial and credit data. By identifying patterns and correlations in the data, deep learning algorithms can help banks make more accurate credit risk assessments.

1.4.4.3 Customer service

Deep learning algorithms can analyze customer data to provide personalized recommendations and insights. For example, banks can use deep learning algorithms to analyze customer spending patterns and offer personalized product recommendations.

1.4.4.4 Chatbots

Deep learning algorithms can be used to develop chatbots that can provide customer support and answer common questions. By using natural language processing, deep learning algorithms can provide more personalized and accurate responses to customer inquiries.

1.4.4.5 Investment analysis

Deep learning algorithms can analyze large amounts of financial data to identify investment opportunities and predict market trends. By providing more accurate and timely investment insights, deep learning algorithms can help banks make more informed investment decisions.

So, it has the potential to revolutionize the banking industry by enabling more efficient and personalized services while also improving security and fraud detection. As the technology continues to evolve and new applications are discovered, we can expect to see more banks adopting deep learning to improve their operations and customer experiences.

1.4.5 Fishing industry

Deep learning has the potential to revolutionize the fishing industry by enabling more efficient and sustainable fishing practices while also reducing environmental impacts. Here are some of the key applications of deep learning in the fishing industry.

1.4.5.1 *Fish detection*

Deep learning algorithms can be used to analyze underwater images and video footage to detect and identify fish species. By using computer vision and machine learning (ML) techniques, deep learning algorithms can accurately identify different types of fish, which can help improve fishing practices and reduce catch.

1.4.5.2 *Improving fish stock assessments*

By identifying patterns and correlations in the data, deep learning algorithms can help fisheries make more accurate assessments of fish stocks, which can improve sustainability and reduce overfishing.

1.4.5.3 *Real-time insights for operational optimization*

By providing real-time insights into ocean conditions, deep learning algorithms can help fisheries optimize their operations and reduce environmental impacts.

1.4.5.4 *Aquaculture*

Deep learning algorithms can be used to monitor fish farms and optimize production. By analyzing data on water quality, feeding patterns, and other factors, deep learning algorithms can help fish farmers improve efficiency and reduce waste.

1.4.5.5 *Sustainable fishing practices*

Deep learning algorithms can be used to develop models that optimize fishing practices to reduce bycatch and minimize environmental impacts. By analyzing data on fishing practices and environmental factors, deep learning algorithms can help fisheries develop more sustainable fishing practices.

Deep learning has the potential to transform the fishing industry by enabling more sustainable and efficient fishing practices. As the technology continues to evolve and new applications are discovered, we can expect to see more fisheries adopting deep learning to improve their operations and reduce environmental impacts

1.4.6 Textile industry

Deep learning has the potential to revolutionize the textile industry by enabling more efficient and sustainable production processes while also improving product quality and customization. Here are some of the key applications of deep learning in the textile industry.

1.4.6.1 Quality control

Deep learning algorithms can be used to analyze images of textile products to detect defects and inconsistencies in the fabric. By using computer vision and ML techniques, deep learning algorithms can accurately identify defects such as holes, snags, or color variations, which can help improve product quality and reduce waste.

1.4.6.2 Product design

Deep learning algorithms can be used to analyze customer preferences and develop personalized textile products. By using data on customer preferences and previous purchases, deep learning algorithms can help textile companies develop customized products that meet individual customer needs.

1.4.6.3 Textile manufacturing

Deep learning algorithms can be used to optimize textile manufacturing processes and reduce waste. By analyzing data on production processes, such as machine settings and environmental factors, deep learning algorithms can help manufacturers improve efficiency and reduce environmental impacts.

1.4.6.4 Enhancing textile recycling with deep learning

By using computer vision and ML techniques, deep learning algorithms can accurately identify different types of fabrics, which can help improve the efficiency and effectiveness of textile recycling.

1.4.6.5 Optimizing supply chains for environmental efficiency

By analyzing data on suppliers, transportation, and other factors, deep learning algorithms can help textile companies identify inefficiencies and reduce environmental impacts.

So we can say, deep learning has the potential to transform the textile industry by enabling more efficient and sustainable production processes while also improving product quality and customization. As the technology continues to evolve and new applications are discovered, we can expect to see more textile companies adopting deep learning to improve their operations and reduce environmental impacts.

1.4.7 Agriculture industry

Deep learning has the potential to revolutionize agriculture by enabling more efficient and sustainable farming practices while also improving crop yields and reducing environmental impacts. Here are some of the key applications of deep learning in agriculture.

1.4.7.1 *Crop monitoring*

Deep learning algorithms can be used to analyze satellite imagery and other data to monitor crop growth and detect potential issues. By providing real-time insights into crop conditions, deep learning algorithms can help farmers optimize their farming practices and improve crop yields.

1.4.7.2 *Pest and disease detection*

Deep learning algorithms can be used to analyze images of crops to detect pest infestations and diseases. By using computer vision and ML techniques, deep learning algorithms can accurately identify signs of pests or diseases, which can help farmers take proactive measures to prevent crop damage.

1.4.7.3 *Soil monitoring*

Deep learning algorithms can be used to analyze soil data to optimize crop growth. By analyzing data on soil moisture, temperature, and other factors, deep learning algorithms can help farmers optimize irrigation and fertilization practices, which can improve crop yields and reduce waste.

1.4.7.4 *Precision farming*

Deep learning algorithms can be used to develop models that optimize farming practices for specific crops and growing conditions. By analyzing data on weather patterns, soil conditions, and other factors, deep learning algorithms can help farmers develop more precise and efficient farming practices.

1.4.7.5 *Harvesting and sorting*

Deep learning algorithms can be used to analyze images of harvested crops and sort them based on quality and other factors. By using computer vision and ML techniques, deep learning algorithms can accurately identify the quality of individual crops, which can help farmers optimize their harvesting and sorting processes.

Overall, deep learning has the potential to transform agriculture by enabling more efficient and sustainable farming practices while also improving crop yields and reducing environmental impacts. As the technology continues to evolve and new applications are discovered, we can expect to see more farmers adopting deep learning to improve their operations and reduce environmental impacts.

1.4.8 Food industry

Deep learning has the potential to transform the food industry by enabling more efficient and sustainable production processes while also improving food quality and safety. Here are some of the key applications of deep learning in the food industry.

1.4.8.1 Quality control

Deep learning algorithms can be used to analyze images of food products to detect defects and inconsistencies in the product. By using computer vision and ML techniques, deep learning algorithms can accurately identify defects such as discoloration, spoilage, or foreign objects, which can help improve product quality and safety.

1.4.8.2 Enhancing food manufacturing efficiency

By analyzing data on processing times, temperatures, and other factors, deep learning algorithms can help food manufacturers improve efficiency and reduce environmental impacts.

1.4.8.3 Food safety

Deep learning algorithms can be used to analyze data on food safety and identify potential hazards. By analyzing data on foodborne illnesses, contamination incidents, and other factors, deep learning algorithms can help food manufacturers take proactive measures to prevent foodborne illnesses and ensure food safety.

1.4.8.4 Recipe development

Deep learning algorithms can be used to analyze data on consumer preferences and develop new food products. By using data on consumer preferences and previous purchases, deep learning algorithms can help food manufacturers develop customized products that meet individual consumer needs.

1.4.8.5 Supply chain

Deep learning algorithms can be used to optimize food supply chains and improve transparency. By analyzing data on suppliers, transportation, and other factors, deep learning algorithms can help food manufacturers identify inefficiencies and reduce environmental impacts.

Deep learning has the potential to revolutionize the food industry by enabling more efficient and sustainable production processes while also improving food quality and safety. As the technology continues to evolve and new applications are discovered, we can expect to see more food manufacturers adopting deep learning to improve their operations and reduce environmental impacts.

1.4.9 Construction industry

Deep learning can have a significant impact on the construction industry by enabling more efficient and cost-effective building processes while also improving safety and reducing environmental impacts. Here are some of the key applications of deep learning in the construction industry.

1.4.9.1 Safety

Deep learning algorithms can be used to analyze data on construction site safety and identify potential hazards. By analyzing data on accidents, near-misses, and other factors, deep learning algorithms can help construction companies take proactive measures to improve safety and prevent accidents.

1.4.9.2 Quality control

Deep learning algorithms can be used to analyze images of construction projects to detect defects and inconsistencies in the work. By using computer vision and ML techniques, deep learning algorithms can accurately identify defects such as cracks, uneven surfaces, and other issues, which can help improve the quality of construction projects.

1.4.9.3 Project management

Deep learning algorithms can be used to analyze data on construction projects and optimize construction schedules. By analyzing data on labor availability, weather patterns, and other factors, deep learning algorithms can help construction companies plan and execute projects more efficiently.

1.4.9.4 Building information modeling

Deep learning algorithms can be used to analyze data from building information modeling (BIM) systems and identify potential issues in construction projects. By analyzing data on building components, materials, and other factors, deep learning algorithms can help construction companies optimize their designs and construction processes.

1.4.9.5 Equipment maintenance

Deep learning algorithms can be used to predict equipment failures and optimize maintenance schedules. By analyzing data on equipment usage, wear and tear, and other factors, deep learning algorithms can help construction companies prevent downtime and reduce repair costs.

Deep learning has the potential to transform the construction industry by enabling more efficient and cost-effective building processes while also improving safety and reducing environmental impacts. As the technology continues to evolve and new applications are discovered, we can expect to see more construction companies adopting deep learning to improve their operations and reduce environmental impacts.

1.4.10 Transportation industry

Deep learning has the potential to revolutionize the transportation industry by enabling more efficient and sustainable transportation systems while also improving safety and reducing environmental impacts. Here are some of the key applications of deep learning in the transportation industry.

1.4.10.1 Autonomous vehicles

Deep learning algorithms can be used to enable autonomous vehicles to navigate roads and avoid obstacles. By analyzing data from cameras, LIDAR, and other sensors, deep learning algorithms can help vehicles make real-time decisions about how to maneuver through traffic and avoid accidents.

1.4.10.2 Traffic optimization

Deep learning algorithms can be used to optimize traffic flow and reduce congestion. By analyzing data from traffic cameras, sensors, and other sources, deep learning algorithms can help transportation planners make real-time decisions about how to route traffic and manage congestion.

1.4.10.3 Predictive maintenance

Deep learning algorithms can be used to predict equipment failures and optimize maintenance schedules. By analyzing data on equipment usage, wear and tear, and other factors, deep learning algorithms can help transportation companies prevent downtime and reduce repair costs.

1.4.10.4 Energy efficiency

Deep learning algorithms can be used to optimize vehicle performance and reduce fuel consumption. By analyzing data on driving patterns, traffic conditions, and other factors, deep learning algorithms can help transportation companies optimize their fleets for maximum efficiency.

1.4.10.5 Public transportation

Deep learning algorithms can be used to optimize public transportation systems and improve passenger experience. By analyzing data on passenger flow, bus and train schedules, and other factors, deep learning algorithms can help transportation planners make real-time decisions about how to optimize service and improve the overall passenger experience.

It has the potential to transform the transportation industry by enabling more efficient and sustainable transportation systems while also improving safety and reducing environmental impacts. As the technology continues to evolve and new applications are discovered, we can expect to see more transportation companies adopting deep learning to improve their operations and reduce environmental impacts.

1.4.11 Health industry

Deep learning has the potential to revolutionize the healthcare industry by enabling more accurate diagnoses, personalized treatment plans, and improved patient outcomes. Here are some of the key applications of deep learning in the healthcare industry [20].

1.4.11.1 Medical imaging

Deep learning algorithms can be used to analyze medical images such as CT scans, X-rays, and MRIs to detect abnormalities and diagnose diseases. By analyzing large amounts of image data, deep learning algorithms can learn to recognize patterns and identify early signs of disease, improving accuracy and reducing the risk of misdiagnosis.

1.4.11.2 Drug discovery

Deep learning algorithms can be used to accelerate drug discovery by predicting the efficacy of new compounds and identifying potential side effects. By analyzing large amounts of data on molecular structures and drug interactions, deep learning algorithms can help researchers develop new treatments more quickly and efficiently.

1.4.11.3 Electronic health records

Deep learning algorithms can be used to analyze electronic health records to identify patterns and predict patient outcomes. By analyzing data on patient demographics, medical histories, and other factors, deep learning algorithms can help doctors and healthcare providers make more informed decisions about patient care.

1.4.11.4 Personalized medicine

Deep learning algorithms can be used to develop personalized treatment plans based on an individual's genetic profile and medical history. By analyzing large amounts of data on genetic mutations, disease pathways, and treatment outcomes, deep learning algorithms can help doctors tailor treatments to the specific needs of each patient.

1.4.11.5 Medical robotics

Deep learning algorithms can be used to power medical robots and assistive devices, enabling more precise and efficient surgeries and treatments. By analyzing data from sensors and cameras, deep learning algorithms can help medical robots navigate complex environments and perform delicate procedures with greater accuracy.

So it has the potential to transform the healthcare industry by enabling more accurate diagnoses, personalized treatment plans, and improved patient outcomes. As the technology continues to evolve and new applications are discovered, we can expect to see more healthcare providers adopting deep learning to improve patient care and outcomes.

1.4.12 Retail industry

Personalized and efficient customer experiences while also improving operational efficiency and reducing costs. Here are some of the key applications of deep learning in the retail industry.

1.4.12.1 Recommendation engines

Deep learning algorithms can be used to analyze customer data such as browsing history, purchase history, and demographic information to provide personalized recommendations. By analyzing large amounts of data, deep learning algorithms can learn to predict which products a customer is likely to be interested in, improving customer satisfaction and increasing sales.

1.4.12.2 Inventory management

Deep learning algorithms can be used to optimize inventory management by predicting demand for different products. By analyzing data on past sales, seasonal trends, and other factors, deep learning algorithms can help retailers better forecast demand and adjust their inventory levels accordingly, reducing waste and improving operational efficiency.

1.4.12.3 Fraud detection

Deep learning algorithms can be used to detect and prevent fraud by analyzing data on customer behavior, transaction history, and other factors. By identifying unusual patterns or behavior, deep learning algorithms can flag potentially fraudulent transactions and help retailers prevent financial losses.

1.4.12.4 Visual search

Deep learning algorithms can be used to enable visual search, allowing customers to search for products based on images rather than keywords. By analyzing images of products, deep learning algorithms can identify similar items and make personalized recommendations to customers, improving the overall customer experience.

1.4.12.5 Supply chain optimization

Deep learning algorithms can be used to optimize supply chain operations by analyzing data on shipping times, delivery routes, and other factors. By identifying areas for improvement, deep learning algorithms can help retailers reduce costs and improve delivery times, improving customer satisfaction.

Overall, deep learning has the potential to transform the retail industry by enabling more personalized and efficient customer experiences while also improving operational efficiency and reducing costs. As the technology continues to evolve and new applications are discovered, we can expect to see more retailers adopting deep learning to improve their operations and better serve their customers.

1.4.13 Mining industry

Deep learning can play a significant role in the mining industry by improving safety, efficiency, and profitability. Here are some of the key applications of deep learning in the mining industry.

1.4.13.1 Autonomous mining equipment

Deep learning algorithms can be used to power autonomous mining equipment, such as trucks and drills. By analyzing data from sensors and cameras, deep learning algorithms can help these vehicles navigate complex terrain and perform their tasks with greater efficiency and accuracy.

1.4.13.2 Predictive maintenance

Deep learning algorithms can be used to predict equipment failures and schedule maintenance proactively. By analyzing data on equipment performance and environmental conditions, deep learning algorithms can identify signs of impending equipment failure and alert maintenance teams before a breakdown occurs, reducing downtime and maintenance costs.

1.4.13.3 Mineral exploration

Deep learning algorithms can be used to analyze geological data and identify potential mineral deposits. By analyzing data from various sources such as remote sensing, geophysics, and geochemistry, deep learning algorithms can help mining companies target their exploration efforts more effectively, reducing the risk of costly drilling and other exploratory work.

1.4.13.4 Safety monitoring

Deep learning algorithms can be used to monitor safety in mining operations by analyzing data from sensors and cameras. By identifying potential hazards such as falls, collisions, or fires, deep learning algorithms can alert safety teams to take action before an incident occurs, improving safety for workers.

1.4.13.5 Energy management

Deep learning algorithms can be used to optimize energy use in mining operations by analyzing data on power consumption and environmental conditions. By identifying areas for improvement, deep learning algorithms can help mining companies reduce energy costs and minimize their environmental impact.

Overall, deep learning has the potential to transform the mining industry by improving safety, efficiency, and profitability. As the technology continues to evolve and new applications are discovered, we can expect to see more mining companies adopting deep learning to improve their operations and better serve their stakeholders.

1.4.14 Marketing

Deep learning has emerged as a powerful tool in the field of marketing, enabling marketers to leverage large amounts of data and extract valuable insights. Here are some key areas where deep learning is being used in marketing.

1.4.14.1 Customer segmentation

Deep learning algorithms can analyze vast amounts of customer data, such as demographics, behaviors, and purchase history, to identify patterns and segment customers into distinct groups. This allows marketers to personalize their campaigns and target specific segments with tailored messaging.

1.4.14.2 Recommendation systems

Deep learning models are used in recommendation systems to provide personalized product or content recommendations to customers. These models analyze user behavior, preferences, and historical data to predict and suggest relevant products or content, enhancing the customer experience and driving sales.

1.4.14.3 Sentiment analysis

Deep learning techniques, such as natural language processing (NLP), enable marketers to analyze customer sentiment from social media posts, reviews, and customer feedback. Sentiment analysis helps companies understand customer opinions, identify trends, and respond to customer needs effectively.

1.4.14.4 Image and video analysis

Deep learning models can analyze visual content, such as images and videos, to extract valuable information. Marketers can use this technology to identify brand logos, objects, and scenes within visual content shared by customers on social media. It can also help in monitoring and analyzing user-generated content to gain insights into brand perception.

1.4.14.5 Predictive analytics

Deep learning algorithms can be used to predict customer behavior, such as purchasing patterns, churn probability, and customer lifetime value. By analyzing historical data, these models can generate accurate predictions, allowing marketers to make data-driven decisions and optimize their marketing strategies.

1.4.14.6 Ad targeting and optimization

Deep learning techniques can improve ad targeting and optimization by analyzing user data, including browsing history, demographics, and online behavior. Marketers can leverage this technology to deliver relevant ads to the right audience at the right time, increasing the effectiveness of their campaigns.

1.4.14.7 Fraud detection

Deep learning algorithms can detect patterns and anomalies in transactional data to identify fraudulent activities. In marketing, this is particularly useful for detecting click fraud, ad fraud, and other forms of online fraud, helping marketers protect their investments and optimize their ad spend.

Overall, deep learning in marketing enables marketers to gain valuable insights, enhance customer experiences, and optimize their marketing strategies by leveraging the power of data and advanced analytics techniques.

1.4.15 Financial services

Deep learning has made significant contributions to the financial services industry, revolutionizing various aspects of operations and decision-making. Here are some key applications of deep learning in financial services.

1.4.15.1 Fraud detection

Deep learning algorithms can analyze large volumes of financial data, including transactional records, user behavior, and historical patterns, to detect fraudulent activities. These models can identify anomalies and patterns that indicate potential fraud, enabling financial institutions to take proactive measures to prevent fraudulent transactions.

1.4.15.2 Credit scoring and underwriting

Deep learning models can analyze a wide range of data, such as credit history, income statements, and social media data, to assess creditworthiness and predict default risks. These models can automate the credit scoring and underwriting process, improving accuracy and efficiency while minimizing human bias.

1.4.15.3 Algorithmic trading

Deep learning techniques are used in algorithmic trading to analyze market data, including historical price movements, news articles, and social media sentiment, to make predictions and generate trading signals. Deep learning models can adapt to changing market conditions and identify complex patterns, helping financial institutions make more informed and profitable trading decisions.

1.4.15.4 Risk assessment

Deep learning models can analyze and process vast amounts of data to assess risk in financial portfolios. These models can identify patterns and correlations in historical market data, helping financial institutions manage and mitigate risks associated with investments, loans, and insurance underwriting.

1.4.15.5 Customer service and chatbots

Deep learning-powered chatbots can provide personalized customer service, addressing customer queries, and providing support 24/7. These chatbots leverage NLP and deep learning algorithms to understand and respond to customer inquiries accurately and efficiently, improving customer satisfaction and reducing operational costs.

1.4.15.6 Market forecasting

Deep learning models can analyze historical financial data, market trends, and macroeconomic indicators to generate forecasts and predictions. These models can provide insights into stock prices, exchange rates, and other financial indicators, assisting investors and financial analysts in making informed decisions.

1.4.15.7 NLP for document analysis

Deep learning algorithms, combined with NLP techniques, can analyze and extract relevant information from financial documents such as annual reports, news articles, and regulatory filings. This helps financial institutions automate tasks like document classification, sentiment analysis, and summarization, improving efficiency and reducing manual effort.

These are just a few examples of how deep learning is transforming the financial services industry. By leveraging the power of advanced algorithms and data analytics, deep learning enables financial institutions to streamline processes, enhance risk management, and provide personalized services to their customers.

1.4.16 Hospitality

Deep learning has found several applications in the hospitality industry, enhancing guest experiences, optimizing operations, and enabling personalized services. Here are some key areas where deep learning is being used in hospitality.

1.4.16.1 Recommendation systems

Deep learning algorithms are used to develop recommendation systems that provide personalized recommendations to guests based on their preferences and behavior. These systems analyze historical data, such as past bookings, reviews, and guest feedback, to suggest tailored options for accommodations, dining, activities, and amenities, improving the guest experience.

1.4.16.2 Image and video analysis

Deep learning models can analyze images and videos to automatically identify and classify objects, scenes, and activities within a hospitality setting. This technology can be used for various purposes, such as monitoring guest areas for safety and security, analyzing crowd flow and occupancy patterns, and even recognizing guests for personalized greetings and services.

1.4.16.3 Chatbots and virtual assistants

Deep learning-powered chatbots and virtual assistants are being used in the hospitality industry to provide 24/7 customer support, handle guest inquiries, and assist with reservations, recommendations, and other guest services. These chatbots leverage NLP and deep learning algorithms to understand and respond to guest queries accurately and efficiently.

1.4.16.4 Demand forecasting and pricing optimization

Deep learning models can analyze historical booking data, market trends, and external factors to forecast demand and optimize pricing strategies. By accurately predicting demand patterns, hotels and resorts can optimize room rates, availability, and revenue management, maximizing profitability while meeting guest expectations.

1.4.16.5 Sentiment analysis

Deep learning techniques, such as NLP, can be applied to analyze guest reviews and social media posts related to hospitality establishments. Sentiment analysis helps hotels and resorts gain insights into guest experiences, identify areas for improvement, and respond to feedback in a timely manner, ultimately enhancing guest satisfaction and reputation management.

1.4.16.6 Voice recognition and voice-activated services

Deep learning algorithms can power voice recognition systems that enable guests to control various aspects of their stay using voice commands. These systems can be integrated into hotel rooms to control lighting, temperature, entertainment systems, and other amenities, providing a seamless and personalized guest experience.

1.4.16.7 Operational efficiency and resource optimization

Deep learning models can analyze operational data, such as occupancy rates, staff schedules, energy consumption, and maintenance records, to optimize resource allocation and improve operational efficiency. This can help hotels and resorts streamline processes, reduce costs, and deliver better services to guests.

Deep learning technologies offer great potential for the hospitality industry, enabling hotels, resorts, and other establishments to provide personalized experiences, optimize operations, and stay ahead in a competitive market. By leveraging the power of data and advanced analytics, deep learning is reshaping the way hospitality services are delivered.

1.4.17 Energy

Deep learning has significant applications in the energy sector, contributing to various aspects of energy generation, distribution, and consumption. Here are some key areas where deep learning is being used in the energy industry.

1.4.17.1 Energy demand forecasting

Deep learning models can analyze historical energy consumption data, weather patterns, economic indicators, and other factors to forecast energy demand accurately. These models can help utilities and grid operators optimize energy generation, storage, and distribution, ensuring a reliable and efficient energy supply.

1.4.17.2 Power grid optimization

Deep learning algorithms can optimize the operation of power grids by analyzing real-time data from sensors, meters, and other devices. These models can identify patterns, anomalies, and predict grid failures, enabling proactive maintenance, load balancing, and grid stability.

1.4.17.3 Renewable energy generation

Deep learning techniques are used to improve the efficiency and output of renewable energy generation systems. For example, deep learning models can analyze weather data, solar radiation levels, and wind patterns to optimize the performance of solar panels and wind turbines, enhancing renewable energy production.

1.4.17.4 Energy trading and pricing

Deep learning algorithms can analyze energy market data, historical prices, and other relevant factors to predict energy market trends and optimize trading strategies. These models can help energy traders and market participants make informed decisions regarding energy pricing, buying, and selling in competitive energy markets.

1.4.17.5 Energy asset management

Deep learning can assist in managing energy assets, such as power plants, by analyzing sensor data and equipment performance. These models can detect anomalies, predict failures, and optimize maintenance schedules, reducing downtime and improving asset efficiency.

1.4.17.6 Smart grid management

Deep learning algorithms can analyze data from smart meters, sensors, and IoT devices to manage and control energy distribution in smart grids. These models can optimize load distribution, identify energy losses, and enable demand response programs for efficient energy utilization.

1.4.17.7 Energy consumption optimization

Deep learning can be used to optimize energy consumption in buildings and industrial processes. By analyzing data from smart meters, sensors, and historical usage patterns, deep learning models can provide insights and recommendations for energy-efficient operations, reducing energy waste and costs.

1.4.17.8 Energy system planning

Deep learning models can assist in long-term energy system planning by analyzing data on energy resources, infrastructure, and demand projections. These models can help policymakers, energy companies, and utilities make informed decisions regarding the development of new energy projects, grid expansion, and resource allocation.

Deep learning technologies have the potential to revolutionize the energy sector by enabling more efficient energy generation, distribution, and consumption. By leveraging advanced analytics and data-driven insights, deep learning can contribute to a sustainable and intelligent energy ecosystem.

1.4.18 Social media

Deep learning plays a crucial role in social media by powering various applications and enhancing user experiences. Here are some key areas where deep learning is utilized in social media.

1.4.18.1 Image and video recognition

Deep learning models enable accurate image and video recognition, allowing social media platforms to automatically tag and categorize user-generated content. These models can identify objects, scenes, faces, and even sentiment within visual content, enhancing search capabilities and content discovery.

1.4.18.2 Content recommendation

Deep learning algorithms analyze user behavior, preferences, and interactions to provide personalized content recommendations on social media platforms. These models can suggest relevant posts, articles, videos, and ads to users, increasing engagement and user satisfaction.

1.4.18.3 Sentiment analysis

Deep learning techniques, such as NLP, enable sentiment analysis of social media posts, comments, and reviews. By analyzing text data, deep learning models can determine the sentiment behind user-generated content, helping companies understand customer opinions, monitor brand perception, and respond to user feedback effectively.

1.4.18.4 User profiling and targeted advertising

Deep learning algorithms can analyze user data, including demographics, interests, and online behaviors, to create detailed user profiles. These profiles enable social media platforms to deliver targeted advertisements, ensuring that ads are shown to the most relevant audience based on their preferences and behaviors.

1.4.18.5 Social network analysis

Deep learning models can analyze social network data, including connections, interactions, and patterns, to uncover valuable insights. These models can identify influential users, detect communities, predict user behavior, and understand the dynamics of social interactions on platforms.

1.4.18.6 Trend analysis and prediction

Deep learning algorithms can analyze large volumes of social media data to identify emerging trends, topics, and conversations. These models can help social media platforms and businesses understand user interests, anticipate market trends, and adapt their strategies accordingly.

1.4.18.7 Content moderation

Deep learning is employed in content moderation systems to automatically identify and filter out inappropriate or offensive content, such as hate speech, harassment, and explicit images. These models assist social media platforms in maintaining a safe and inclusive environment for users.

1.4.18.8 Real-time event detection

Deep learning algorithms can analyze real-time social media data to detect and track events as they unfold. These models can identify breaking news, natural disasters, public gatherings, and other significant events, allowing platforms to provide timely updates and deliver relevant content to users.

Deep learning continues to advance the capabilities of social media platforms, enabling them to deliver personalized content, understand user behavior, and maintain a safe and engaging environment. By leveraging the power of data and advanced analytics, deep learning enhances the overall user experience on social media.

1.4.19 Tertiary sector

The tertiary sector, also known as the service sector, encompasses a wide range of industries, including education, healthcare, finance, hospitality, transportation, and more. Deep learning has applications in various areas within the tertiary sector, providing opportunities for improved services, operational efficiency, and data-driven decision-making. Here are some examples of how deep learning is used in the tertiary sector.

1.4.19.1 Personalized education

Deep learning algorithms can be employed in educational settings to personalize learning experiences. These algorithms analyze student data, such as performance, learning styles, and preferences, to recommend tailored learning materials, adaptive assessments, and individualized instruction. This promotes personalized learning and helps students achieve better educational outcomes.

1.4.19.2 Healthcare diagnosis and treatment

Deep learning models are used in healthcare for medical image analysis, disease diagnosis, and treatment planning. These models can analyze medical images like X-rays, MRIs, and CT scans to detect abnormalities, assist in the diagnosis of diseases, and provide treatment recommendations. Deep learning can also aid in predicting patient outcomes and supporting precision medicine.

1.4.19.3 Financial services and risk assessment

Deep learning algorithms are used in the financial sector for risk assessment, fraud detection, and investment analysis. These models analyze large amounts of financial data, market trends, and customer behavior to predict risks, identify fraudulent activities, and generate investment insights. This helps financial institutions make informed decisions and optimize their services.

1.4.19.4 Customer service and chatbots

Deep learning-powered chatbots and virtual assistants are used in the service industry to enhance customer service experiences. These chatbots can understand natural language, answer customer queries, provide recommendations, and assist with transactions. Deep learning enables them to learn from interactions and improve over time, delivering more accurate and efficient customer support.

1.4.19.5 Transportation and logistics

Deep learning algorithms can optimize transportation and logistics operations. For example, they can be used for route optimization, demand forecasting, and predictive maintenance in transportation systems. Deep learning models can analyze historical data, traffic patterns, and other relevant factors to make real-time recommendations and improve operational efficiency.

1.4.19.6 Hospitality and customer experience

Deep learning is used in the hospitality sector to enhance guest experiences. For instance, deep learning models can analyze customer preferences, booking data, and reviews to provide personalized recommendations, improve service offerings, and optimize operations in hotels and resorts.

1.4.19.7 Data analytics and decision-making

Deep learning techniques enable data-driven decision-making in various tertiary sector industries. By analyzing large volumes of data, deep learning models can uncover insights, detect patterns, and make predictions. This helps organizations in planning, resource allocation, marketing strategies, and overall operational efficiency.

Deep learning has the potential to transform the tertiary sector by enabling personalized services, enhancing operational efficiency, and delivering improved customer experiences. By harnessing the power of data and advanced analytics, deep learning drives innovation and optimization across various industries within the tertiary sector.

1.4.20 Service sector

Deep learning has significant applications in the service sector, encompassing industries such as hospitality, healthcare, transportation, finance, customer support, and more. Here are some key areas where deep learning is being used in the service sector.

1.4.20.1 *Personalized customer experiences*

Deep learning algorithms enable businesses to provide personalized experiences to their customers. By analyzing customer data, preferences, and behavior, deep learning models can recommend tailored products, services, and content, enhancing customer satisfaction and loyalty.

1.4.20.2 *Chatbots and virtual assistants*

Deep learning-powered chatbots and virtual assistants are extensively used in customer support and service interactions. These bots can understand natural language, answer queries, assist with transactions, and provide personalized recommendations, improving customer service and reducing response times.

1.4.20.3 *Sentiment analysis*

Deep learning techniques, such as NLP, enable sentiment analysis of customer feedback, reviews, and social media posts. This helps businesses gauge customer satisfaction, identify potential issues, and take proactive measures to improve service quality.

1.4.20.4 *Fraud detection and security*

Deep learning algorithms can be employed in the service sector to detect fraudulent activities and enhance security. Whether it's identifying fraudulent transactions, detecting cybersecurity threats, or monitoring suspicious user behavior, deep learning models provide advanced capabilities to safeguard businesses and their customers.

1.4.20.5 *Service optimization and resource allocation*

Deep learning models can analyze data on service demand, resource availability, and operational constraints to optimize service delivery. For instance, in transportation, deep learning can optimize routes and schedules for improved efficiency. In healthcare, it can assist with patient flow management and resource allocation.

1.4.20.6 *Predictive maintenance*

Deep learning is used in the service sector to predict and optimize maintenance activities. By analyzing sensor data, equipment performance, and historical patterns, deep learning models can detect anomalies, predict failures, and schedule maintenance proactively, minimizing downtime and improving operational efficiency.

1.4.20.7 Recommendation systems

Deep learning algorithms can be employed to develop recommendation systems that suggest relevant services, products, or content to customers. This helps in cross-selling, upselling, and enhancing the overall customer experience by providing personalized recommendations based on their preferences and behavior.

1.4.20.8 Data analytics and insights

Deep learning techniques enable businesses in the service sector to analyze large volumes of data and extract valuable insights. By uncovering patterns, trends, and correlations, deep learning models help businesses make data-driven decisions, optimize operations, and identify opportunities for improvement.

Deep learning plays a vital role in transforming the service sector by enabling personalized experiences, enhancing operational efficiency, and delivering improved customer service. By leveraging the power of advanced analytics and data-driven insights, businesses in the service sector can gain a competitive edge and provide exceptional services to their customers.

1.4.21 Secondary sector

The secondary sector, also known as the industrial sector, includes industries involved in manufacturing, construction, and other industrial activities. Deep learning has several applications in the secondary sector, contributing to process optimization, quality control, predictive maintenance, and more. Here are some key areas where deep learning is utilized in the secondary sector.

1.4.21.1 Quality control

Deep learning algorithms can be used for quality control in manufacturing processes. By analyzing sensor data, images, and other production data, deep learning models can identify defects, anomalies, and deviations from quality standards. This helps in ensuring product quality and reducing waste.

1.4.21.2 Predictive maintenance

Deep learning models can analyze sensor data from machinery and equipment to predict maintenance needs. By identifying patterns and anomalies, these models can detect potential equipment failures and recommend proactive maintenance, minimizing downtime and optimizing production efficiency.

1.4.21.3 Supply chain optimization

Deep learning algorithms can optimize supply chain processes, such as inventory management, demand forecasting, and logistics. By analyzing historical data, market trends, and customer behavior, deep learning models can improve production planning, reduce inventory costs, and enhance overall supply chain efficiency.

1.4.21.4 Process optimization

Deep learning techniques can be used to optimize various industrial processes. By analyzing process data and historical patterns, deep learning models can identify areas for improvement, optimize parameters, and increase production efficiency. This leads to cost savings, higher productivity, and better resource utilization.

1.4.21.5 Computer vision and object recognition

Deep learning algorithms enable computer vision capabilities in the secondary sector. These models can analyze images and videos to identify objects, monitor production lines, and ensure adherence to safety protocols. Computer vision can also be used for automated product inspection and quality control.

1.4.21.6 Energy efficiency

Deep learning can contribute to energy efficiency in the secondary sector. By analyzing energy consumption data, operational patterns, and environmental factors, deep learning models can identify opportunities for energy optimization and recommend strategies to reduce energy waste and costs.

1.4.21.7 Robotics and automation

Deep learning algorithms are employed in robotic systems to enable automation and enhance industrial processes. These models can analyze sensor data, control robotic movements, and enable robots to perform complex tasks with precision and adaptability.

1.4.21.8 Fault detection and diagnosis

Deep learning techniques can be used for fault detection and diagnosis in industrial systems. By analyzing sensor data and operational parameters, deep learning models can detect abnormalities, pinpoint the root cause of faults, and facilitate quick troubleshooting and repairs.

Deep learning offers tremendous potential for the secondary sector, driving advancements in manufacturing, construction, and industrial processes. By leveraging advanced analytics and data-driven insights, deep learning enhances process efficiency, quality control, and predictive maintenance, leading to increased productivity and competitiveness in the secondary sector.

1.4.22 Telecommunication

Deep learning plays a crucial role in the telecommunications industry, enabling improved network performance, efficient resource management, customer experience enhancement, and more. Here are some key areas where deep learning is utilized in telecommunications.

1.4.22.1 Network optimization

Deep learning algorithms can analyze large volumes of network data, including network logs, performance metrics, and user behavior, to optimize network performance. These models can identify network bottlenecks, predict network congestion, and optimize network resources for better reliability and quality of service.

1.4.22.2 Predictive maintenance

Deep learning models can analyze data from network equipment and sensors to predict and prevent network failures. By detecting patterns and anomalies in the data, these models can anticipate equipment malfunctions, proactively schedule maintenance, and minimize downtime.

1.4.22.3 Fraud detection and security

Deep learning algorithms are used in telecommunications to detect fraudulent activities and enhance network security. These models can analyze call records, network traffic, and user behavior to identify suspicious patterns, detect fraudulent calls or transactions, and mitigate security threats.

1.4.22.4 Customer experience management

Deep learning techniques enable telecommunications companies to analyze customer data, including call records, service usage patterns, and feedback, to improve customer experience. By understanding customer preferences and behavior, deep learning models can personalize service offerings, provide targeted recommendations, and enhance customer satisfaction.

1.4.22.5 Network planning and design

Deep learning algorithms can assist in network planning and design by analyzing historical data, traffic patterns, and geographic information. These models can optimize network deployment, determine optimal cell tower locations, and predict network capacity requirements, ensuring efficient network coverage and capacity management.

1.4.22.6 NLP and customer support

Deep learning-powered NLP models are used in telecommunications for customer support and service interactions. These models can understand and respond to customer queries, provide automated assistance, and enable more efficient and personalized customer support experiences.

1.4.22.7 Network traffic management

Deep learning algorithms can analyze network traffic patterns and predict future traffic demands. This information helps telecommunication providers optimize network resources, allocate bandwidth effectively, and ensure smooth network operations, particularly during peak usage periods.

1.4.22.8 Service recommendations and marketing

Deep learning models can analyze customer behavior, usage patterns, and preferences to provide personalized service recommendations and targeted marketing campaigns. By understanding individual customer needs, deep learning enables telecommunications companies to offer tailored services, promote relevant products, and increase customer engagement.

Deep learning continues to advance the telecommunications industry by enabling intelligent network management, personalized services, and enhanced customer experiences. By leveraging the power of data analytics and advanced algorithms, telecommunications companies can optimize their networks, improve security, and deliver innovative services to meet the evolving needs of their customers.

1.4.23 Petroleum

Deep learning has various applications in the petroleum industry, helping to improve exploration and production processes, enhance operational efficiency, and optimize decision-making. Here are some key areas where deep learning is utilized in the petroleum sector.

1.4.23.1 Seismic interpretation

Deep learning algorithms can analyze seismic data, which provides information about subsurface structures and potential hydrocarbon reservoirs. Deep learning models can enhance the accuracy of seismic interpretation by automatically detecting patterns, identifying anomalies, and generating detailed geological models, assisting in exploration and reservoir characterization.

1.4.23.2 Reservoir characterization

Deep learning techniques can be applied to reservoir characterization tasks, such as rock property prediction, faces classification, and fluid identification. By analyzing well logs, core data, and production history, deep learning models can improve the understanding of reservoir properties and aid in optimizing production strategies.

1.4.23.3 Production optimization

Deep learning algorithms can optimize production processes by analyzing production data, well performance, and operational parameters. These models can predict production rates, identify production anomalies, and optimize well placement and completion techniques, ultimately improving production efficiency and maximizing hydrocarbon recovery.

1.4.23.4 Equipment health monitoring

Deep learning models can be used for equipment health monitoring, such as predicting equipment failures, detecting anomalies, and optimizing maintenance schedules. By analyzing sensor data, operational parameters, and historical patterns, these models can provide early warnings, reduce downtime, and enhance the reliability of critical equipment in the petroleum industry.

1.4.23.5 Drilling operations

Deep learning techniques can improve drilling operations by analyzing real-time drilling data, including drilling parameters, geotechnical data, and wellbore stability information. Deep learning models can provide insights for optimizing drilling efficiency, detecting drilling hazards, and reducing non-productive time.

1.4.23.6 *Data analytics and decision-making*

Deep learning enables data-driven decision-making by analyzing vast amounts of data in the petroleum industry. These models can uncover hidden patterns, correlations, and trends, helping in reservoir management, production forecasting, and investment decision-making.

1.4.23.7 *Asset integrity and safety*

Deep learning algorithms can be employed for asset integrity management and safety monitoring in the petroleum sector. These models can analyze sensor data, inspection reports, and maintenance records to detect anomalies, predict potential risks, and ensure the safety and reliability of assets and infrastructure.

1.4.23.8 *Environmental impact assessment*

Deep learning techniques can assist in environmental impact assessment by analyzing data related to emissions, environmental monitoring, and regulatory compliance. These models can help in predicting and mitigating environmental risks associated with petroleum operations.

It holds significant potential to revolutionize the petroleum industry by improving exploration outcomes, enhancing production efficiency, and optimizing operational processes. By leveraging advanced analytics and ML algorithms, deep learning enables petroleum companies to make informed decisions, reduce costs, and maximize the value of their assets.

1.5 LIMITATIONS AND CHALLENGES

Deep learning has gained significant attention and success in various domains, including industrial applications. However, it is important to recognize that deep learning also faces limitations and challenges when applied to industrial settings. Some of these limitations and challenges include.

Data requirements: Deep learning models typically require a large amount of labeled data to achieve high performance. Obtaining sufficient and high-quality labeled data in industrial environments can be challenging, as it often requires manual annotation, which can be time-consuming and expensive. Additionally, some industrial processes may have limited data available, making it difficult to train deep learning models effectively.

1.5.1 Generalization

Deep learning models are known for their ability to learn complex patterns and make accurate predictions. However, they can struggle with generalizing well to new, unseen data. In industrial applications, where conditions and environments can vary significantly, the ability of deep learning models to generalize across different scenarios can be limited.

1.5.2 Interpretability

Deep learning models are often considered black boxes, meaning it can be challenging to understand how they arrive at their predictions. This lack of interpretability can be problematic in industrial applications, where stakeholders may require explanations and insights into the decision-making process. For safety-critical applications, interpretability becomes even more crucial for ensuring transparency and trust.

1.5.3 Computational resources

Deep learning models, especially deep neural networks, are computationally intensive and require significant computational resources for training and inference. Industrial settings may not always have access to the necessary computing infrastructure, which can hinder the adoption and deployment of deep learning models in these environments.

1.5.4 Real-time constraints

Some industrial applications require real-time or near real-time decision-making. Deep learning models, particularly complex ones, may not be able to meet these strict time constraints. In such cases, model simplification or alternative approaches may be required to ensure timely and efficient decision-making.

1.5.5 Robustness and reliability

Deep learning models can be sensitive to adversarial attacks or noisy input data. In industrial environments, where there may be environmental disturbances, sensor noise, or deliberate attacks, the robustness and reliability of deep learning models can be a concern. Ensuring that the models are resilient to such challenges becomes essential for their practical deployment.

1.5.6 Ethical and legal considerations

Industrial applications often involve sensitive data, and decisions made by deep learning models can have significant implications. Ensuring fairness, avoiding biases, and addressing legal and ethical concerns are crucial considerations when applying deep learning in industrial contexts.

1.5.7 Integration with existing systems

Many industrial processes and systems are already established and may have legacy technologies in place. Integrating deep learning models seamlessly with existing systems can be a complex task, requiring careful planning and coordination to ensure compatibility, reliability, and scalability.

1.5.8 Continuous learning and adaptation

Industrial environments are dynamic and subject to changes over time. Deep learning models may struggle to adapt and continuously learn from evolving data streams. Re-training and updating models in real time can be challenging, and methods for continual learning need to be developed to address this limitation.

1.5.9 Cost considerations

Implementing deep learning solutions in an industrial setting may involve significant costs related to data acquisition, infrastructure, expertise, and maintenance. These costs need to be carefully evaluated and justified against the expected benefits and outcomes.

Addressing these limitations and challenges requires interdisciplinary efforts, including collaborations between domain experts, data scientists, and ML researchers. Advancements in research and the development of novel algorithms and techniques can help overcome these obstacles and enable the successful application of deep learning in industrial contexts.

1.6 CONCLUSION

In conclusion, this chapter aims to provide a comprehensive understanding of the fundamentals of deep learning for industrial applications. Through an extensive review of existing literature and practical examples, we have highlighted the key components and principles that underpin the successful implementation of deep learning in industrial settings. We began by discussing the fundamental concepts of deep learning, including neural networks, activation functions, and backpropagation. These concepts serve as the building blocks for training deep learning models and understanding their inner workings. We explored different types of neural networks, such as convolutional neural networks (CNNs) for image processing and recurrent neural networks (RNNs) for sequential data analysis, and highlighted their relevance in industrial applications. Moreover, we examined the importance of data preprocessing and augmentation techniques in deep learning. Proper preprocessing ensures the quality and suitability of data for training deep learning models.

Techniques like data normalization, dimensionality reduction, and feature engineering play a crucial role in enhancing model performance and generalization. Furthermore, we addressed the challenges associated with training deep learning models in industrial settings, including the need for large labeled datasets, computational resources, and model interpretability. We discussed various strategies to mitigate these challenges, such as transfer learning, semi-supervised learning, and model compression techniques. Furthermore, we highlighted several real-world industrial applications where deep learning has demonstrated its effectiveness. These include predictive maintenance, quality control, anomaly detection, NLP, and robotics. We discussed how deep learning models have revolutionized these domains, enabling more accurate predictions, efficient operations, and enhanced decision-making. In conclusion, deep learning has emerged as a powerful tool in industrial applications, with the potential to revolutionize various sectors. However, successful implementation requires a deep understanding of its fundamentals, including neural networks, data preprocessing, model training, evaluation, and validation. By embracing these principles and addressing the associated challenges, industries can unlock the full potential of deep learning and pave the way for a more intelligent and automated future.

REFERENCES

[1] Molnar, C., Bennett, S., Berendt, B., Burrell, J., Friedman, B., & Wallach, H. (2020). Interpretable Machine Learning: Definitions, Methods, and Applications. arXiv preprint arXiv:2009.11669.

[2] Wang, X., Zhang, S., Zhang, Y., & Yan, R. (2020). Deep learning-based fault diagnosis for rotating machinery. *Journal of Intelligent Manufacturing*, 31(1), 197–206.

[3] Zhang, Y., Zhao, L., Zhang, J., & Yu, Y. (2020). Machine learning for supply chain management: A review. *Journal of Cleaner Production*, 256, 120464.

[4] Xu, Z., Chen, W., Zhou, L., & Li, Y. (2021). Deep learning-based remaining useful life prediction for equipment maintenance in a manufacturing plant. *Journal of Manufacturing Systems*, 59, 294–303. https://doi.org/10.1016/j.jmsy.2020.12.013

[5] Li, Q., Jiang, L., & Huang, X. (2021). A deep learning-based approach for early diagnosis of Alzheimer's disease using MRI images. *BMC Medical Imaging*, 21(1), 1–9. https://doi.org/10.1186/s12880-021-00603-1

[6] Zheng, X., Li, J., & Zhou, Y. (2021). Portfolio optimization with deep reinforcement learning. *Applied Soft Computing*, 102, 107002. https://doi.org/10.1016/j.asoc.2021.107002

[7] Chen, Y., Liu, W., Su, X., Gao, Y., & Zheng, Y. (2021). A novel deep learning-based speech recognition system for noisy industrial environments. *Applied Sciences*, 11(6), 2722. https://doi.org/10.3390/app11062722

[8] Tang, Y., Xu, X., Liu, Y., Wang, L., & Wang, B. (2022). Anomaly detection for IoT-enabled manufacturing systems based on deep learning. *IEEE Access*, 10, 23298–23309.

[9] Xu, J., Liu, J., Chen, Y., & Wang, J. (2022). Edge computing and machine learning-based predictive maintenance system for manufacturing. *IEEE Access*, 10, 25813–25821.
[10] Sethi, R., Bhushan, B., Sharma, N., Kumar, R., & Kaushik, I. (2020). Applicability of industrial IoT in diversified sectors: Evolution, applications and challenges. In: Kumar, R., Sharma, R., & Pattnaik, P. K. (eds) *Studies in Big Data Multimedia Technologies in the Internet of Things Environment*, pp. 45–67. Springer, Singapore. https://doi.org/10.1007/978-981-15-7965-3_4
[11] Kakandwar, S., Bhushan, B., & Kumar, A. (2023). Integrated machine learning techniques for preserving privacy in internet of things (IOT) systems. In: Bharat, B., Sudhir, K. S., Muzafer, S., & Azedine, B. (eds) *Blockchain Technology Solutions for the Security of IoT-Based Healthcare Systems*, pp. 45–75. Academic Press, London, UK. https://doi.org/10.1016/b978-0-323-99199-5.00012-4
[12] Kumar, A., Bhushan, B., Shristi, S., Chaganti, R., & Soufiene, B. O. (2023). Blockchain-based decentralized management of IOT devices for preserving data integrity. In: Bharat, B., Sudhir, K. S., Muzafer, S., & Azedine, B. (eds) *Blockchain Technology Solutions for the Security of IoT-Based Healthcare Systems*, pp. 263–286. Academic Press, London, UK. https://doi.org/10.1016/b978-0-323-99199-5.00009-4
[13] Han, J., Pei, J., & Tong, H. (2022). *Data Mining: Concepts and Techniques.* Morgan Kaufmann, Burlington, MA.
[14] Verma, A., Bhattacharya, P., Madhani, N., Trivedi, C., Bhushan, B., Tanwar, S., Sharma, G., Bokoro, P. N., & Sharma, R. (2022). Blockchain for industry 5.0: Vision, opportunities, key enablers, and future directions. *IEEE Access*, 10, 69160–69199. https://doi.org/10.1109/access.2022.3186892
[15] Pathak, L. K., Mahato, S., & Sweta, S. (2022). Technological dimension of a smart city. *Smart City Infrastructure: The Blockchain Perspective*, 13, 247–268.
[16] "Role of AI in Industrial 4.0" https://www.virtualpaper.pro/index.php?route=mag/article&tracking=5ff2be89b6d0a&a_id=594 (Accessed on Feb 14, 2024).
[17] Chen, T., Sampath, V., May, M. C., Shan, S., Jorg, O. J., Aguilar Martín, J. J., Stamer, F., Fantoni, G., Tosello, G., Calaon, M. (2023). Machine learning in manufacturing towards industry 4.0: From 'for now' to 'four-know'. *Applied Sciences*, 13, 1903. https://doi.org/10.3390/app13031903
[18] "Machine Learning for Industrial Applications: A Comprehensive Literature Review" https://www.sciencedirect.com/science/article/abs/pii/S095741742100261X (Accessed on Feb 19, 2023).
[19] Tercan, H., & Meisen, T. (2022). Machine learning and deep learning based predictive quality in manufacturing: A systematic review. *Journal of Intelligent Manufacturing*, 33(7), 1879–1905.
[20] Mahato, S., Pathak, L. K., Sweta, S., & Choubey, D. K. (2024). Wearable smart technologies: Changing the future of healthcare. In: Pranav, P., Patel, A., & Jain, S. (eds) *Machine Learning in Healthcare and Security*, pp. 130–148. CRC Press, Boca Raton, FL, 2024.

Chapter 2

Application of machine learning/deep learning in healthcare

Rabab Anjum, Mohit Kumar, Zeeshan Rafi, Kunal Nagarkoti, and Nimish Srivastava

2.1 INTRODUCTION

Artificial intelligence (AI) can be defined as the capability of a machine to solve different problems using its self-intelligence. It refers to the simulation of human intelligence so that machines could be programmed to perceive their environment, reasoning, learning, and making decisions like humans. AI is powered by various techniques and methodologies, including machine learning, deep learning, natural language processing, computer vision, and robotics. These technologies enable AI systems to acquire knowledge, adapt to new information, and improve their performance over time.

AI and similar technologies are becoming more and more common in business and society, and they are starting to be used in healthcare. Many facets of patient care might be changed by this technology, as well as internal administrative procedures at payer, provider, and pharmaceutical organisations.

Numerous studies have already shown that AI is capable of doing important healthcare jobs including illness diagnosis as well as or better than humans. Today, algorithms already surpass radiologists in identifying cancerous tumours and advising researchers on how to create cohorts for expensive clinical trials. However, it has been predicted long time before that AI completely replaces humans in large medical process areas for a variety of reasons [1].

In recent times, AI has found applications in several fields including agriculture, finance, transportation, education, customer service and natural language processing, environmental conservation, entertainment and most importantly in healthcare by enhancing the procedure and improving outcomes. In health care, AI has already been used widely in health equipment that can collect healthcare data effectively to be used to improve health outcomes in patients by disease prediction, drug discovery, diagnosis, and personalised treatment [2].

The widely used term "Artificial Intelligence" can be further categorised into several subfields, such as machine learning, deep learning, cognitive

 DOI: 10.1201/9781003534761-2

computing, and computer vision. Machine learning (ML) is however an important part of AI that provides machines the ability to learn without human involvement [3]. ML is the application of AI, which can create algorithms that generate predictions and analyse data [4].

There are three types of ML algorithms.

2.1.1 Supervised

Supervised learning is a method in which we provide labelled training data that has one or more inputs and labelled outputs. The ML model uses the labelled given set of data and access the data during training and provides the prediction of a new set of data. Usually, supervised learning models focus on methods for classification and regression. Several models have been discovered based on function separating the data into multiple categorical classes in classification. The input set of data is categorised under labels given in the input set of labelled data. In regression, model or function is developed for distinguishing the data into continuous real values rather than using categorical classes which helps to identify adequate data depending on the historical data [3,5]. Supervised learning can be explained with the help of Figure 2.1 which shows the learning model using input data and providing supervised output data.

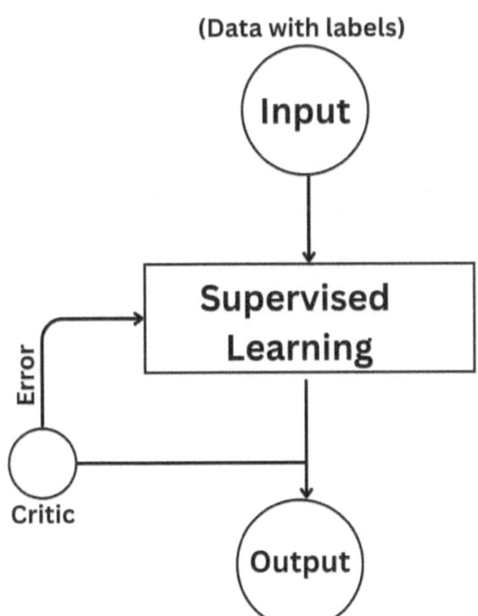

Figure 2.1 Learning model of supervised learning.

2.1.2 Unsupervised

An unsupervised ML algorithm has no labelled data set as shown in Figure 2.2, and these algorithms discover or find hidden patterns in data without human intervention. It mainly consists of three tasks: clustering, association, and dimensionality reduction. In clustering, algorithm groups similar data experiences together based on similarities like age, location, and similar parameters. Whereas in association, the algorithm looks for a relationship between variables. In dimensional reduction, the algorithm reduces the number of variables in the data. This technique is also used in the pre-processing stage of data to have quality data.

The most common and known algorithm for unsupervised learning is K-mean clustering. It has many variations, such as K-medians and K-medoids, but the main principle of the algorithm is to find the nearest mean for all the assumed clusters from the data. Then it assigns the current

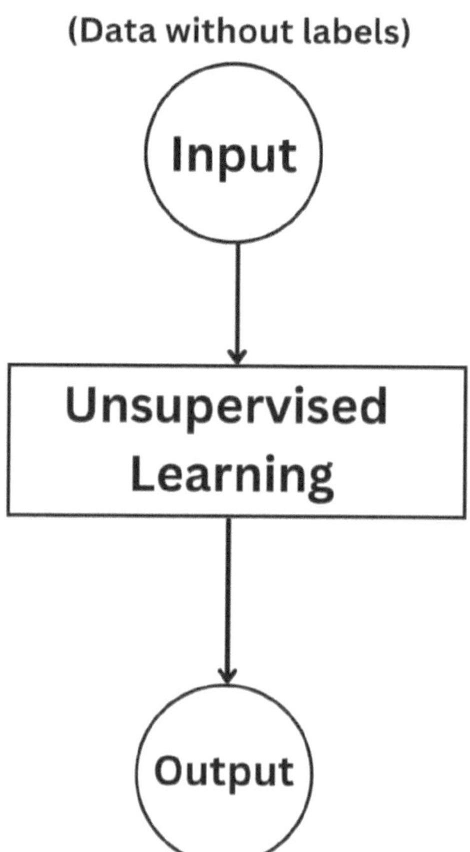

Figure 2.2 Learning model of unsupervised learning.

data point to a cluster, recalculates the mean for the cluster, and updates it for the following data point. There is one drawback it initialises with an expected number of means, and proper selection of the number of clusters can result in better clustering and more accuracy [3].

2.1.3 Reinforcement learning

Another type of ML is reinforcement learning, wherein initially, labelled data is used and further, it tries to predict from the unlabelled data. The main feature of reinforcement learning is to find optimal and prestige design-making strategies and learning by the feedback given to the prediction in the form of rewards and punishments for the prediction. It tries to improve its ability and make the model more accurate [6]. Figure 2.3 shows the learning model of reinforcement learning.

2.2 ARTIFICIAL NEURAL NETWORK AND DEEP LEARNING

An advanced and widely used model is artificial neural network (ANN) which simulates the human brain's working of processing information by identifying the patterns, and relations in the data [1,7]. As shown in Figure 2.4, the block diagram of ANN is inspired and simulates the neural

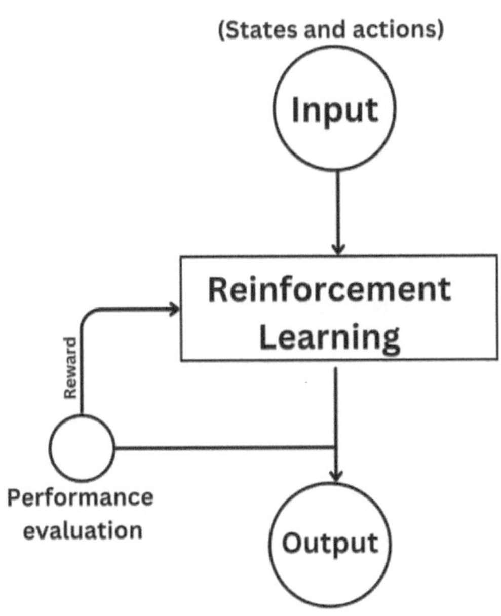

Figure 2.3 Learning model of reinforcement learning.

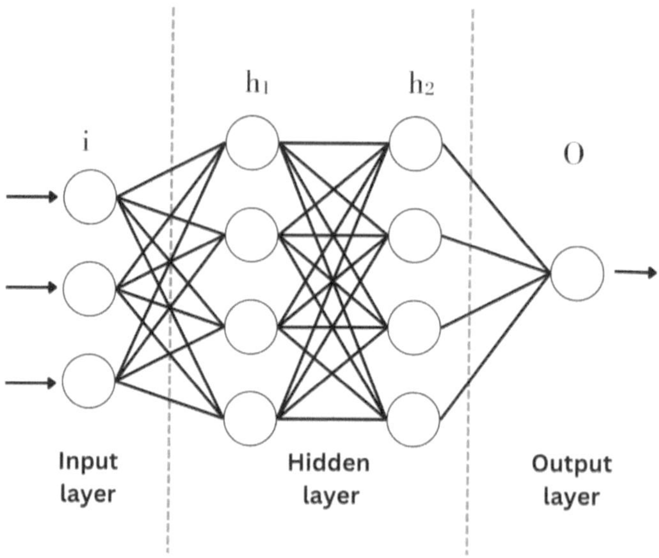

Figure 2.4 Working model of artificial neural network.

network of the human brain and processes the data and gives output. ANN are weighted directed graphs that are massive computing systems that work parallelly. These are typically inspired by biological networks, and the cell in this network is called a neuron, which is used to process information. In ANN, the neuron output and input are connected with the help of edges and nodes indicated as artificial neurons [7,8].

Equation represents the basic working of ANN.

X1.......... *Xn*: input
W1.......... *Wn*: Weights
b: bias

$$z = ((X1 * W1) + (X2 * W2) + \ldots + (Xn * Wn)) + b$$

$$Z = \left(\sum_{i=1}^{n} Xi * Wi \right) + b \tag{2.1}$$

Another important type of ML is deep learning which is an advanced phase of ML and mimics the neural network in which the human brain works; it consists of three or more layers of neural network. It uses deep processing layers connected with weights [2].

2.3 APPLICATION OF AI/ML IN HEALTHCARE

Artificial intelligence and machine learning (AI/ML) applications in healthcare are many and are transforming the sector. A few significant uses as depicted in Figure 2.5 also are explained below.

i. **Medical Imaging Analysis:** AI/ML algorithms can accurately identify and classify a wide range of illnesses and ailments by analysing medical pictures such as X-rays, CT scans, and MRI scans.
ii. **Diagnosis and Disease Prediction:** AI/ML can evaluate patient data, including genetic data and electronic health records (EHRs), to help diagnose diseases and forecast patient outcomes, enabling physicians to make better choices.
iii. **Personalised Treatment Plans:** AI/ML algorithms may suggest individualised treatment strategies depending on the particular traits of a patient, such as genetics, medical history, and lifestyle elements.
iv. **Discovery of New Drugs and Drug Development:** AI/ML are used to forecast outcomes, discover viable medication candidates, and optimise drug development procedures.

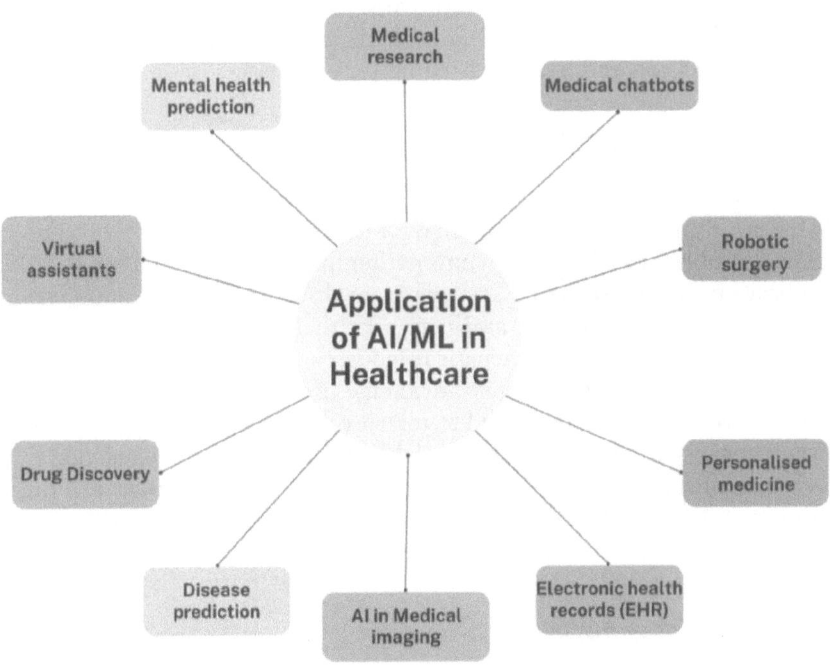

Figure 2.5 Several applications of AI/ML in healthcare.

v. **Virtual Health Assistants:** AI-driven virtual health assistants may provide patients with individualised health advice, respond to questions, make appointments, and help with the management of chronic diseases.
vi. **Wearable Technology and Remote Monitoring:** AI/ML can evaluate data from wearable technology and remote monitoring systems to track patient health in real time, allowing for earlier illness identification and improved health care.
vii. **Natural Language Processing (NLP):** NLP algorithms assist in structuring unstructured healthcare data—such as notes from physicians and patient records—for additional analysis and decision-making.
viii. **Clinical Decision Support:** By offering evidence-based advice and treatment guidelines, AI/ML systems may help healthcare practitioners make better judgements.
ix. **Fraud Detection and Security:** AI/ML algorithms are used to identify fraud and safeguard patient data.
x. **Administrative Automation:** AI/ML automates administrative processes including billing, claims processing, and appointment scheduling, which lowers administrative strain and boosts productivity.
xi. **Population Health Management:** To support public health activities, AI/ML may evaluate population-level data to find patterns, risk factors, and possible disease outbreaks.
xii. **Surgical Robotics:** Powered by AI/ML, surgical robots improve surgical precision and lessen procedure invasiveness.

2.3.1 Drug discovery

As data digitisation in the pharmaceutical sector is increasing AI typically ML algorithms could be used to analyse the medical data to solve complex clinical problems and enhance automation in the files healthcare. Another application of AI/ML-based systems in healthcare involves using various advanced tools, algorithms, and networks that can mimic human intelligence and help in the pharmaceutical industry.

However, the most important advantage of AI/ML is in drug discovery wherein ML algorithms are used to interpret and learn from the input data from previous experiments on wet labs and make independent decisions for accomplishing specific objectives like finding the target molecule, target validation, lead identification, etc. AI/ML can also be used to minimise wet lab efforts and automate the drug discovery procedure.

2.3.1.1 AI in drug discovery

Recent discoveries in the field of healthcare show the large availability of adequate data about target and lead compounds from the experiments performed in web laboratories. However, due to the intricate procedures involved in the process of drug discovery, it is a time-consuming process and expensive task. Therefore, AI/ML algorithms could be used to use that data to predict the target, lead compound, and optimise the drug. ML algorithms also help in the 3D structure prediction of a target protein and the outcome of drug-protein prediction and determining drug activity and many similar applications [9].

AI/ML algorithms can also be frequently used for drug screening, like predicting toxic compounds and side effects of lead compounds. It can save many resources and time simultaneously with the help of permutation and combinations done by the ML model.

Despite its several advantages, there are many significant data challenges. One of them is the uncertainty of the data. The previously available data by pharmaceutical companies for drug development needs to be more sufficient and accurate, and the compound behaves differently in different people and conditions. There are many attributes/affecting factors for a single possibility, and traditional ML tools might not be able to analyse this type of data.

2.3.1.2 AI yields new antibiotic

AI-generated computer models can be used to discover many antibiotic compounds. An AI model can quickly screen and simulate more than billions of data and pick out potential lead compounds or antibiotics that kill bacteria. At the Massachusetts Institute of Technology (MIT), researchers identified a drug called Halicin, a c-Jun N-terminal protein kinase (JNK) inhibitor that kills many strains of bacteria. In their research, they found that this drug is capable of killing the world's most problematic infectious disease-causing bacteria and it also prevents the development of antibodies resistance in *E. coli* [10].

2.3.1.3 AI in drug repurposing (repositioning)

Drug repurposing is most commonly used practice in recent times which involves using drugs already approved, passed through clinical trials, and approved by regulatory agencies to be used for therapeutic use. It allows the development of faster and more cost-efficient drugs. With the help of AI, this process can be easily enhanced and accelerated to multiply its applications in healthcare. In this era of big data, we can utilise the medicine network data and drug target interaction data and find the most effective molecule with the help of computational power. Later, this process could be utilised to discover relevant compounds in a wet laboratory [10].

2.3.2 Disease prediction

AI and ML have increased productivity of drugs as well as decreased medical errors. Nowadays, MRI, CT, and X-ray images are used by doctors along with AI-assisted analysis, which is used to diagnose and predict diseases. In Radiology and Cardiology, AI uses image-based disease diagnosis, and due to financial issues and accuracy, it is applied in ophthalmology with the help of a Deep learning algorithm. ML classification has excellent potential in predicting chronic diseases including fundus bleeding, microaneurysms, and neovascularisation that could be identified with AI-assisted medical diagnosis as well as the prediction of Diabetic Retinopathy [11].

ML uses past data for disease prediction, so that it could be effectively used to track patient health. So, by using this, doctors could make accurate decisions, and good treatment could be applied to the patient. The K-nearest neighbour protocols could be used in the health care system to predict diseases and target labels by defining the nearest neighbour class where Euclidean distances are used. If $K = 1$, the assigned case is to the category of its nearest neighbour. The help of a decision tree, where a large collection of data is divided into small sets, could be used to calculate the probability of a given record to each category and classify the record most likely in class or category [12,13].

ML is also used in predicting human heart diseases. In auscultation techniques, the production of sound could be distinguished by physicians between normal and abnormal cardiac sounds. Even though this method of predicting heart disease has many drawbacks, in spite, several ML methods have been used nowadays to detect heart diseases. Several algorithms, such as support vector machine, Vote, and K-nearest neighbour could be also used to detect cardiac diseases [14].

2.3.3 AI in medical imaging

The application of ML protocols has been increasing exponentially in the past few years, particularly in healthcare. Some relevant applications currently used include AI image recognition technique wherein the computer uses AI and ML algorithms to identify objects, places, and humans). Many medical imaging tools like CT scans, X-rays, MRIs & ultrasounds use those images and extract features from those images and assist in identifying the patterns and helping to make predictions in the future [39,40].

The main advancements are creating a deep learning model consisting of various layers that better evaluate the input image, produce more accurate disease prediction, and enhance patient care. In deep learning, a medical image is given as input, a deep feature is extracted from the picture by a pre-trained convolutional neural network (CNN), and then different layers analyse the data, and then the model predicts the disease (AI applications to medical images: From ML to deep learning).

2.3.4 Electronic health records

EHR could manage as well as store data related to patient health. It includes patient health history like tests, diagnosis, medicine, allergies, immunisations, personalised medical care, treatment plans, and medical quality and safety improvement. Data from EHR could be used in AI-based new drug development. AI could also accurately classify diseases and reclassify them according to the individuals with a quick analysis of images and medical data. It could also provide new required services such as improved medical quality and real-time monitoring. This could provide personalised or customised healthcare services as it could also affect future diseases [15].

2.3.5 Personalised medicine

Personalised medicine practice can be done in the development of drugs as well as health devices. The first step involves diagnosis or finding out the risk of developing disease; the second step involves pathophysiologic processes identification; In the third step, the appropriate intervention identification is made, which was earlier identified in the previous two steps; fourth step involves intervention testing on the individual going through diagnosis and pathobiology assessment; fifth step involves warehousing result in suitable databases so as patient information could be used in other patients assessment to find a meaningful pattern [16].

Currently, there is no sufficient infrastructure, and AI is resource-intensive for collecting, annotating, and integrating data to treat patients. An alternative to personalised medicine is provided by phenotypic personalised medicine. Its objective is to choose the best combination of drugs and doses with building relationships between stimuli and responses. It forms a feedback system control platform that includes neural networks for multi-signal combinations with their effects through their learning, leading to a relation between doses and treatment efficacy. The CURATE. AI finds the relationship between stimuli (input) and responses (output), described by quadratic polynomial equations [17].

2.3.6 Virtual assistants

Initially, the chatbot could provide tasks such as reminding or scheduling to assist patients. With the help of the Alleviate chatbot, patients with health challenges could get personalised treatment and care to assist clinicians in understanding their patients more accurately [18].

Smart healthcare was introduced by IBM in 2009. Smart healthcare is a health service system that uses sensors to get and transform information through the Internet of Things, which consists of participants like doctors, patients, hospitals, and research institutions. It has several dimensions, such as diagnosis as well as treatment of diseases, health decision formation, management of hospitals, medical research, and virtual assistance. Virtual assistants communicate with the help of speech recognition and respond according to the patient. Virtual assistants have a role in making a connection between doctors and patients. Virtual assistants can save doctors time by replying based on the patient's basic information. It could also help in treatment as well as improve mental health [19].

2.3.7 Mental health prediction

Nowadays, ML techniques are used for the diagnosis of mental health diseases including depression, anxiety, schizophrenia, post-traumatic stress, bipolar disorders, and children's mental health problems.

It has been found in the study that to predict schizophrenia with the help of the random forest method accuracy of 73.7% was achieved, which is relatively less accurate than others. According to Jo et al., the paper uses ML and network analysis to recognise 48 patients and 24 healthy controls with the help of data based on probability on the brain by tractography and after using ML. It was found that the highest accuracy of 68.6% was achieved by the random forest method [17].

For Depression and Anxiety, according to the paper of Ahmed et al. [20], a model is created using physiological assessment and ML algorithms, including a support vector machine, a convolutional neural network (CNN) commonly referred to as a Convolution network, linear discriminant analysis as well as K-nearest neighbour. CNN, which is nearly 96% accurate for anxiety while 96.8% accurate for depression; support vector, which is nearly 95% accurate for anxiety while 95.8% accurate for depression; linear discriminant analysis which is nearly 93% accurate for anxiety while 87.9% accurate for depression and K-nearest neighbour which is nearly 70.96% accurate for anxiety while 81.82% accurate for depression. These algorithms are used to categorise based on strength related to anxiety and depression, consisting of two data sets [20].

ML could also be used for predicting bipolar disorders as stated in a research paper by Rocha-Rego et al., who have applied ML algorithms to find out that the K-nearest neighbours algorithm mean *F*1 score of about 0.67 and support vector machine, the mean *F*1 score is about 0.72, while with the random forest algorithm, the mean *F*1 score is about 0.73 could be used to predict the bipolar diseases in two different populations [21].

ML is also used in estimating Post-traumatic stress disorder (PTSD); according to a paper by Reece et al., they used random forest ML algorithm to determine PTSD among Twitter users where more than 243,000 posed were analysed, and it was found to be AUC score of 0.89. According to a research paper [7], ML algorithm among military forces of the UK and collected 13,690 subjects from 2004 to 2009. ML algorithms such as random forest, which have the highest accuracy of 97% in prediction, while artificial neural networks have the least accuracy of 89% [22]. ML algorithms could be used in estimating mental health problems among children where several features, as well as attributes, are considered for the categorising process [23]. The accuracy of different ML algorithms in predicting mental health disorders is shown in Figure 2.6 [23].

Some major applications of AI in health care can be categorised in several domains and have been listed in Table 2.1.

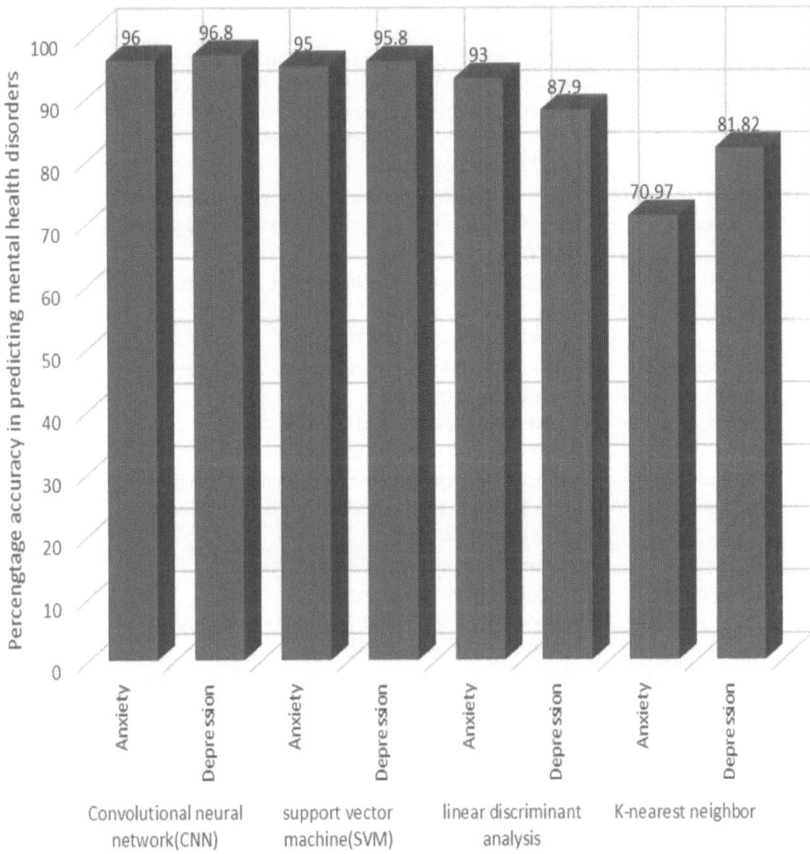

Figure 2.6 **Comparative analysis of accuracy of different machine learning algorithms in predicting mental health disorders.**

2.4 FUTURE ASPECTS OF AI/ML IN HEALTHCARE

In the near future, AI will have a great impact on extensive healthcare options. It can be extensively used in the development of precision medicine, so as to further improve healthcare options. There is a high possibility that AI will eventually be used in several fields of healthcare including radiology and pathology pictures to be further reviewed by a computer given the rapid advancements in AI for imaging analysis. In future, AI can also be used for recognising speech and text for purposes like patient communication and clinical note transcription etc.

The primary difficulty facing AI in various healthcare fields involves ensuring that the technologies are used in routine clinical practice, not determining if they will be capable enough to be beneficial. In order for

Table 2.1 Several applications of artificial intelligence in healthcare

S. No.	*Field*	*Role of AI*	*Reference*
1	Personalised medicine	CURATE.AI could be used to find the association between stimuli (input) and responses (output), which are expressed by quadratic polynomial equations and selecting the optimum combination of drugs.	[5]
2	Virtual assistants	With the help of sensors, smart healthcare could get information and transformation through the Internet.	[3]
3	Mental health prediction	With the help of clustering which is used to categorise the group of data which could be used in mental health prediction.	[24,25]
4	Electronic health records (EHR)	EHR was used for early detection of cancer such as pancreatic cancer. By analysing data from EHR, new drug development could take place.	[26,27]
5	AI in medical imaging	With the help of AI in medical imaging, more accurate disease prediction can be achieved and patient care can be enhanced.	[28]
6	Disease prediction	Currently for predicting disease AI uses image-based disease diagnosis. ML algorithms such as the K-nearest algorithm are used for disease diagnosis.	[11,12]
7	Drug discovery	With the help of ML algorithm, the 3D Structure of protein could be easily predicted which could help in prediction as well as determining the activity of Drugs.	[9]
8	Surgery	AI could enhance the surgery by more precision with high efficiency in the desired task which could ultimately enhance the decision-making.	[29]
9	Maternal care	AI could also enhance the maternal care by first collecting data through sensors. Then Fog layer could mediate in between IoT devices and the cloud to enhance response time. Data collected could be stored in the cloud and an emergency alert could be generated.	[30]
10	Hospital Robots	Recently, robotic nursing has been used more frequently. During the COVID-19 pandemic, they show importance as they are not vulnerable to viruses.	[31]
11	SmartScope	It is a device to detect Cardiopulmonary Diseases with the help of AI. It could have an accuracy of more than 90% and could avoid other procedures such as CT and MRI.	[32]
12	Clinical trial	Clinical Trail Matching (CTM) predicts degree of eligibility. CMT uses NLP (natural language processing) which provides us 96% accuracy.	[33]

AI systems to be widely adopted, they need to be endorsed by regulatory bodies, integrated with EHR systems, sufficiently standardised so that similar products function similarly, taught to clinicians, paid for by public or private payer organisations, and improved over time in the field. These difficulties will eventually be resolved, but it will take considerably longer than it will for the technology to advance. We thus anticipate modest AI usage in clinical practice in 5 years and more widespread use within 10 years.

Nowadays, with the advent of IoT real-time monitoring gives rise to smart health care. Due to increase of older adults is seen and as maximum old adults prefers to stay home, so smart home system is also produced. The smart health care could be implemented using devices like smartphones, but the main problem using smart healthcare is different sensor used for integrating data. Therefore, in future, various techniques related to data fusion so as to get integrated information from devices which would result in low bandwidth and streamlined signal. In future, blockchain could be used to avoid security issues as secure framework and protocol.

2.4.1 Nanorobots

These are small robotic devices or machines whose components are nearly in nanometer. It is an emerging technology that could help in precision medicine including surgery, delivery, diagnosis as well as in medical imaging [34].

2.4.2 Brain–computer interfaces for diagnosis and treatment

Brain–computer interfaces (BCI) is a technology that receives the signal and after receiving it analyses that signal and in the end transfers the signal to real world from brain. BCI uses brain signal for controlling external device which helps disabled people to communicate [35].

2.4.3 Human-AI symbiosis in medical research

There is increase of Human-AI symbiosis which shows division of work between humans and machines. There are human and AI which could help in the betterment of various research fields with the help of brute force for retrieving as well as analysing large amount of data and making complex decisions [36].

2.4.4 AI-enhanced telepathology

With the help of AI, the diagnosis could be done faster which would result in fast and accurate prediction. It could also help in the prediction of disease for a large number of patients with more accuracy with that pathologists could work from distance. Currently, the cost could be very high for incorporating AI in Pathology [37].

2.4.5 Healthcare "blockchain" with AI verification

Medical crucial information could be verified using blockchain technology. Image data could be verified more accurately. This technology guarantees the security of crucial information of healthcare like medical transaction and other important data. The security of block is very different from cryptography and the individual block stores information [19,38].

2.5 CONCLUSION

AI is a great tool that includes both ML and deep learning algorithms. The applications of AI are wide and are gaining much popularity in health care. Several applications of ML have been discussed in this chapter which include drug discovery, disease prediction with the help of image-based disease diagnosis, medical imaging, EHRs which include storing patient health history, to provide personalised medicine. These applications further improve the patient health outcomes by providing personalised treatment as well as personalised medicines. Nowadays, different application-based systems as well as tools are used such as CURATE. AI and virtual assistants could be provided by Alleviate chatbot and smart healthcare. Moreover, AI tools particularly ML algorithms have proved beneficial in the prediction of mental health disorders such as anxiety, depression, schizophrenia, bipolar disorders, PTSD as well as the prediction of children's mental health problems. It can help in the timely formulation of treatment and health plans to overcome these problems.

REFERENCES

[1] Davenport, T., & Kalakota, R. (2019). The potential for artificial intelligence in healthcare. *Future Healthcare Journal*, 6(2), 94–98. https://doi.org/10.7861/futurehosp.6-2-94

[2] Panesar, A. (2020). *Machine Learning and AI for Healthcare: Big Data for Improved Health Outcomes*, pp. 1–368. Apress, Berkeley, CA. https://doi.org/10.1007/978-1-4842-6537-6

[3] Toh, C., Brody, J. P., Toh, C., & Brody, J. P. (2021). Applications of machine learning in healthcare. *Smart Manufacturing - When Artificial Intelligence Meets the Internet of Things*, 65, 92297. https://doi.org/10.5772/INTECHOPEN.92297

[4] Singh, A. M., & Haju, W. B. (2022). Artificial intelligence. *International Journal for Research in Applied Science and Engineering Technology*, 10(7), 1210–1220. https://doi.org/10.22214/IJRASET.2022.44306

[5] Blasiak, A., Khong, J., & Kee, T. (2020). CURATE. AI: Optimizing personalized medicine with artificial intelligence. *SLAS Technology*, 25(2), 95–105. https://doi.org/10.1177/2472630319890316

[6] Erickson, B. J., Korfiatis, P., Akkus, Z., & Kline, T. L. (2017). Machine learning for medical imaging. *Radiographics*, 37(2), 505–515. https://doi.org/10.1148/rg.2017160130

[7] Agatonovic-Kustrin, S., & Beresford, R. (2000). Basic concepts of artificial neural network (ANN) modeling and its application in pharmaceutical research. *Journal of Pharmaceutical and Biomedical Analysis*, 22(5), 717–727. https://doi.org/10.1016/S0731-7085(99)00272-1

[8] Jain, A. K., Jianchang, M., & Mohiuddin, K. M. (1996). Artificial neural networks: A tutorial. *Computer*, 29(3), 31–44. https://doi.org/10.1109/2.485891

[9] Paul, D., Sanap, G., Shenoy, S., Kalyane, D., Kalia, K., & Tekade, R. K. (2021). Artificial intelligence in drug discovery and development. *Drug Discovery Today*, 26(1), 80–93. https://doi.org/10.1016/J.DRUDIS.2020.10.010

[10] Farghali, H., Canová, N. K., & Arora, M. (2021). The potential applications of artificial intelligence in drug discovery and development. *Physiological Research*, 70(Suppl4), 715–722. https://doi.org/10.33549/PHYSIOLRES.934765

[11] Liu, C., Jiao, D., & Liu, Z. (2020). Artificial intelligence (AI)-aided disease prediction. *BIO Integration*, 1(3), 130–136. https://doi.org/10.15212/BIOI-2020-0017

[12] Mishra, N., Kaitholia, K., Srivastava, B., Shah, N. K., Narayan, J. P., Dev, V., Phookan, S., Anvikar, A. R., Rana, R., Bharti, R. S., Sonal, G. S., Dhariwal, A. C., & Valecha, N. (2014). Declining efficacy of artesunate plus sulfadoxine-pyrimethamine in northeastern India. *Malaria Journal*, 13(1), 284.

[13] Patil, J., & Varsha Pimprale, M. (n.d.). Disease classification and prediction based on symptoms using machine learning. *European Chemical Bulletin*, 12(4), 9.

[14] Saboor, A., Usman, M., Ali, S., Samad, A., Abrar, M. F., & Ullah, N. (2022). A method for improving prediction of human heart disease using machine learning algorithms. *Mobile Information Systems*, 2022, 1–9. https://doi.org/10.1155/2022/1410169

[15] Lee, S., & Kim, H. S. (2021). Prospect of artificial intelligence based on electronic medical record. *Journal of Lipid and Atherosclerosis*, 10(3), 282–290. https://doi.org/10.12997/JLA.2021.10.3.282

[16] Stephanopoulos, N., Tong, G. J., Hsiao, S. C., & Francis, M. B. (2010). Dual-surface modified virus capsids for targeted delivery of photodynamic agents to cancer cells. *ACS Nano*, 4(10), 6014–6020. https://doi.org/10.1021/NN1014769

[17] Jo, Y. T., Joo, S. W., Shon, S. H., Kim, H., Kim, Y., & Lee, J. (2020). Diagnosing schizophrenia with network analysis and a machine learning method. *International Journal of Methods in Psychiatric Research*, 29(1), 1818. https://doi.org/10.1002/MPR.1818

[18] Roy, K., Khandelwal, V., Goswami, R., Dolbir, N., & Malekar, J. (2023). Demo alleviate: Demonstrating artificial intelligence enabled virtual assistance for telehealth: The mental health case. *Proceedings of the AAAI Conference on Artificial Intelligence*, 37(13), 16479–16481. www.aaai.org

[19] Tian, S., Yang, W., Grange, J. M. L., Wang, P., Huang, W., & Ye, Z. (2019). Smart healthcare: Making medical care more intelligent. *Global Health Journal*, 3(3), 62–65. https://doi.org/10.1016/J.GLOHJ.2019.07.001

[20] Ahmed, A., Sultana, R., Ullas, M. T. R., Begom, M., Rahi, M. M. I., & Alam, M. A. (2020). A machine learning approach to detect depression and anxiety using supervised learning. *2020 IEEE Asia-Pacific Conference on Computer Science and Data Engineering (CSDE), Gold Coast, Australia*, pp. 1–6. https://doi.org/10.1109/CSDE50874.2020.9411642

[21] Rocha-Rego, V., Jogia, J., Marquand, A. F., Mourao-Miranda, J., Simmons, A., & Frangou, S. (2014). Examination of the predictive value of structural magnetic resonance scans in bipolar disorder: A pattern classification approach. *Psychological Medicine*, 44(3), 519–532.

[22] Reece, A. G., Reagan, A. J., Lix, K. L. M., Dodds, P. S., Danforth, C. M., & Langer, E. J. (2017). Forecasting the onset and course of mental illness with Twitter data. *Scientific Reports*, 7(1), 1–11. https://doi.org/10.1038/s41598-017-12961-9

[23] Chung, J., & Teo, J. (2022). Mental health prediction using machine learning: Taxonomy, applications, and challenges. *Applied Computational Intelligence and Soft Computing*, 2022, 1–19. https://doi.org/10.1155/2022/9970363

[24] Srivastava, A., Nagpal, B. N., Saxena, R., & Subbarao, S. K. (2001). Predictive habitat modelling for forest malaria vector species An. dirus in India – A GIS-based approach. *Current Science*, 80(9), 1129–1134.

[25] Mahato, S., Paul, S., Goyal, N., Mohanty, S. N., & Jain, S. (2023). 3EDANFIS: Three channel EEG-based depression detection technique with hybrid adaptive neuro fuzzy inference system. *Recent Patents on Engineering*, 17(6), 32–48.

[26] Kenner, B. J., Abrams, N. D., Chari, S. T., Field, B. F., Goldberg, A. E., Hoos, W. A., Klimstra, D. S., Rothschild, L. J., Srivastava, S., Young, M. R., & Go, V. L. W. (2021). Early detection of pancreatic cancer: Applying artificial intelligence to electronic health records. *Pancreas*, 50(7), 916–922. https://doi.org/10.1097/MPA.0000000000001882

[27] Hu, Z. S., Hung, F. Y., Chang, S. J., & Chen, K. J. (2012). The crystallization characteristics and photoluminescence properties of ZnO/Ag nanoflower arrays. *Technical Digest -25th International Vacuum Nanoelectronics Conference, IVNC 2012, Jeju, Korea*, pp. 176–180. https://doi.org/10.1109/IVNC.2012.6316893

[28] Shin, W.-K., Cho, J., Kannan, A. G., Lee, Y.-S., & Kim, D.-W. (2016). Cross-linked composite gel polymer electrolyte using mesoporous methacrylate-functionalized SiO2 nanoparticles for lithium-ion polymer batteries. *Scientific Reports*, 6(1), 26332. https://doi.org/10.1038/srep26332

[29] Zhou, X. Y., Guo, Y., Shen, M., & Yang, G. Z. (2020). Application of artificial intelligence in surgery. *Frontiers of Medicine*, 14(4), 417–430. https://doi.org/10.1007/S11684-020-0770-0

[30] Gulzar Ahmad, S., Iqbal, T., Javaid, A., Ullah Munir, E., Kirn, N., Ullah Jan, S., & Ramzan, N. (2022). Sensing and artificial intelligent maternal-infant health care systems: A review. *Sensors*, 22(12), 4362. https://doi.org/10.3390/s22124362

[31] Kyrarini, M., Lygerakis, F., Rajavenkatanarayanan, A., Sevastopoulos, C., Nambiappan, H. R., Chaitanya, K. K., Babu, A. R., Mathew, J., & Makedon, F. (2021). A survey of robots in healthcare. *Technologies*, 9(1), 8.

[32] Prabu, A. (2021). *SmartScope: An AI-Powered Digital Auscultation Device to Detect Cardiopulmonary Diseases.* TechRxiv. Preprint. https://doi.org/10.36227/TECHRXIV.14921268.V1

[33] Alexander, M., Solomon, B., Ball, D. L., Sheerin, M., Dankwa-Mullan, I., Preininger, A. M., Jackson, G. P., & Herath, D. M. (2020). Evaluation of an artificial intelligence clinical trial matching system in Australian lung cancer patients. *JAMIA Open*, 3(2), 209–215. https://doi.org/10.1093/jamiaopen/ooaa002

[34] Soto, F., Wang, J., Ahmed, R., & Demirci, U. (2020). Medical micro/nanorobots in precision medicine. *Advanced Science*, 7(21), 2002203. https://doi.org/10.1002/advs.202002203

[35] Mudgal, S. K., Sharma, S. K., Chaturvedi, J., & Sharma, A. (2020). Brain computer interface advancement in neurosciences: Applications and issues. *Interdisciplinary Neurosurgery*, 20, 100694. https://doi.org/10.1016/J.INAT.2020.100694

[36] Jarrahi, M. H. (2018). Artificial intelligence and the future of work: Human-AI symbiosis in organizational decision making. *Business Horizons*, 61(4), 577–586. https://doi.org/10.1016/j.bushor.2018.03.007

[37] Zarella, M. D., McClintock, D. S., Batra, H., Gullapalli, R. R., Valante, M., Tan, V. O., Dayal, S., Oh, K. S., Lara, H., Garcia, C. A., & Abels, E. (2023). Artificial intelligence and digital pathology: Clinical promise and deployment considerations. *Journal of Medical Imaging (Bellingham, Wash.)*, 10(5), 051802. https://doi.org/10.1117/1.JMI.10.5.051802

[38] Kim, S. K., & Huh, J. H. (2020). Artificial neural network blockchain techniques for healthcare system: Focusing on the personal health records. *Electronics (Switzerland)*, 9(5), 763. https://doi.org/10.3390/electronics9050763

[39] Andrew, J., Deva Priya, I., Sagayam, K. M., Bharat, B., Yuichi, S., & Jennifer, E. (2023). Blockchain for healthcare systems: Architecture, security challenges, trends and future directions. *Journal of Network and Computer Applications*, 215, 103633. https://doi.org/10.1016/j.jnca.2023.103633.

[40] Khadidos, A. O., Alyoubi, K. H., Mahato, S., Khadidos, A. O. and Mohanty, S. N.(2023). Machine learning and electroencephalogram signal based diagnosis of depression. *Neuroscience letters*, 809, 137313. https://doi.org/10.1016/j.neulet.2023.137313

Chapter 3

Implementing machine learning technology for smart agriculture based on data collection systems and predictive analytics use cases

Raviduth Ramful and Yogesh Beeharry

3.1 INTRODUCTION

The agriculture sector worldwide is at a turning point in time with the pressing issue of climate change and loss in biodiversity [1,2]. Despite living in an era of global challenges ranging from scarcity of resources and surge in resistant pests and diseases, the urging need to produce more to meet the growing demand of an increasing worldwide population is further straining agricultural production practice [3]. As specified by the second Sustainable Development Goals (SDG 2) of the United Nations, more efforts should be centered towards ending hunger, achieving food security while maintaining its nutritional value and promoting sustainable agricultural practice [4]. The realization of this goal through large-scale agricultural production is however impacted by climate change. The expansion of agricultural land leads to deforestation and contributes to Green House Gas (GHG) emissions while adversely affecting biodiversity [2,5,6]. To minimize the extensive impacts of conventional agricultural practice, new approach of growing more with less resources are being developed such as vertical farming in a controlled environment [7,8]. Such alternative agricultural practice will also address SDG 13 about climate change notably by limiting the impact of urbanization and land conversion on the environment and on its biodiversity [9].

3.1.1 Smart agriculture

The inclusion of smart agriculture practice, which enables automatic production, optimal control, intelligent management, and logistic control, can notably improve the use of agricultural resources while minimizing the adverse impacts on the ecological environment [10–12]. Moreover, the need to generate and collect data in smart agriculture has been substantiated by numerous factors. The operational models of modern farms have been drastically changed given the data-based opportunities which are radically different from previous generations [13]. Smart agricultural practice has been reported to be

DOI: 10.1201/9781003534761-3

data intensive in contrast to traditional farming methods which were known to be both labor and capital-intensive [13]. To improve agricultural production and operational efficiency, agricultural data can be captured by the use of appropriate agricultural sensors and cloud computing for real-time monitoring followed by suitable analysis techniques such as data mining and other relevant technologies [10,11,14,15]. The low quality of collected data has often been reported to be one of the main shortcomings in modern agricultural practice, for instance in the accurate estimation of fertilizer application rates [16,17].

3.1.1.1 Improved agriculture practice

One popular method that can significantly boost crop yield in modern farming practice is via the use of fertilizers [18,19]. Fertilizers have long been used in conventional farming practice to boost crop yield. Nitrogen-based fertilizer has significantly increased the crop yield and overall productivity of farming operations since the beginning of the 20th century. On the downside, intensive agricultural practice has led to a gradual degradation on the resilience and fertility of soil [20]. Precision agriculture is indispensable to maximize the use of agricultural resources such as fertilizers for maximum productivity and reduced waste in modern farming practice [16,19]. Fertilizers, which generally originate from either natural or synthetic materials, have to be applied in controlled amounts to soil or to plant tissues for optimum nutrient supply [20]. Additionally, another way to maintain good harvest and prevent yield loss in smart agricultural practice is via the prevention and control of pests and diseases [21,22].

3.1.2 Data collection in modern agriculture practice

In sustainable agricultural practice, namely for the precise application of fertilizers, real-time monitoring via adequate sensor systems and network technologies is intrinsically required in order to acquire diverse and comprehensive agricultural data ranging from soil nutrients to climatic variations [16,18,19,23]. The advent of Agriculture 4.0 has revolutionized modern agriculture practice with the surge in utilization of sensors to capture a wide variety of plant and soil information. Conventional rain gauges and soil moisture probes have been modified for enhanced connectivity to provide real-time data of cultivated lands [24]. The collected data on yield for intensive cropping systems such as rice and maize for a wide range of landscape positions, irrigation method, and cropping intensities, is further promoting investment in agricultural research and development [25].

3.1.2.1 Application of smart sensors

With the emergence of data-driven agriculture, notable improvement has been observed in the development of smart sensors for remote data collection [26]. Smart sensors, which incorporate digital technology, are enabling better monitoring and visualization of crops and farming processes respectively [27,28]. The implementation of smart sensors in modern agricultural practice was found to provide numerous benefits, namely in the precise control and application of resources, quality monitoring, and enhanced productivity. Sensors in precision agriculture vary in terms of their principle of operation and mode of application [19,29]. The most commonly used sensors in agriculture are the position sensors, optical sensors, electrochemical sensors, mechanical sensors, and dielectric sensors which provide vital information such as precise positioning, plant color data, soil nutrient levels, soil compactness, and soil moisture level, respectively [30–34].

In specific agricultural operations such as in hydroponics or soil less farming concept, whereby it is essential to monitor the level of nutrients in plants, the EC (Electrical conductivity) sensor, pH sensor, light intensity sensor, and humidity sensor were found to greatly alleviate the laborious task of manually monitoring the plant physiological parameters [30,35]. Other sensors such as thermostats for instance are used in nursery farming to control the temperature of seedling mat whereby the optimal control of temperature is required in the rooting area so as to maximize the chance of germination. More advanced applications of sensors for instance in the field of plant growth monitoring have made the use of 3D and multi-spectral imaging to detect plant height, the projected leaf area, leaf area index, and digital biomass among other essential plant growth parameters [30,35].

3.1.3 Remote sensing technology in agriculture

Another widely used technique to gather data in modern agriculture is via remote sensing technology [36,37]. Remote sensing technology, which consists of the acquisition of information from distance via remote sensors and airborne platforms, has been widely applied in the field of agriculture to estimate and predict corresponding crop and soil indices [38]. From a review study, remote sensing was found to be a necessity in modern farming processes owing to its numerous benefits such as collecting data about crop yield, weed control, assessing nutritional requirements in plants and in soil as well as determining the water requirement for optimum growth [36]. In one study, remote sensing data from image sensing instruments including satellite-derived images have been extracted and further analyzed to predict soil organic carbon in agricultural soils by considering the soil-adjusted vegetable and green ratio vegetation indices [38]. In the irrigation process for instance, the use of remote sensing in recent years was found to be an indispensable tool in various locations around the world. The synoptic coverage generated by remote sensing for irrigated fields subjected to a variety of environmental conditions enabled the assessment of vegetation growth, maturity, and crop yield [39,40]. Moreover, remote sensing technology was also found to be useful in secondary aspects of agricultural operations such as weed control. Weed control, which is indirectly linked to plant physiological parameters, does have an impact on the overall crop yield and necessitates proper monitoring. The common remote sensing methods used to detect weed infestation are via the ground, aircraft, and satellite-based platforms which make use of specific imagery techniques, namely the plant canopy spectral response [41].

3.1.4 UAV technology in agriculture

One customarily used airborne platform for remote sensing technology is via unmanned aerial vehicles (UAVs). The implementation of such technology is proving to be a game-changer in modern agricultural practice and livestock farming. The use of UAV in agriculture can provide numerous benefits to farmers ranging from improved crop health monitoring to precise application of inputs [42–48]. UAVs can be fitted and customized with high-resolution cameras and sensors to capture detailed image of crops enabling the detection of plant diseases and growth deficiencies. Addressing plant deficiencies such as lack of nutrients or pest infestations from an early stage would minimize crop loss and improve overall crop health as farmers are able to take swift action. Moreover, highly advanced and configured UAV platforms with high payload capacity could additionally be used in the precise dispensing of plant inputs such as fertilizers and pesticides for weeds control.

3.1.5 IoT in agriculture

New research findings have shown that the implementation of modern technologies is indispensable to assist farmers to achieve sustainable agriculture by improving crop yield, reduce waste and make better management decision [49]. To optimize the use of resources, improve crop yield and minimize waste, a selection of digital technologies, namely in precision farming, Internet of Things (IoT) automation in agro industry and UAVs-based imagery, have been proposed to assist the farming communities by generating real-time farm data [30]. IoT for instance, has been identified as a potential solution to address the present challenges faced in agriculture, such as climate change and biodiversity crisis [20]. The implementation of IoT in conjunction with remote sensing technologies and smart sensors was found to yield satisfactory results in various data-driven agricultural concepts ranging from data analytics platforms to aerial imaging [30]. Moreover, the implementation of agricultural technologies such as IoT and remote sensing has led to a surge in big data in modern farming processes [50,51].

3.1.6 Datasets in agriculture

With this surge in datasets, new data analytics platforms are being explored to extract valuable information which will support the decision-making process in farming operations [52]. Additionally, data-based analytics were reported to have the potential to optimize the agricultural production systems, value chains, and food systems [53]. From a recent review study, the types of data-analytics objectives were mainly found to be focused on increasing production, controlling the environmental field and reducing consumption of resources [54]. In fertilizer application for instance, real-world agricultural dataset such as climate, soil nutrients, crop yield and fertilization records were found to be crucial for accurate estimation of fertilizer application rates by considering machine learning techniques such as the multi-task learning (MTL) model [16].

3.2 REVIEW OF AI TECHNIQUES USED IN AGRICULTURE

The advent of artificial intelligence (AI) technology such as machine learning and computer vision are revolutionizing the way data are processed in agriculture given their high flexibility, high performance, accuracy, and cost-effectiveness [55–60]. Other specific types of machine learning technology such as the Deep Learning (DL) technique have been employed to forecast yields in wheat and barley cereal crops which were subjected to nitrogen treatment for increased yield variability [61]. For accurate prediction of forecasted yields, the model training, which is a prerequisite stage, was conducted by considering a wide variety of datasets, namely yield data from the harvester, in-season weather data from sowing to growth stage, Normalized Difference Vegetation Index (NDVI) and the Normalized Difference Red-Edge (NDRE) [61]. Furthermore, Few-Shot Learning (FSL), which is another type of Machine Learning framework, has been successfully applied to the classification of agricultural image datasets for pest recognition and plant disease identification in smart agriculture [21]. One of the limitations of such a data-analytics approach in the field of image classification in smart agriculture is the quality of collected data and data labeling [21].

With the rapid evolution of technology and fast adoption of the IoT as well as the Internet of Flying Things (IoFT) in several sectors, the agricultural sector has not been left behind. In addition to the sensorial data collection mechanisms, AI-based techniques are being incorporated for the predictive analytics aspect and to enable stakeholders make informed decisions. Depending on the predictive problem being considered, the categorization can be made in terms of supervised learning and unsupervised learning. Supervised learning essentially deals with the training of a model on datasets comprising input features and corresponding outputs. It can be further categorized into classification and regression problems. Unsupervised learning deals with the analysis and clustering of unlabeled datasets and helps in discovering hidden patterns or data groupings in the corresponding datasets.

3.2.1 Supervised learning

The supervised learning approach trains models from labeled datasets for achieving the desired predictive outputs. The two main types of problems in which supervised learning can be categorized are Classification and Regression.

3.2.1.1 Classification problem

The classification problem comprises datasets with feature variables and a finite integer number of categories to which each of the feature sets can be mapped. The minimum number of categories is two whereby a binary classification problem is being considered. Problems consisting of more than two classes are referred to as multiclassification ones. Some common examples of classification algorithms are support Vector Machine (SVM), Decision Trees, K-Nearest Neighbor (K-NN), and Random Forest.

3.2.1.1.1 Support Vector Machine

In Support Vector Machines, the dataset is represented in an n-dimensional space, where n represents the number of features. The classification is then performed by finding the optimal hyper-plane that differentiates the different classes. An example of a random dataset comprising two features (feature x and feature y) and labels with two different categories (Label 1 and Label 2). It can be observed from Figure 3.1 that the hyper-plane separates the data points representing the two different labels.

3.2.1.1.2 Decision Trees

A decision tree is a tree-structured classifier comprising the following:

Internal nodes that represent the features of a dataset;
Branches which represent the decision rules; and
Leaf nodes which represent the different outcomes.

A general graphical structure of a decision tree is shown in Figure 3.2.

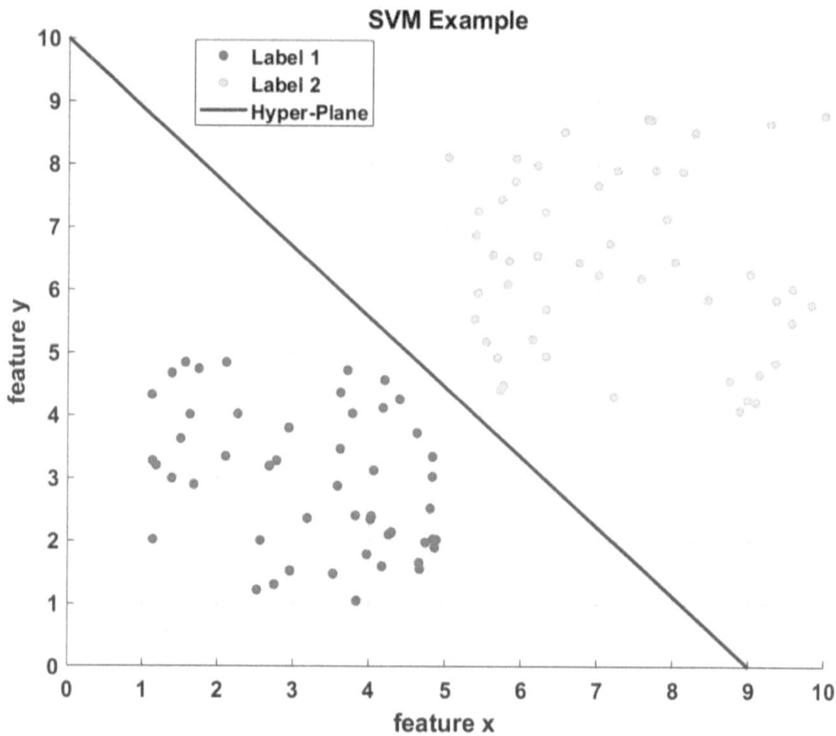

Figure 3.1 Example for support vector machine (SVM).

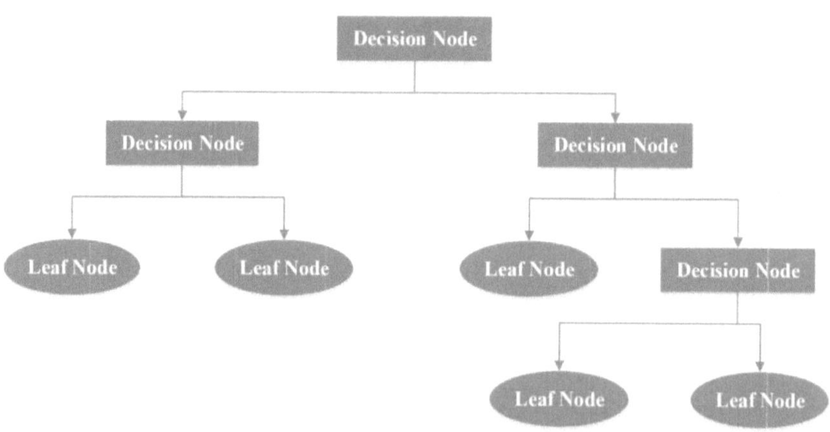

Figure 3.2 General graphical structure of a decision tree.

Considering the previous sample dataset and applying the decision tree algorithm to it yields the tree as shown in Figure 3.3. Clearly from Figure 3.3, it can be observed that the tree comprises:

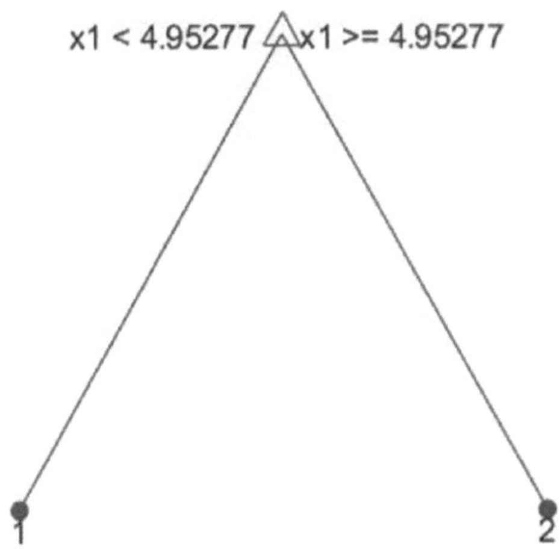

Figure 3.3 Decision tree for the sample dataset.

One decision node
Two branches and
Two leaf nodes pertaining to the classes in the dataset.

The split at the decision node is performed on the first feature such that if its value is less than 4.95277, the association is made to label 1, otherwise, the association is made to label 2.

Figure 3.4 demonstrates the classification of the new data point using the above decision tree algorithm. It can be observed that the data point has been assigned to label 2.

3.2.1.1.3 K-Nearest Neighbors (K-NN)

K-NN is one of the simplest supervised machine learning algorithms which essentially parses the existing dataset to spot the pre-defined N-neighbors based on a distance measurement and thereby categorizing the new input data. Consider the same sample dataset used in Section 3.2.1.1.1. Figure 3.5 shows the plot for the corresponding labels based on the feature values. The two categories in the dataset are represented in red and green colors. A new data point is included (blue cross) and the K-NN algorithm would be used to categorize it into either of the classes.

The algorithm works as per the following:

- The number of neighbors (K) is selected to be 5.
- The Euclidean distances from all data points in the dataset are computed.

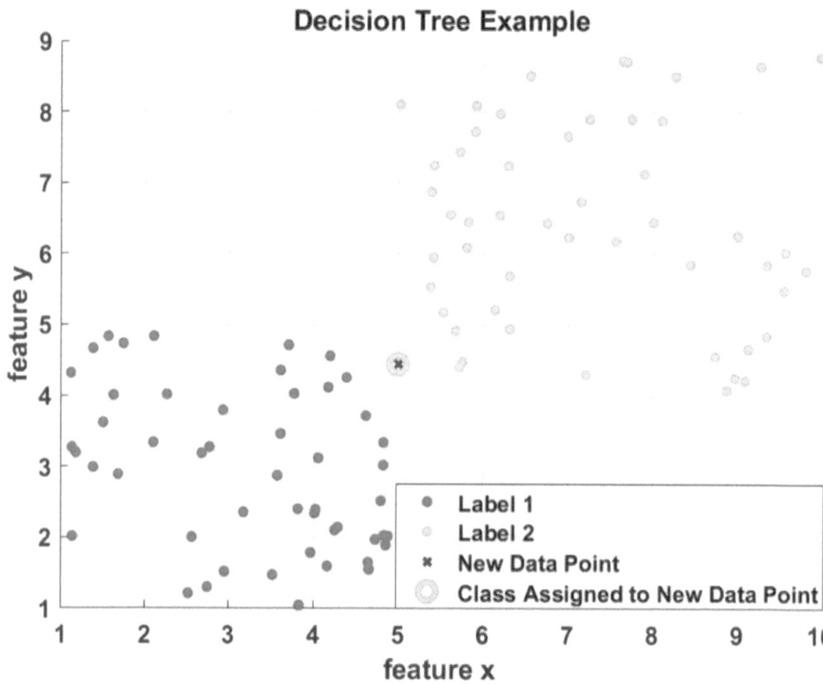

Figure 3.4 Example for decision tree.

- The K nearest neighbors are selected as per the computed Euclidean distances in ascending order.
- The number of data points in each category is counted among the K neighbors.
- The new data point is assigned to the category for which the number of neighbor is maximum.

Figure 3.6 demonstrates the classification of the new data point using the above K-NN algorithm. It can be observed that the data point has been assigned to label 1.

3.2.1.1.4 Random Forest

The random forest algorithm is a classifier comprising several decision trees on various subsets of the dataset and takes the average in order to improve the predictive accuracy. A voting mechanism is thus setup which takes the majority of votes and outputs the corresponding predicted label. A general graphical structure of a random forest is shown in Figure 3.7.

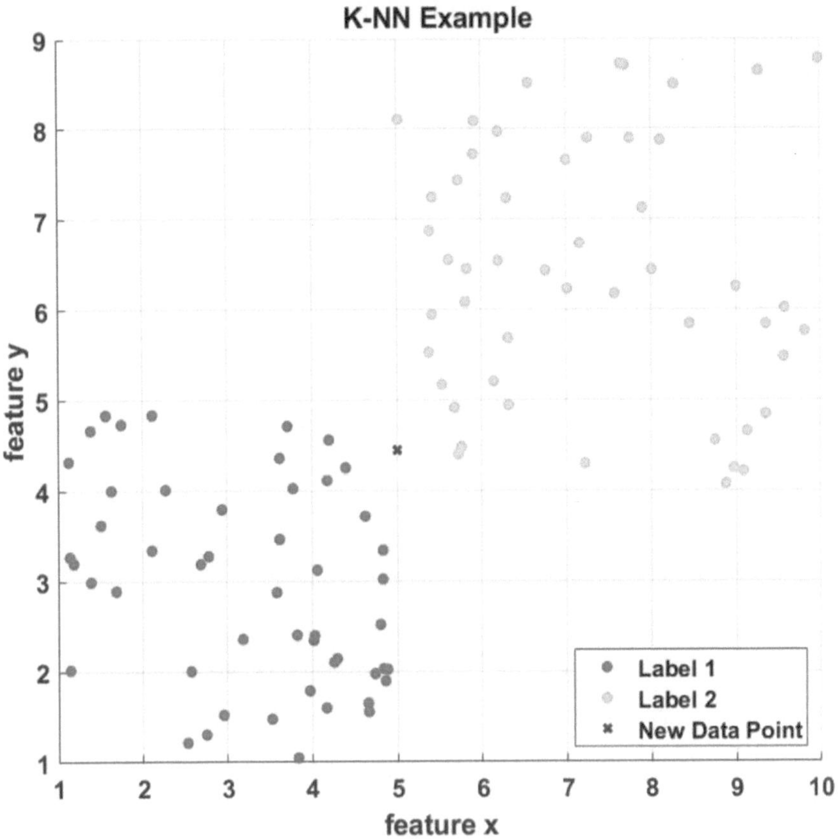

Figure 3.5 Example for K-NN before classification.

Figure 3.8 demonstrates the classification of the new data point using the above Random Forest algorithm. It can be observed that the data point has been assigned to label 2.

3.2.1.2 Regression problem

The regression problem comprises datasets comprising of dependent and independent variables for which the mathematical relationship is sought. This is commonly faced in cases where projections need to be made. Some popular algorithms are linear regression, logistic regression, and polynomial regression.

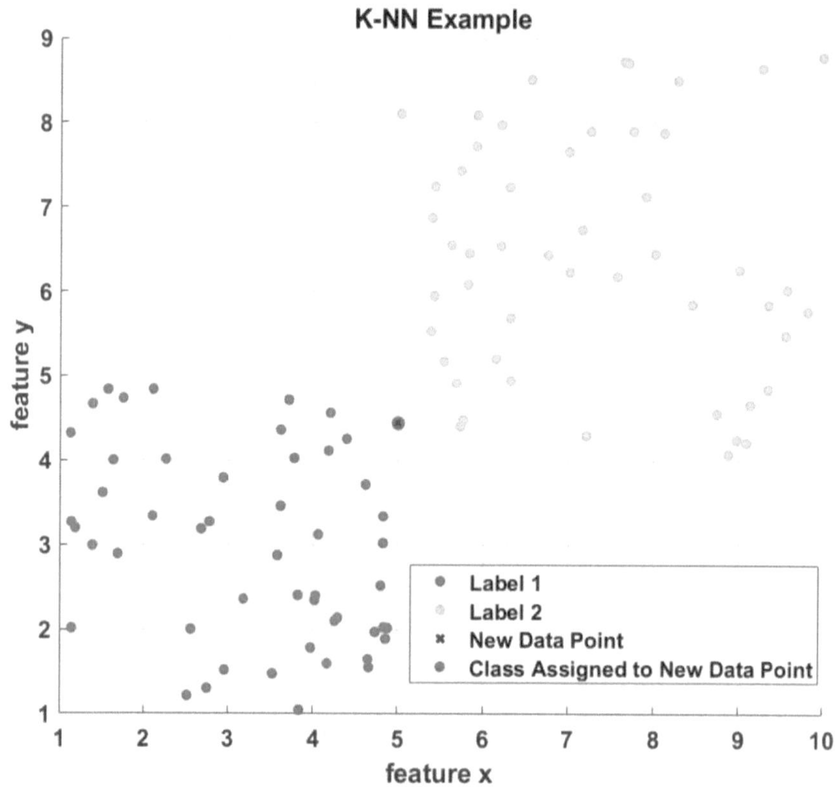

Figure 3.6 Example for K-NN after classification.

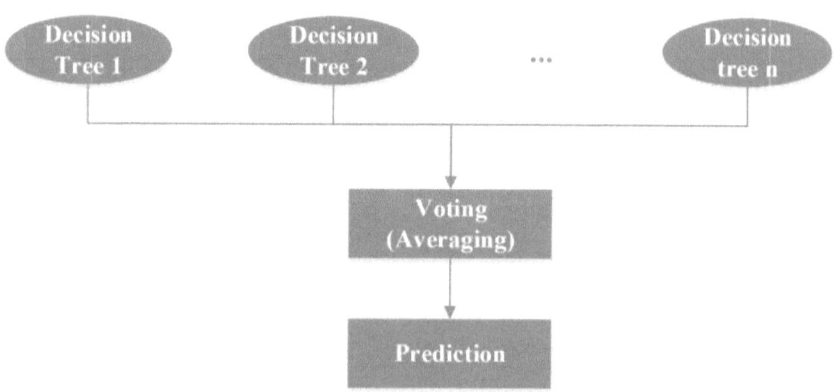

Figure 3.7 General graphical structure of a random forest.

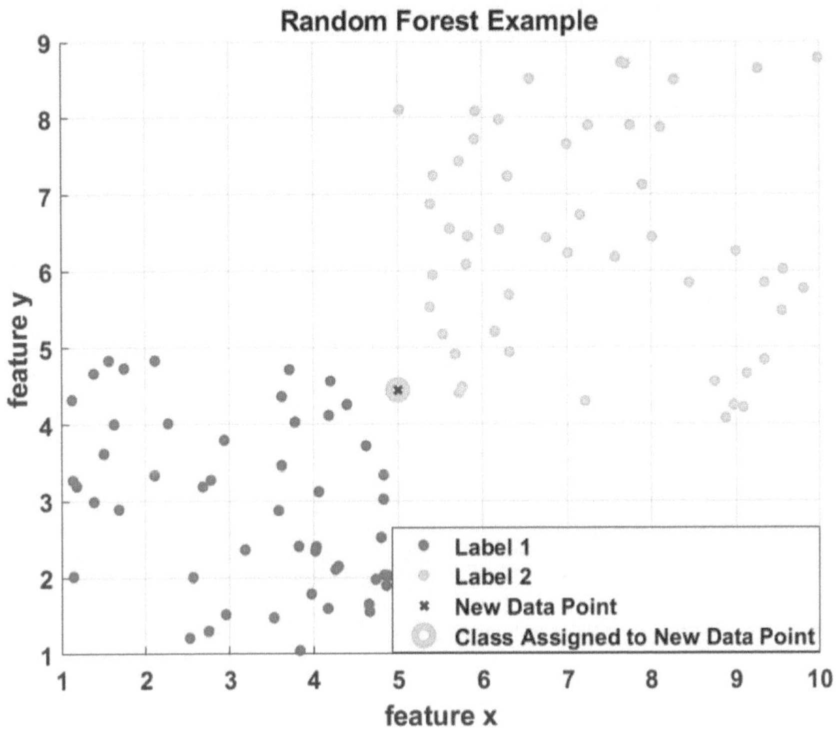

Figure 3.8 Example for random forest after classification.

3.2.1.2.1 Linear Regression

Linear regression model is a popular statistical technique for prediction. It demonstrates the linear relationship between a dependent variable and one or more independent variables. It can thus be categorized into simple linear regression and multiple linear regression based on the number of independent variables. For an illustration of the multiple linear regression, the 'carsmall.dat' dataset is used in MATLAB®.

The predictive model for the number of miles per gallon as a function of the weight and acceleration of the car is obtained using the built-in 'fitlm' function. The general equation for the relationship sought with this method is:

$$\text{MPG} = 45.155 - (0.0082475 \times \text{Weight}) + (0.19694 \times \text{Acceleration}) \quad (3.1)$$

The plot for the original dataset as well as a predicted data point with sample values of weight and acceleration are shown in Figure 3.9.

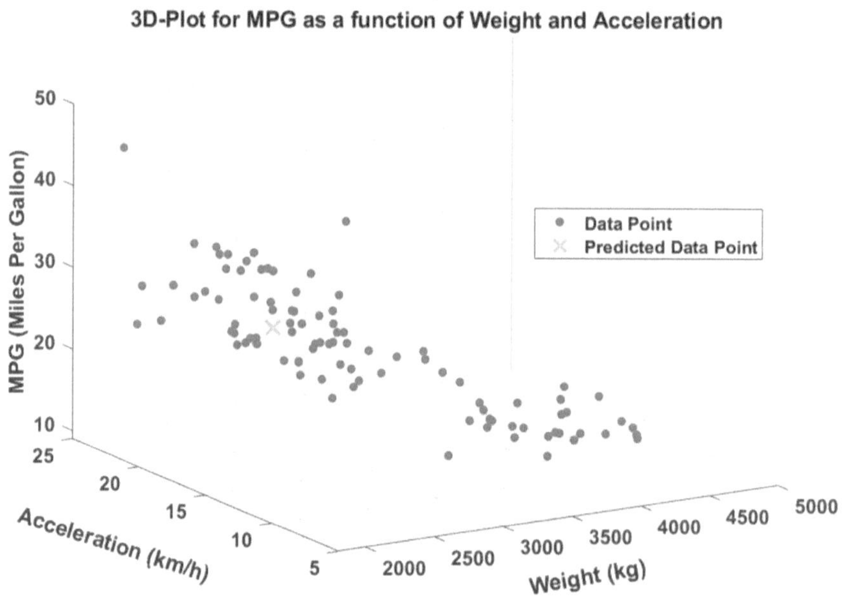

Figure 3.9 3D-plot for MPG as a function of weight and acceleration.

3.2.1.2.2 Logistic Regression

Logistic regression operates in a similar way as linear regression with the exception that a sigmoid function is fitted in view to address classification problems. Consider a random dataset generated in MATLAB®. The binary logistic regression model is obtained using the 'fitglm' function and the plots or the original and predicted data are shown in Figure 3.10.

3.2.1.2.3 Polynomial Regression

Polynomial regression models the relationship between dependent and independent variables as a polynomial of nth degree. A sample example of population data over the years can be generated and fitted to a polynomial of the 5th degree. Using the 'polyfit' and 'polyval' functions in MATLAB®, the following equation is obtained:

$$Y = (-0.5943X^5 - 0.1876X^4 + 2.6074X^3 + 2.5973X^2 + 0.6848X + 1.2361) \times 10^9 \quad (3.2)$$

The sample data points and predicted model can be observed in Figure 3.11.

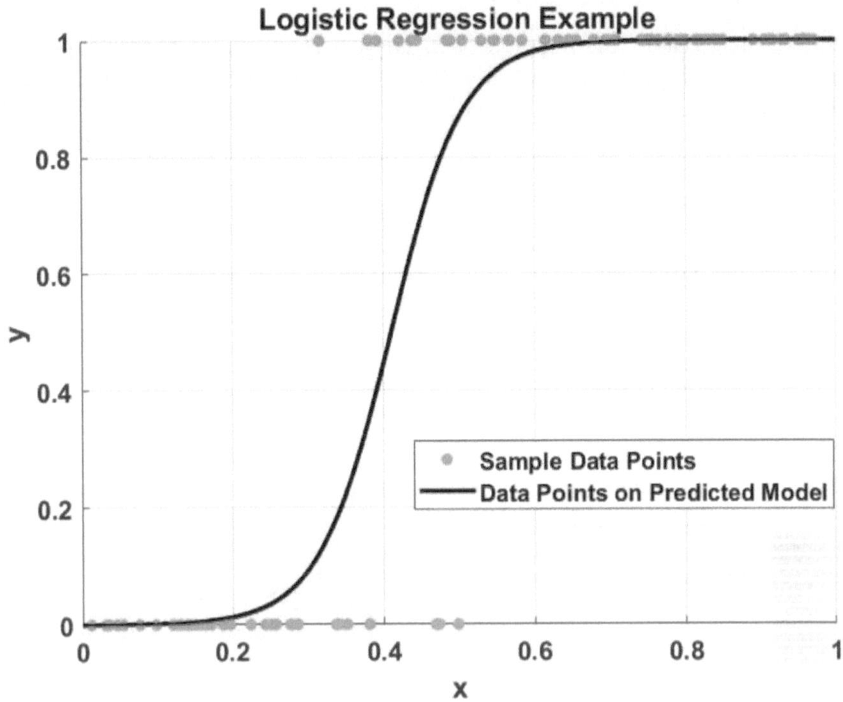

Figure 3.10 Example for logistic regression model.

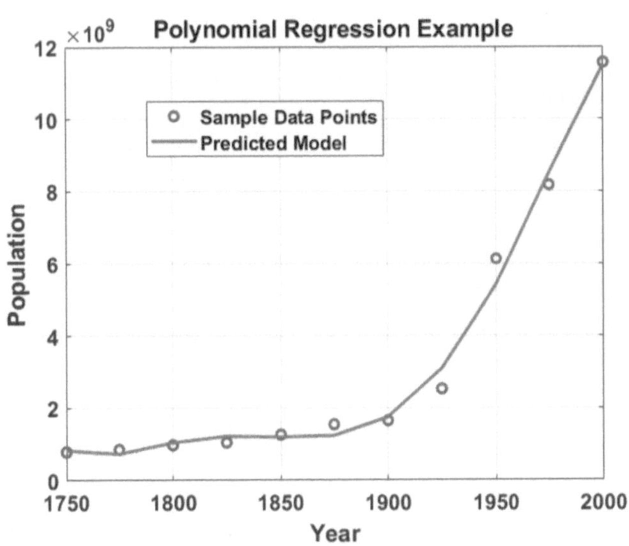

Figure 3.11 Example for polynomial regression model.

3.2.2 Unsupervised learning

The unsupervised learning approach comprises datasets without specific labels and the algorithms have the ability to discover hidden patterns or data groupings without the requirement for human intervention. Unsupervised learning is used for three main tasks, namely clustering, association, and dimensionality reduction.

Some common algorithms used for these tasks are:

1. **Clustering:** K-Means clustering and Gaussian Mixture Models.
2. **Association Rules:** A-priori algorithms.
3. **Dimensionality Reduction:** Principle Component Analysis (PCA) and Auto encoders.

3.2.2.1 Clustering

Clustering is a technique that processes raw and unclassified data into different groups based on similarities, differences, and even patterns in the dataset. Some common algorithms used are K-Means and Gaussian Mixture Models (GMM).

3.2.2.1.1 K-Means Clustering

K-Means algorithm performs an exclusive clustering on unlabeled data points and assigns them into K different groups, where K is representative of the number of clusters and is based on the distance from each group's centroid. The prediction of new data in this case will measure the distance from the different centroids and assign to the group which is closest.

3.2.2.1.2 Gaussian Mixture Models

In the case of Gaussian Mixture Models, the classification is done in terms of an unspecified number of probability distribution functions. New data points are then tested statistically to determine the model it belongs to.

3.2.2.2 Association rules

As the name suggests, Association rule is a rule-based mechanism that aids in finding the corresponding links between different variables in a given dataset. The most common use-case for association rules are market basket analysis which can help the companies understand the relationship and thereby enable them to perform cross selling. Amongst the few existing algorithms, the a-priori algorithm is most commonly used.

3.2.2.2.1 A-Priori Algorithms

The a-priori algorithm helps in studying the frequency of common collections of items in transactional datasets of companies. These insights help companies to have better control and management of products as well as increase sales.

3.2.2.3 Dimensionality reduction

Despite the fact that more data is desired in view to build more accurate models, it can also negatively affect the performance by causing overfitting to happen. With a number of features present in a dataset, it is necessary to understand the level of correlations between the different features and reduce the dimensionality at the preprocessing stage itself. Some widely used techniques used in machine learning problems are described in the next sub-sections.

3.2.2.3.1 Principle Component Analysis (PCA)

PCA operates on the normalized dataset and computes the covariance Eigen values and Eigen vectors. Following the computation, a sorting mechanism is performed to create new features and removing the redundant ones.

3.2.2.3.2 Auto-Encoders

Auto-encoders leverage neural networks in principle in view to compress the dataset and creating a new representation of the original input data. The auto-encoder operates in the region between the input and hidden layers in the architecture of neural networks.

3.2.3 ML algorithms for use cases in smart agriculture

In view to optimize the way food gets to the people and revolutionize agriculture which is a very critical sector for the economy, ML technology has emerged as an important tool. There exist different use cases whereby ML technology helps to increase efficiency and accuracy in smart decision-making while reducing the costs and risks associated with the operations in the agricultural sector. The data types involved can be sensorial information as well as imaging data from which insights can be extracted using appropriate and adapted ML algorithms.

3.2.3.1 *Crop management*

Crop management is one of the most crucial and challenging stages in the agricultural lifecycle. With the effects of climate change, the world is experiencing a non-uniform frequency and variation of droughts, temperature, rainfall as well as wetting and drying cycles. This has a drastic effect on the crops' resistance which eventually affects the yield. With all these challenges and variations at hand, the emergence of ML techniques has demonstrated that they have the ability to mitigate these issues. The approach to adopt in this case, is to collect field data on plant behaviors and use it to develop the corresponding probabilistic model(s). Some major parameters that can be of prime importance for this problem are environmental temperature, soil type, rainfall as well as actual crop information. Based on these data, models such as multiple linear regression or neural networks can be implemented to provide insights as to which crops would be most suitable and provide high yields under varying conditions.

3.2.3.2 *Precision spraying*

In fields spanning over several hectares, spraying needs to be done in order to prevent infestation of pests and diseases. In so doing, over-spraying cannot be avoided which drastically impacts on the expenses, operational costs, soil quality, and environment. To combat this problem, it is essential to have insights on plant sizes, types of disease affecting the crops, types of pests present in the field as well as specific locations where the spraying needs to be performed. All this information would help with understanding the type(s) and quantity of spraying that needs to be performed at specific locations. This would result in smart spraying while reducing the expenses, operational costs as well as impacts on the environment without impacting the yield. In this case scenario, computer-vision algorithms such as convolutional neural networks (CNNs) operating on drone-based imagery are mostly suitable.

3.2.3.3 *Insect detection*

With high percentages of global crop production being destroyed by pests, farmers have resorted to the usage of pesticides which kills smaller non-threatening insects living around the fields in the process. Considering this aspect and the popularity of drones, ML techniques are being increasingly used for insect detection on live drone imagery/video. This allows for the real-time detection and tracking of pests/insects which can affect the crops drastically and enable timely actions to be taken by the farmers. In this case, computer vision-based ML models are suitable.

More specifically, properly designed and trained CNN models can be used for the identification of bees, flies, mosquitoes, rats as well as fruit flies with significantly high accuracies of the orders of 90.

3.2.3.4 *Field conditions management*

Among several important aspects of farming, water and soil management in the fields are more crucial. Depending on the areas of the fields being considered, it can be very difficult to manage. IoT is essentially deployed in these cases to collect and monitor the soil and water conditions in micro-sections of the plantation. Feeding these data together with weather conditions to the appropriate ML models can provide insights into the rate of fertilizer application at the different locations, irrigation schedules and amount as well as the best timings for harvesting. All these together contribute to better planning of resources, cost reductions, and improved yields.

3.2.3.5 *Yield mapping*

One of the ultimate goals of every farmer is to get the most out of their crops, contribute to feeding the global population as well as earn decent profits. Advanced ML techniques have been introduced to operate on satellite images as well as multi-spectral imaging from drones in order to predict crop yields. In addition to image processing and predictions of current plantations, it is important to have information on the weather and soil condition profiles to enable a better understanding of which crops would be most suitable and provide higher yields. This in turn results in a reduction in the costs, usage of pesticides, and impacts on the environment. In this case, time series predictive models can be used in tandem with digital image processing and prediction models.

3.3 ANALYSIS OF DATASETS FOR SMART AGRICULTURE

A number of sample datasets in the context of smart agriculture are available for modeling. Some of the existing datasets and the corresponding references are shown in Table 3.1.

Table 3.1 Existing sample datasets in the context of smart agriculture

	Dataset name	*Reference*
Dataset 1	Smart Agricultural Production Optimization Engine	[62]
Dataset 2	Global Wheat Head Dataset	[63]
Dataset 3	2023 EY Open Science Data Challenge	[64]

3.3.1 Dataset 1

Dataset 1 is available from the Kaggle community and comprises agricultural data comprising of the following features, namely nitrogen content of soil, phosphorus content of soil, potassium content of soil, temperature, humidity, pH level of soil, and rainfall.

The data also consists of labels pertaining to the crop suitable for the different values attributed to the features above. The 22 different crops in the dataset are {'rice', 'maize', 'chickpea', 'kidneybeans', 'pigeonpeas', 'mothbeans', 'mungbean', 'blackgram', 'lentil', 'pomegranate', 'banana', 'mango', 'grapes', 'watermelon', 'muskmelon', 'apple', 'orange', 'papaya', 'coconut', 'cotton', 'jute', and 'coffee'}.

There are 100 rows of feature values for each label. Thus, the dataset comprises 2,200 rows in total. A summarization of the features in the dataset is given in Table 3.2.

3.3.1.1 *Visualization plots of the different features*

The visualization plots for the different features in the dataset are shown in Figures 3.12–3.18. Each figure contains the histogram with 20 bins, the violin plot, and the corresponding boxplot.

Table 3.2 Summary of the features in dataset 1

	Nitrogen	*Phosphorus*	*Potassium*	*Temperature*	*Humidity*	*pH*	*Rainfall*
Count	2,200	2,200	2,200	2,200	2,200	2,200	2,200
Mean	50.55	53.36	48.15	25.61	71.48	6.469	103.5
STD DEV	36.92	32.98	50.65	5.064	22.26	0.7739	54.96
Minimum	0.0	5.0	5.0	8.826	14.26	3.505	20.21
Maximum	140.0	145.0	205.0	43.67	99.98	9.935	298.6

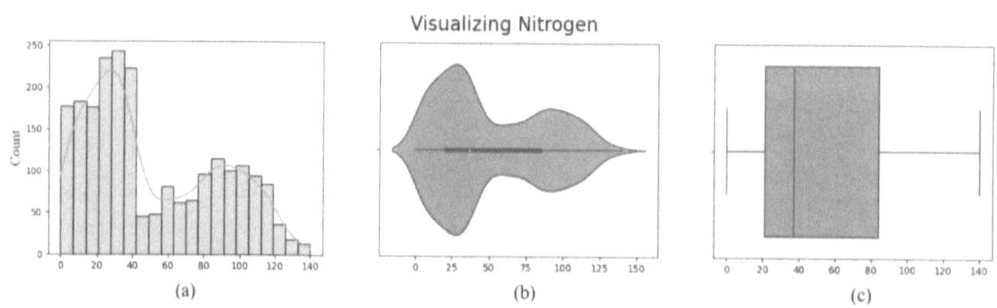

Figure 3.12 Visualizations for nitrogen: (a) Histogram for nitrogen, (b) Violin plot for nitrogen, (c) Box plot for nitrogen.

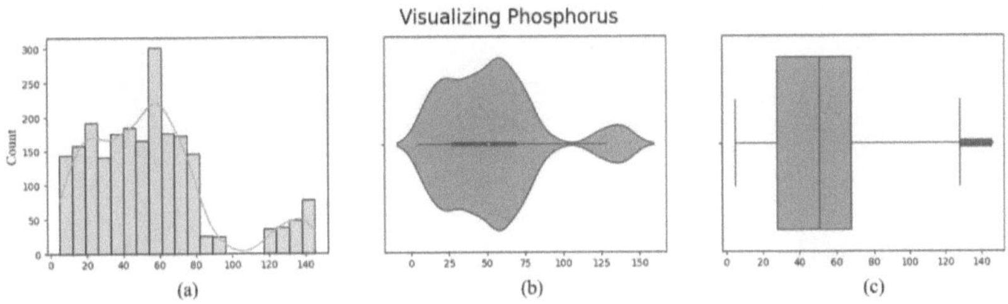

Figure 3.13 Visualizations for phosphorus: (a) Histogram for phosphorus, (b) Violin plot for phosphorus, (c) Box plot for phosphorus.

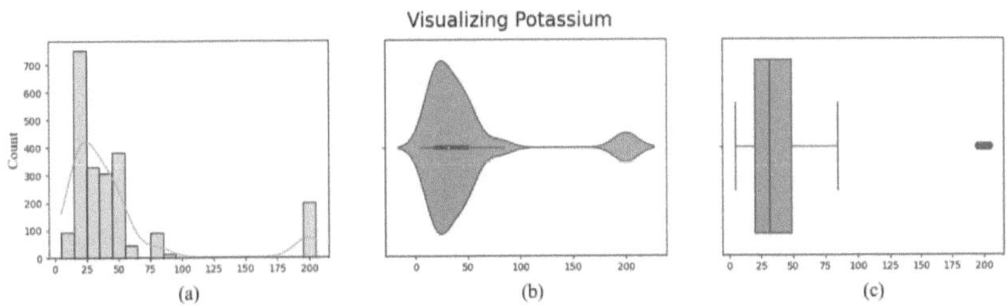

Figure 3.14 Visualizations for potassium: (a) Histogram for potassium, (b) Violin plot for potassium, and (c) Box plot for potassium.

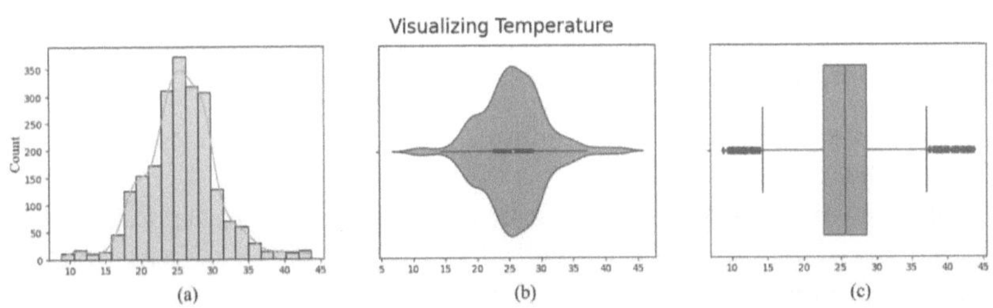

Figure 3.15 Visualizations for temperature: (a) Histogram for temperature, (b) Violin plot for temperature, and (c) Box plot for temperature.

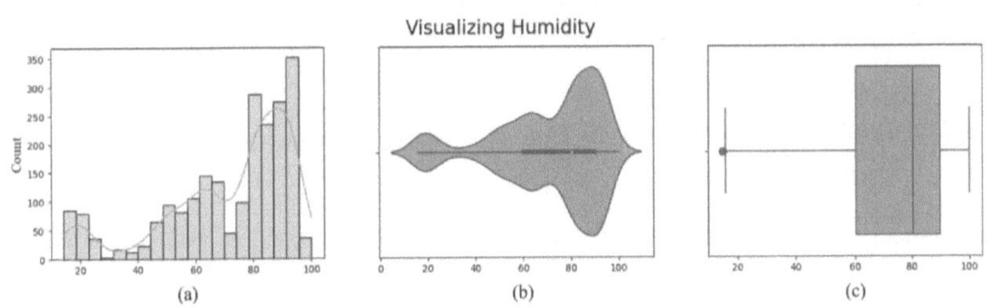

Figure 3.16 Visualizations for humidity: (a) Histogram for humidity, (b) Violin plot for humidity, and (c) Box plot for humidity.

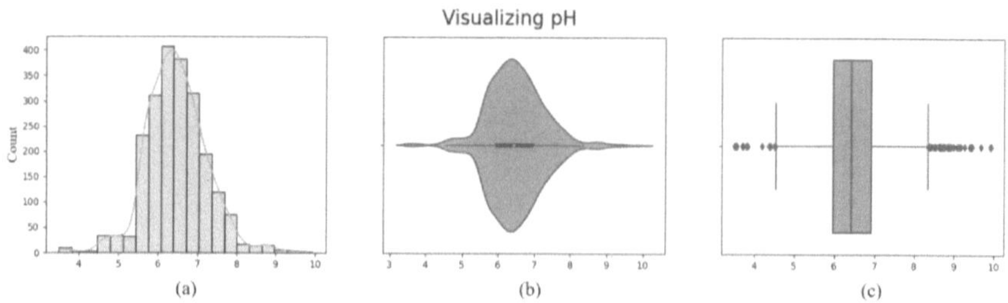

Figure 3.17 Visualizations for pH: (a) Histogram for pH, (b) Violin plot for pH, and (c) Box plot for pH.

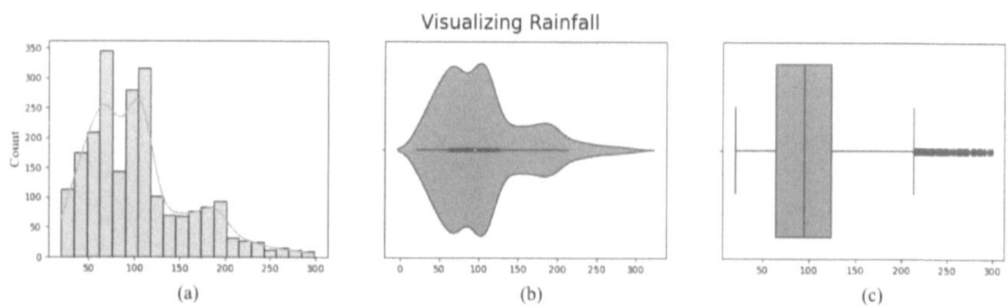

Figure 3.18 Visualizations for rainfall: (a) Histogram for rainfall, (b) Violin plot for rainfall, (c) Box plot for rainfall.

3.3.1.2 Comparison of the different classes

The comparison of the mean attributes for the different classes is depicted in Figure 3.19.

From the figure above, a sorting of the crops in descending order of the different features can be performed. The top five and least five crops with dependence on each feature are shown in Table 3.3.

A number of observations can be made. For example, if 'lentil' is considered, it shows high dependency on phosphorus and pH while least dependency on nitrogen, potassium, and rainfall. Similarly, if 'orange' is considered, it has a high dependency on humidity and pH and least dependency on nitrogen, phosphorus, and potassium.

3.3.1.3 Decomposition using the principle component analysis algorithm

The principle component analysis (PCA) algorithm forms part of the dimensionality reduction mechanisms as discussed in previous Section 3.2.2.3. For the dataset under consideration, newer datasets comprising two or three features only. The corresponding scatter plots can be used for visualization as demonstrated in Figures 3.20 and 3.21. Figures 3.20 and 3.21

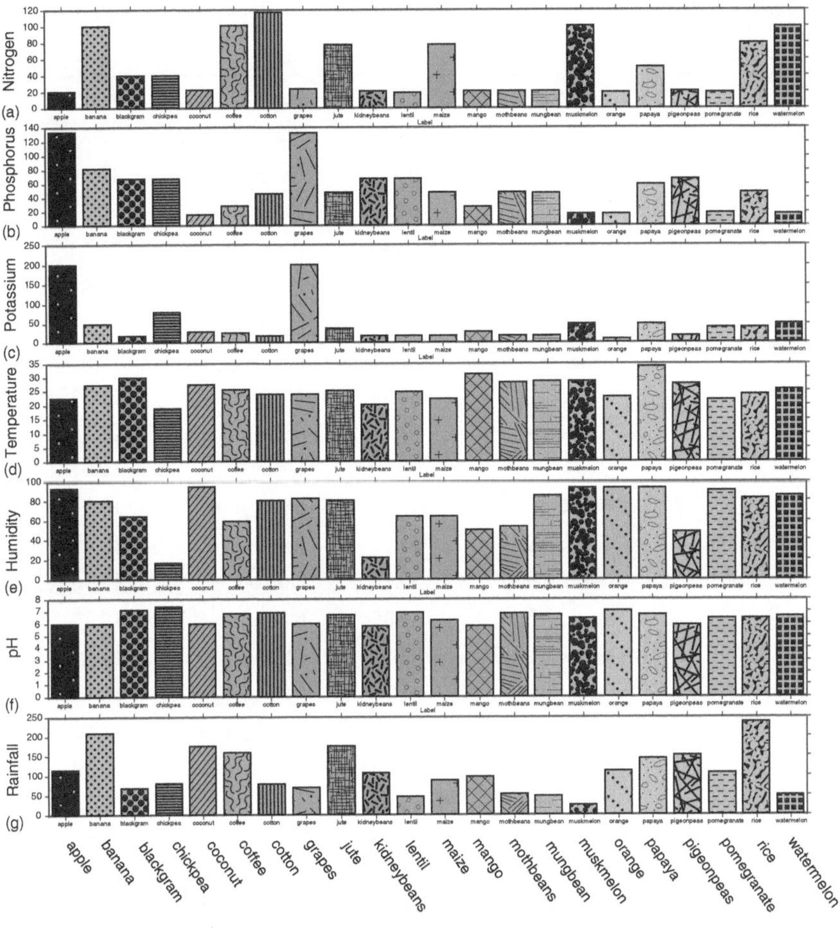

Figure 3.19 Comparison of the mean attributes of the different classes: (a) Nitrogen, (b) Phosphorus, (c) Potassium, (d) Temperature, (e) Humidity, (f) pH, and (g) Rainfall.

represent the two-dimensional and three-dimensional plots, respectively, for the decomposition using the PCA algorithm.

3.3.2 Dataset 2

Another dataset that was developed in agriculture is the Global Wheat Head Dataset [63], which was led by nine different institutions from seven countries. The dataset constitutes of images of wheat fields in different locations around the world. In each image of the training set, there are also bounding boxes around each identified wheat head. In addition to the set of images, there is also a CSV file constituting of the filename of

Table 3.3 Groupings of top 5 and least 5 dependent crops on each feature

Feature	*Top 5 crops*	*Value*	*Least 5 Crops*	*Value*
	Cotton	117.77	Lentil	18.77
	Coffee	101.2	Pomegranate	18.87
Nitrogen	Muskmelon	100.32	Orange	19.58
	Banana	100.23	Mango	20.07
	Watermelon	99.42	Pigeon peas	20.73
	Apple	134.22	Orange	16.55
Phosphorus	Grapes	132.53	Coconut	16.93
	Banana	82.01	Watermelon	17.0
	Lentil	68.36	Muskmelon	17.72
	Chickpea	67.79	Pomegranate	18.75
	grapes	200.11	Orange	10.01
	Apple	199.89	Black gram	19.24
Potassium	Chickpea	79.92	Lentil	19.41
	watermelon	50.22	Cotton	19.56
	muskmelon	50.08	Maize	19.79
	papaya	33.72	Chickpea	18.87
Temperature	Mango	31.21	Kidney beans	20.11
	Blackgram	29.97	Pomegranate	21.83
	Muskmelon	28.66	Maize	22.38
	Mungbean	28.52	Apple	22.63
Humidity	Coconut	94.84	Chickpea	16.86
	Papaya	92.40	Kidney beans	21.60
	Muskmelon	92.34	Pigeon peas	48.06
	Apple	92.33	Mango	50.15
	Orange	92.17	Mothbeans	53.16
pH	Chickpea	7.336	Kidney beans	5.749
	Blackgram	7.133	Mango	5.766
	Orange	7.016	Pigeon peas	5.794
	Lentil	6.927	Apple	5.929
	Cotton	6.912	Coconut	5.976
Rainfall	Rice	236.1	Muskmelon	24.68
	Coconut	175.6	Lentil	45.68
	Jute	174.7	Mung bean	48.40
	Coffee	158.0	Watermelon	50.78
	Pigeon peas	149.4	Moth beans	51.19

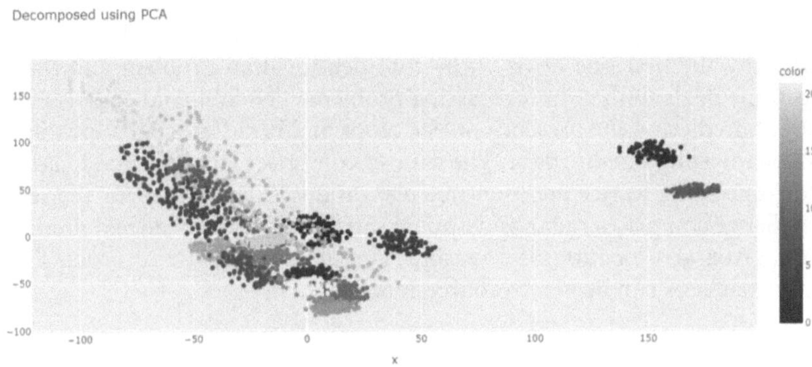

Figure 3.20 Two-dimensional plot for decomposition using PCA.

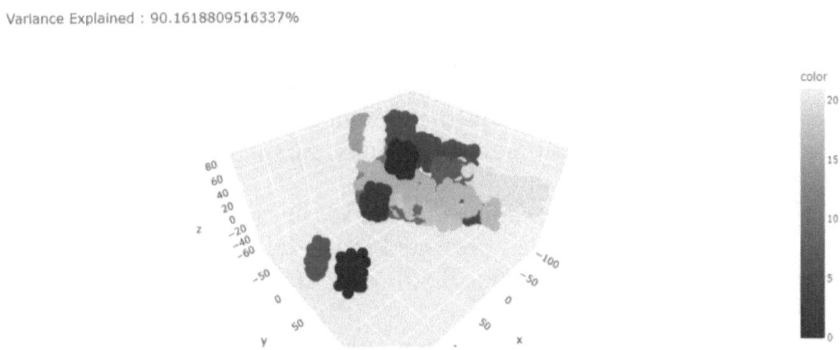

Figure 3.21 Three-dimensional plot for decomposition using PCA.

the images, width and height of the image, and details pertaining to the bounding boxes present in the images. This dataset is important in smart farming as it can potentially help farmers with health assessment and maturity levels of the wheat crops in view to take management decisions in the fields. Real-case scenarios for wheat detection can be very complicated with different environmental challenges. One example of such a challenge is the processing of blurred images due to the constant movement of the crops in a windy environment.

3.3.3 Dataset 3

Recently, Ernst and Young Company Ltd. in collaboration with Cornell University and Microsoft Ltd launched the 2023 EY Open Science Data

Challenge [64] in view to using AI techniques and harvesting data to help solve world hunger. The challenge is broken down into two different levels whereby the first one deals with crop identification problem and the second one deals with crop forecasting problem. The first challenge is mainly about predicting the presence of rice crops at 250 different geo locations in Vietnam using satellite data. The dataset comprises latitudes and longitudes corresponding to rice and non-rice plantations. The dataset for the second challenge comprises radar and optical satellite data for rice yield prediction at a given geo location and season. The manipulation and processing of these datasets can be very resource-intensive.

3.4 SIMULATIONS OF SYSTEM MODELS AND ANALYSIS OF RESULTS

In this section, Dataset 1 is considered for the demonstration of the application of different ML models. The simulations of the ML models are done using the Python programming language with different corresponding libraries. The programming has essentially been performed using the Jupyter notebooks on the Google Colab platform. The main libraries used are, namely numpy, pandas, matplotlib, seaborn, plotly and sci-kit learn.

The dataset is split into training and test sets in the ratio of 70:30 in this chapter. Different ML models presented in the following sub-sections are the K-NN Classification, Logistic Regression, and Random Forest.

3.4.1 K-NN Classification

To use the K-NN classification mechanism, a suitable value of K needs to be determined prior to using the algorithm. This can be done by computing a cross-validation score for a range of values of K and perform the selection based on the highest score. Using the training dataset at hand and a range of values of K from 1 to 20 in steps of 1, the different scores can be obtained using the cross-validation score library from the sci-kit learn package in Python. The resulting plot is shown in Figure 3.22.

From Figure 3.22, it can be observed that the highest score is attributed to $K=1$. The K-NN classifier can thus be implemented using a single neighbor as parameter and fed with the training set for learning. The model is then tested on the separate data and an accuracy of 98% is obtained from the classification report. The corresponding classification matrix can be presented in the form of a heat map as shown in Figure 3.23.

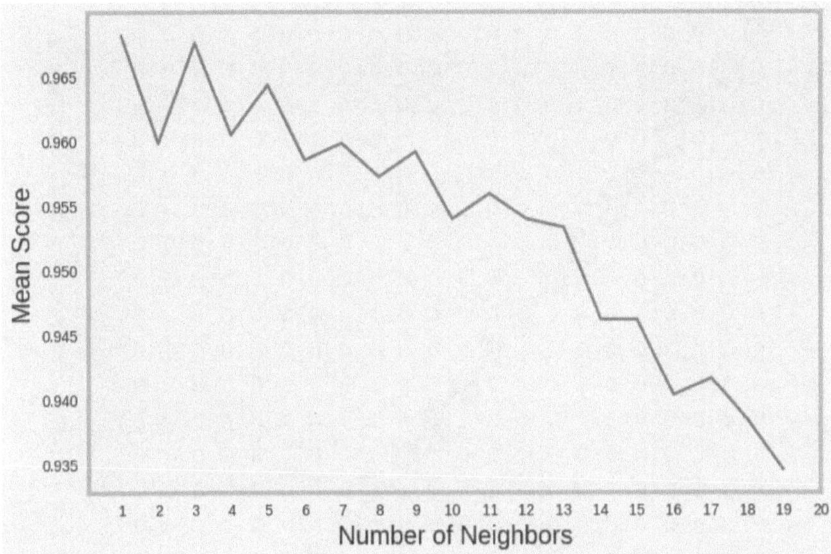

Figure 3.22 Cross-validation score for different values of *K* nearest neighbors.

3.4.2 Logistic regression based classification

The parameters used for the logistic regression model in Python are as follows:

Set of maximum iterations: {50, 75, 100, 200, 300, 400, 500, 700}
Solver: 'lbfgs'(low-memory Broyden Fletcher Goldfarb Shanno)

Following the training process of the model, it is then tested on the separate data and an accuracy of 99% is obtained from the classification report. The corresponding classification matrix can be presented in the form of a heat map as shown in Figure 3.24.

3.4.3 Random forest classification

The parameters used for the random forest model in Python are as follows:

Number of trees in the forest - n estimators: {50, 75, 100, 200, 300, 400, 500, 700}
Random state hyperpameter: 42

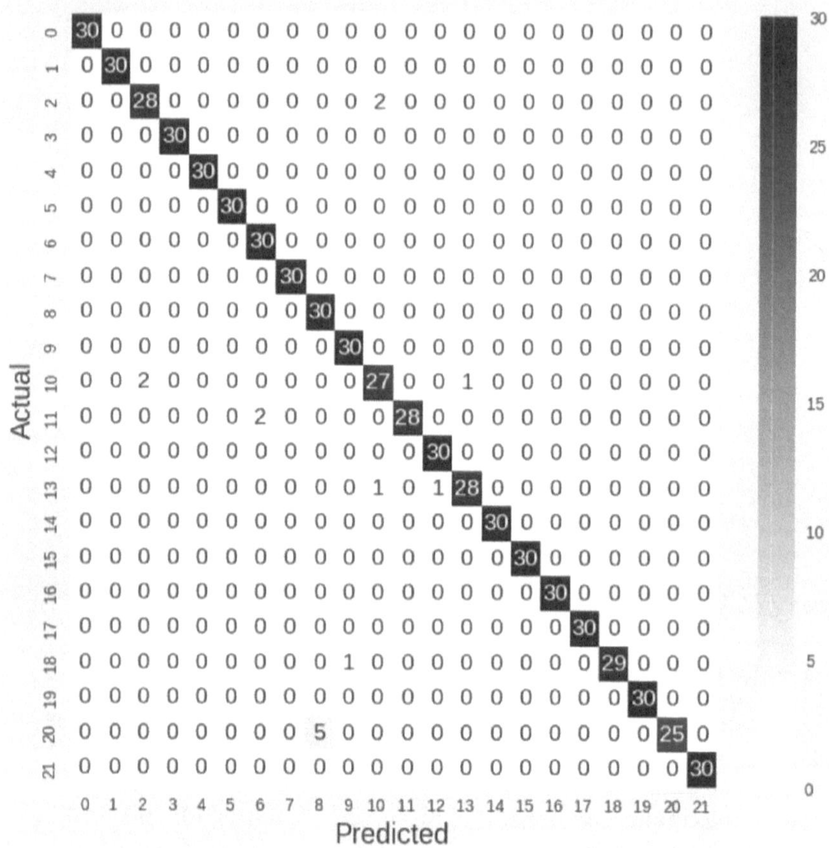

Figure 3.23 Heat map for the confusion matrix with K-NN.

Following the training process of the model, it is then tested on the separate data and an accuracy of 99% is obtained from the classification report. The corresponding classification matrix can be presented in the form of a heat map as shown in Figure 3.25.

3.4.4 Gradient boosting classifier

The parameters used for the gradient boosting classifier model in Python are as follows:

Learning rate: {0.05, 0.075, 0.1, 0.25, 0.5, 0.75, 1}
Number of trees in the forest - n estimators: {50, 75, 100, 200, 300, 400, 500, 700}
Random state hyperpameter: 42

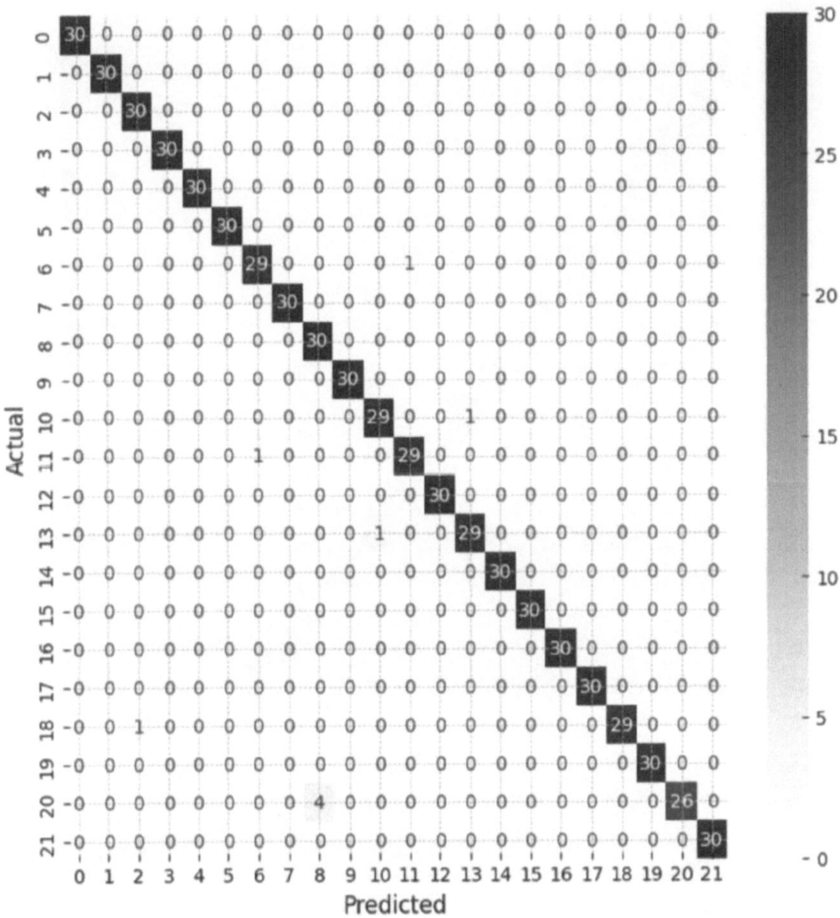

Figure 3.24 Heat map for the confusion matrix with logistic regression.

Following the training process of the model, it is then tested on the separate data and an accuracy of 99% is obtained from the classification report. The corresponding classification matrix can be presented in the form of a heat map as shown in Figure 3.26.

3.5 DISCUSSIONS

From the results presented above, it can be observed that the different classification models based on the soil and environmental parameters as features provide predictive models with accuracies higher than 95%. These models have applications in crop suggestion mechanisms based on

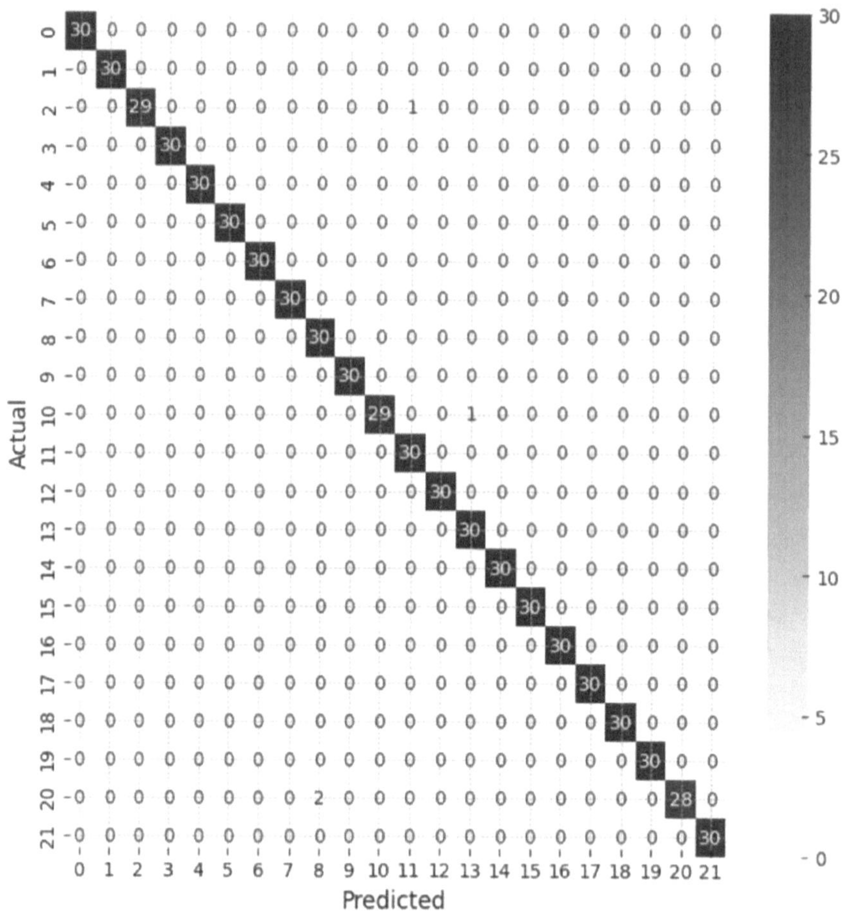

Figure 3.25 Heat map for the confusion matrix with random forest.

the different surrounding conditions which can eventually help farmers obtain better yields and save costs. In addition to these types of predictive models, time-series profiling models of the different parameters can also be implemented prior to feeding to the crop prediction models. Moreover, several other features can be incorporated into the above models in view to have yield forecasts so as to help the farmers plan their harvesting effectively.

3.6 CONCLUSION

In this chapter, we have reviewed existing non-exhaustive datasets and AI techniques used in the context of smart agriculture. The dataset used

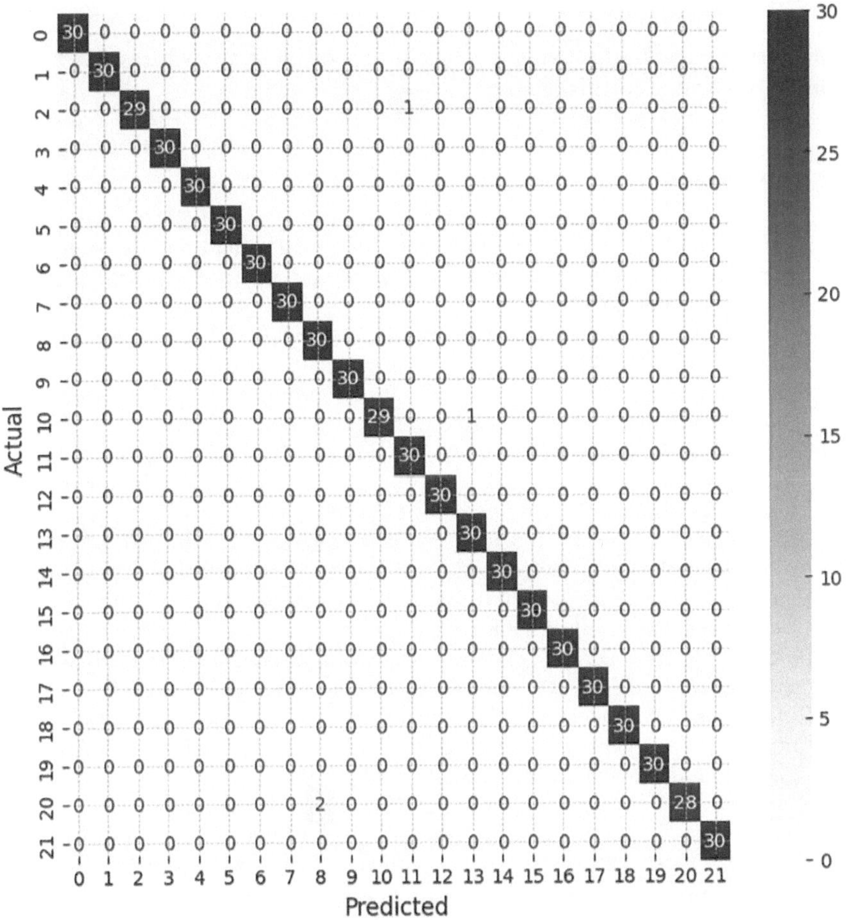

Figure 3.26 Heat map for the confusion matrix with gradient boosting classifier.

comprises soil and environmental parameters and corresponding labels of most suitable crops. The predictive models, which are presented in this chapter, take the soil and environmental parameters as input features to predict the corresponding crop at the output. The four different models presented are K-NN, logistic regression, random forest, and gradient boosting classification. From the different confusion matrices presented, it can be observed that the accuracies of the different models are above 95%.

ACKNOWLEDGMENTS

The authors would like to thank the University of Mauritius for the necessary facilities provided.

DECLARATIONS

The authors declare that there is no conflict of interest with respect to the content presented in this chapter.

REFERENCES

[1] Aydinalp, C., Cresser, M.S.: The effects of global climate change on agriculture. *American-Eurasian Journal of Agricultural & Environmental Sciences* 3(5), 672–676 (2008).

[2] Dudley, N., Alexander, S.: Agriculture and biodiversity: A review. *Biodiversity* 18(2–3), 45–49 (2017).

[3] Transnational Corporations: Agricultural production and development. World Investment Report, UN, New York and Geneva (2009).

[4] Gil, J.D.B., Reidsma, P., Giller, K., Todman, L., Whitmore, A., Ittersum, M.: Sustainable development goal 2: Improved targets and indicators for agriculture and food security. *Ambio* 48(7), 685–698 (2019).

[5] Balafoutis, A., Beck, B., Fountas, S., Vangeyte, J., Wal, T., Soto, I., G´omezBarbero, M., Barnes, A., Eory, V.: Precision agriculture technologies positively contributing to GHG emissions mitigation, farm productivity and economics. *Sustainability* 9(8), 1339 (2017).

[6] Soule, J., Carr´e, D., Jackson, W., et al.: Ecological impact of modern agriculture. *Ecological Impact of Modern Agriculture* 91, 165–188 (1990).

[7] Beacham, A.M., Vickers, L.H., Monaghan, J.M.: Vertical farming: A summary of approaches to growing skywards. *The Journal of Horticultural Science and Biotechnology* 94(3), 277–283 (2019).

[8] Birkby, J.: Vertical farming. *ATTRA Sustainable Agriculture* 2, 1–12 (2016).

[9] Campbell, B.M., Hansen, J., Rioux, J., Stirling, C.M., Twomlow, S., et al.: Urgent action to combat climate change and its impacts (SDG 13): Transforming agriculture and food systems. *Current Opinion in Environmental Sustainability* 34, 13–20 (2018).

[10] Su, Y., Wang, X.: Innovation of agricultural economic management in the process of constructing smart agriculture by big data. *Sustainable Computing: Informatics and Systems* 31, 100579 (2021).

[11] Schuster, J.: Big data ethics and the digital age of agriculture. *Resource Magazine* 24(1), 20–21 (2017).

[12] Lipper, L., Thornton, P., Campbell, B.M., Baedeker, T., Braimoh, A., Bwalya, M., Caron, P., Cattaneo, A., Garrity, D., Henry, K., et al.: Climate-smart agriculture for food security. *Nature Climate Change* 4(12), 1068–1072 (2014).

[13] Wysel, M., Baker, D., Billingsley, W.: Data sharing platforms: How value is created from agricultural data. *Agricultural Systems* 193, 103241 (2021).

[14] Gondchawar, N., Kawitkar, R., et al.: IoT based smart agriculture. *International Journal of Advanced Research in Computer and Communication Engineering* 5(6), 838–842 (2016).

[15] Lee, M., Hwang, J., Yoe, H.: Agricultural production system based on IoT. In: *2013 IEEE 16Th International Conference on Computational Science and Engineering*, pp. 833–837 (2013). IEEE, New York.
[16] Zhang, Y., Wang, X., Liu, T., Wang, R., Li, Y., Xue, Q., Yang, P.: Sustainable fertilisation management via tensor multi-task learning using multi-dimensional agricultural data. *Journal of Industrial Information Integration* 34, 100461 (2023).
[17] He, J., Wang, J., He, D., Dong, J., Wang, Y.: The design and implementation of an integrated optimal fertilization decision support system. *Mathematical and Computer Modelling* 54(3–4), 1167–1174 (2011).
[18] Calabi-Floody, M., Medina, J., Rumpel, C., Condron, L.M., Hernandez, M., Dumont, M., La Luz Mora, M.: Smart fertilizers as a strategy for sustainable agriculture. *Advances in Agronomy* 147, 119–157 (2018).
[19] Agrahari, R.K., Kobayashi, Y., Tanaka, T.S.T., Panda, S.K., Koyama, H.: Smart fertilizer management: The progress of imaging technologies and possible implementation of plant biomarkers in agriculture. *Soil Science and Plant Nutrition* 67(3), 248–258 (2021).
[20] Krause, P.J., Bokinala, V.: A tutorial on data mining for bayesian networks, with a specific focus on IoT for agriculture. *Internet of Things*, 22, 100738 (2023).
[21] Yang, J., Lan, G., Li, Y., Gong, Y., Zhang, Z., Ercisli, S.: Data quality assessment and analysis for pest identification in smart agriculture. *Computers and Electrical Engineering* 103, 108322 (2022).
[22] Pimentel, D.: Pest control in world agriculture. *Agricultural Science* 2, 272–293 (2009).
[23] Lavanya, G., Rani, C., Ganesh Kumar, P.: An automated low cost IoT based fertilizer intimation system for smart agriculture. *Sustainable Computing: Informatics and Systems* 28, 100300 (2020).
[24] Vermesan, O., Friess, P.: *Digitising the Industry Internet of Things Connecting the Physical, Digital and VirtualWorlds*. CRC Press, Boca Raton, FL (2022).
[25] Rizzo, G., Agus, F., Batubara, S.F., Andrade, J.F., Edreira, J.I.R., Purwantomo, D.K., Anasiru, R.H., Marbun, O., Ningsih, R.D., Ratna, B.S., et al.: A farmer data-driven approach for prioritization of agricultural research and development: A case study for intensive crop systems in the humid tropics. *Field Crops Research* 297, 108942 (2023).
[26] Bronson, K., Knezevic, I.: Big data in food and agriculture. *Big Data & Society* 3(1), 2053951716648174 (2016).
[27] Meijer, G.: *Smart Sensor Systems*. John Wiley & Sons, Hoboken, NJ (2008).
[28] Khanal, S., Kc, K., Fulton, J.P., Shearer, S., Ozkan, E.: Remote sensing in agriculture-accomplishments, limitations, and opportunities. *Remote Sensing* 12(22), 3783 (2020).
[29] Srbinovska, M., Gavrovski, C., Dimcev, V., Krkoleva, A., Borozan, V.: Environmental parameters monitoring in precision agriculture using wireless sensornetworks. *Journal of Cleaner Production* 88, 297–307 (2015).
[30] Paul, K., Chatterjee, S.S., Pai, P., Varshney, A., Juikar, S., Prasad, V., Bhadra, B., Dasgupta, S.: Viable smart sensors and their application in data driven agriculture. *Computers and Electronics in Agriculture* 198, 107096 (2022).

[31] Schriber, S.: *Smart Agriculture Sensors: Helping Small Farmers and Positively Impacting Global Issues, Too.* Mouser Electronics, Mansfield, TX (2019).
[32] Law, C.S., Marsal, L.F., Santos, A.: Electrochemically engineered nanoporous photonic crystal structures for optical sensing and biosensing. In: Hussain C.M. (ed) *Handbook of Nanomaterials in Analytical Chemistry*, pp. 201–226. Elsevier, Amsterdam, The Netherlands (2020).
[33] Ali, M.A., Dong, L., Dhau, J., Khosla, A., Kaushik, A.: Perspective-electrochemical sensors for soil quality assessment. *Journal of the Electrochemical Society* 167(3), 037550 (2020).
[34] Manoj, H., Udupa, N.: Application of soil moisture sensor in mixed farming. *International Research Journal of Engineering and Technology (IRJET)* 2(4), 311–314 (2015).
[35] Banerjee, A., Paul, K., Varshney, A., Nandru, R., Badhwar, R., Sapre, A., Dasgupta, S.: Soilless indoor smart agriculture as an emerging enabler technology for food and nutrition security amidst climate change. In: Kumar, V., Srivastava, A. K. & Suprasanna, P. (eds) *Plant Nutrition and Food Security in the Era of Climate Change*, pp. 179–225. Elsevier, Amsterdam, The Netherlands (2022).
[36] W′ojtowicz, M., W′ojtowicz, A., Piekarczyk, J., et al.: Application of remote sensing methods in agriculture. *Communications in Biometry and Crop Science* 11(1), 31–50 (2016).
[37] Shanmugapriya, P., Rathika, S., Ramesh, T., Janaki, P.: Applications of remote sensing in agriculture-a review. *International Journal of Current Mic robiology and Applied Sciences* 8(01), 2270–2283 (2019).
[38] Abdoli, P., Khanmirzaei, A., Hamzeh, S., Rezaei, S., Moghimi, S.: Use of remote sensing data to predict soil organic carbon in some agricultural soils of Iran. *Remote Sensing Applications: Society and Environment* 30, 100969 (2023).
[39] Ozdogan, M., Yang, Y., Allez, G., Cervantes, C.: Remote sensing of irrigated agriculture: Opportunities and challenges. *Remote Sensing* 2(9), 2274–2304 (2010).
[40] Kurz, B., Seelan, S.: 7 use of remote sensing to map irrigated agriculture in areas overlying the ogallala aquifer, United States. *Remote Sensing of Global Croplands for Food Security* 2, 199 (2009).
[41] Thorp, K., Tian, L.: A review on remote sensing of weeds in agriculture. *Precision Agriculture* 5(5), 477–508 (2004).
[42] Vong, C.N., Conway, L.S., Zhou, J., Kitchen, N.R., Sudduth, K.A.: Early corn stand count of different cropping systems using UAV-imagery and deep learning. *Computers and Electronics in Agriculture* 186, 106214 (2021).
[43] Zhang, L., Han, W., Niu, Y., Chavez, J.L., Shao, G., Zhang, H.: Evaluating the sensitivity of water stressed maize chlorophyll and structure based on UAV derived vegetation indices. *Computers and Electronics in Agriculture* 185, 106174 (2021).
[44] Shao, G., Han, W., Zhang, H., Liu, S., Wang, Y., Zhang, L., Cui, X.: Mapping maize crop coefficient KC using random forest algorithm based on leaf area index and UAV-based multispectral vegetation indices. *Agricultural Water Management* 252, 106906 (2021).

[45] Sofonia, J., Shendryk, Y., Phinn, S., Roelfsema, C., Kendoul, F., Skocaj, D.: Monitoring sugarcane growth response to varying nitrogen application rates: a comparison of UAV slam lidar and photogrammetry. *International Journal of Applied Earth Observation and Geoinformation* 82, 101878 (2019).

[46] Calou, V.B.C., Santos Teixeira, A., Moreira, L.C.J., Lima, C.S., Oliveira, J.B., Oliveira, M.R.R.: The use of UAVS in monitoring yellow sigatoka in banana. *Biosystems Engineering* 193, 115–125 (2020).

[47] Cao, Y., Li, G.L., Luo, Y.K., Pan, Q., Zhang, S.Y.: Monitoring of sugar beet growth indicators using wide-dynamic-range vegetation index (WDRVI) derived from UAV multispectral images. *Computers and Electronics in Agriculture* 171, 105331 (2020).

[48] Miranda, V., Pina, P., Heleno, S., Vieira, G., Mora, C., Schaefer, C.E.: Monitoring recent changes of vegetation in Fildes Peninsula (King George Island, Antarctica) through satellite imagery guided by UAV surveys. *Science of the Total Environment* 704, 135295 (2020).

[49] Arora, N.K., Mishra, I.: Current scenario and future directions for sustainable development goal 2: A roadmap to zero hunger. *Environmental Sustainability* 5(2), 129–133 (2022).

[50] Wolfert, S., Ge, L., Verdouw, C., Bogaardt, M.-J.: Big data in smart farming-a review. *Agricultural Systems* 153, 69–80 (2017).

[51] Prathibha, S., Hongal, A., Jyothi, M.: IoT based monitoring system in smart agriculture. In: *2017 International Conference on Recent Advances in Electronics and Communication Technology (ICRAECT)*, pp. 81–84 (2017). IEEE, New York.

[52] Perakis, K., Lampathaki, F., Nikas, K., Georgiou, Y., Marko, O., Maselyne, J.: Cybele-fostering precision agriculture & livestock farming through secure access to large-scale HPC enabled virtual industrial experimentation environments fostering scalable big data analytics. *Computer Networks* 168, 107035 (2020).

[53] Klerkx, L., Jakku, E., Labarthe, P.: A review of social science on digital agriculture, smart farming and agriculture 4.0: New contributions and a future research agenda. *NJAS-Wageningen Journal of Life Sciences* 90, 100315 (2019).

[54] Krisnawijaya, N.N.K., Tekinerdogan, B., Catal, C., Tol, R.: Data analytics platforms for agricultural systems: A systematic literature review. *Computers and Electronics in Agriculture* 195, 106813 (2022).

[55] Eli-Chukwu, N.C.: Applications of artificial intelligence in agriculture: A review. *Engineering, Technology & Applied Science Research* 9(4), 4377–4383 (2019).

[56] Dharmaraj, V., Vijayanand, C.: Artificial intelligence (AI) in agriculture. *International Journal of Current Microbiology and Applied Sciences* 7(12), 2122–2128 (2018).

[57] Liu, S.Y.: Artificial intelligence (AI) in agriculture. *IT Professional* 22(3), 14–15 (2020).

[58] Bannerjee, G., Sarkar, U., Das, S., Ghosh, I.: Artificial intelligence in agriculture: A literature survey. *International Journal of Scientific Research in Computer Science Applications and Management Studies* 7(3), 1–6 (2018).

[59] Patr´ıcio, D.I., Rieder, R.: Computer vision and artificial intelligence in precision agriculture for grain crops: A systematic review. *Computers and Electronics in Agriculture* 153, 69–81 (2018).

[60] Liakos, K.G., Busato, P., Moshou, D., Pearson, S., Bochtis, D.: Machine learning in agriculture: A review. *Sensors* 18(8), 2674 (2018).

[61] Richetti, J., Diakogianis, F.I., Bender, A., Cola¸co, A.F., Lawes, R.A.: A methods guideline for deep learning for tabular data in agriculture with a case study to forecast cereal yield. *Computers and Electronics in Agriculture* 205, 107642 (2023).

[62] Chitrakumari:SmartAgriculturalProductionOptimizingEngine(2022).https://www.kaggle.com/datasets/chitrakumari25/smart-agricultural-production-optimizing-engine?resource=download

[63] Kaggle: Global Wheat Detection (2020). University of Saskatchewan. https://www.kaggle.com/competitions/ global-wheat-detection/overview

[64] Ernst, Young: Welcome to the 2023 EY Open Science Data Challenge (2022). https://challenge.ey.com/

Chapter 4

Empowering learners through generative AI

Human-centric design and machine learning in transforming education

Rishi Raghu, Jayashree J, and Vijayashree J

4.1 INTRODUCTION

Artificial intelligence (AI) is rapidly transforming various sectors, and education is no exception. The advent of AI in education has opened up new avenues for personalized learning, collaborative environments, and enhanced natural language processing, thereby revolutionizing the learning process.

The foundation of modern machine intelligence is deep learning, a subset of AI that has played a key role in the development of artificial intelligence during the past few years. Figure 4.1 illustrates how AI, ML, and deep learning connect to one another and how each of them works. Neural networks, attention mechanisms, and transformer models are a few examples of deep learning techniques that have proved essential in improving the learning process. These methods do have drawbacks, too, such as shortcut learning,

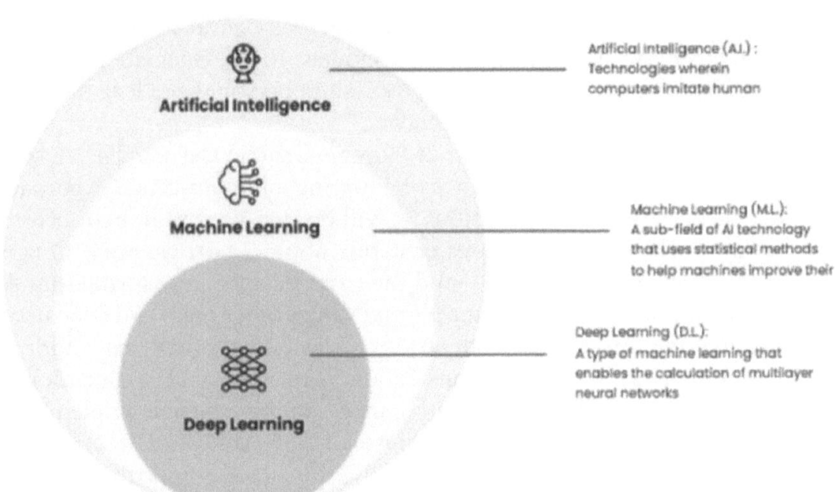

Figure 4.1 How AI, ML, and DL operate and co-relate with each other.

DOI: 10.1201/9781003534761-4

where decision rules work well on common benchmarks but fail to translate to more difficult testing environments, such as real-world scenarios [1].

The use of AI goes beyond only improving learning. Additionally, it is essential in tackling difficult computational science and engineering issues with scant or no physics and high computational demand [2]. Due to its adaptability, AI is a potent tool in the education industry.

Hybrid deep learning models that incorporate a combination of patient information, such as genetic, radiological, and clinical data, have also been developed as a result of the integration of AI in education [3]. These simulations of real-world human behavior capture the essence of AI.

However, integrating AI into education also entails possible dangers and moral questions. It is crucial to adopt a balanced and responsible implementation strategy to ensure that AI applications align with the needs and values of learners and educators.

4.2 OBJECTIVES OF THE CHAPTER

This chapter aims to provide a comprehensive overview of the role of AI in education, with a particular focus on generative AI and human-centric design. The objectives of this chapter are as follows:

i. **Understanding AI and Its Role in Education:** The chapter will begin by providing a background on AI and its significance in the field of education. It will delve into the various ways AI is being utilized to enhance learning experiences and improve educational outcomes.
ii. **Exploring Deep Learning Techniques:** The chapter will explore key aspects of deep learning techniques, such as neural networks, attention mechanisms, and transformer models. It will elucidate their roles in enhancing natural language processing, personalized learning, and collaborative learning environments.
iii. **Highlighting the Importance of Human-Centric Design:** The chapter will emphasize the importance of adopting a human-centric approach in the design of AI applications. It will discuss various human-centric frameworks—including user-centered design, participatory design, and value-sensitive design—and their significance in ensuring that AI applications align with the needs and values of learners and educators.
iv. **Addressing Ethical Considerations:** The chapter will also address potential risks and ethical considerations related to AI in education. It will underscore the need for a balanced and responsible implementation strategy to ensure the ethical use of AI in educational settings.
v. **Providing Recommendations:** Finally, the chapter will offer insights and recommendations for educators, researchers, and developers on harnessing the potential of generative AI to create more inclusive, engaging, and effective learning experiences for all learners.

By achieving these objectives, this chapter aims to contribute to the ongoing discourse on AI in education and provide valuable insights for stakeholders involved in the design and implementation of AI-based educational solutions.

4.3 RELATED WORK

Contributions of AI in the education field are numerous. Some of the contribution are discussed as follows.

4.3.1 Overview of existing research and literature on AI in education

AI has been increasingly recognized as a transformative force in the field of education. The integration of AI in educational settings has been the subject of numerous research studies, each contributing to the understanding of how AI can enhance learning experiences and outcomes.

The paper by Raissi, Perdikaris, and Karniadakis [4], which pioneered the idea of physics-informed neural networks, is one of the ground-breaking works in this field. The general nonlinear partial differential equations that explain the laws of physics are used to train these networks to solve supervised learning tasks while adhering to those rules. This research has important ramifications for AI in education, especially for the teaching and learning of difficult scientific ideas [4].

Atan, Jordon, and Schaar [5] explored the potential of deep learning in creating personalized treatment policies in the medical domain. Their work, titled "Deep-Treat: Learning Optimal Personalized Treatments from Observational Data Using Neural Networks," highlighted the potential of AI in personalizing learning experiences, a concept that can be extended to educational settings [5].

4.3.2 Studies on the impact of human-centric design on learner experiences

The impact of human-centric design on learner experiences has been a focal point of several studies. Schultz and DeMers [6] discussed the transition from emergency remote learning to deep online learning experiences in geography education. Their work emphasized the importance of instructional design in online courses, highlighting the potential of human-centric design in enhancing learner experiences [6].

Gillaspy and Vasilica [7] explored the potential of pedagogical design in fostering the development of the digital self-determined learner. Their findings suggested that with careful scaffolding and adherence to the educational philosophy, this approach has the potential to develop learners who demonstrate key principles of heutagogy, including nonlinear learning, learner agency, capability, self-reflection, and metacognition [7].

Li [8] conducted a study on learner-centered learning tasks in higher education. The results of the study show the value of learner-centered, constructivist-oriented learning tasks because students' enthusiasm to tackle subject material remained constant despite a rise in complexity and workload [8].

4.3.3 Research on the integration of deep learning techniques in educational contexts

Numerous research papers have focused on the integration of deep learning techniques in educational settings. A class-balanced dynamically weighted loss function serves as the foundation for the class rebalancing technique Fernando and Tsokos (2021) presented. With the help of complex minority class samples, this method enables the model to account for instances with varied degrees of difficulty, leading to gradient adjustments. This work has implications for individualized learning, in which the activities and resources can be changed according to the learner's needs and proficiency [9].

The layer depths and level of fine-tuning on transfer learning were assessed by Lee et al. (2020) using a deep convolutional neural network (CNN)-based COVID-19 screening in chest X-ray images.

Their research provided evidence that the right amount of fine-tuning can aid in the building of an effective deep learning model, a concept that can be used in the creation of AI-based educational tools and applications [10].

These papers give a thorough overview of the state of the art in research on the use of AI and human-centered design in education, demonstrating the potential of these tools to improve learning processes and results.

4.4 HUMAN-CENTRIC DESIGN IN AI FOR EDUCATION

Human-centric design, also known as user-centered design, is an approach to designing systems and products that prioritize the needs, preferences, and limitations of end users at each stage of the design process, as shown in Figure 4.2. This approach involves a deep understanding of users, including their tasks, environments, and the entire system within which they operate. The goal is to create solutions that are highly usable and accessible, leading to increased user satisfaction and improved performance.

Holstein and Aleven (2021) emphasize the importance of human-centric design in the context of AI-supported classrooms. They argue for more participatory approaches to research and design in this area, in which practitioners and other stakeholders are deeply, and meaningfully involved throughout the process. Furthermore, they advocate for theory-building and for principled approaches to the study of human-AI decision-making in real-world contexts [11].

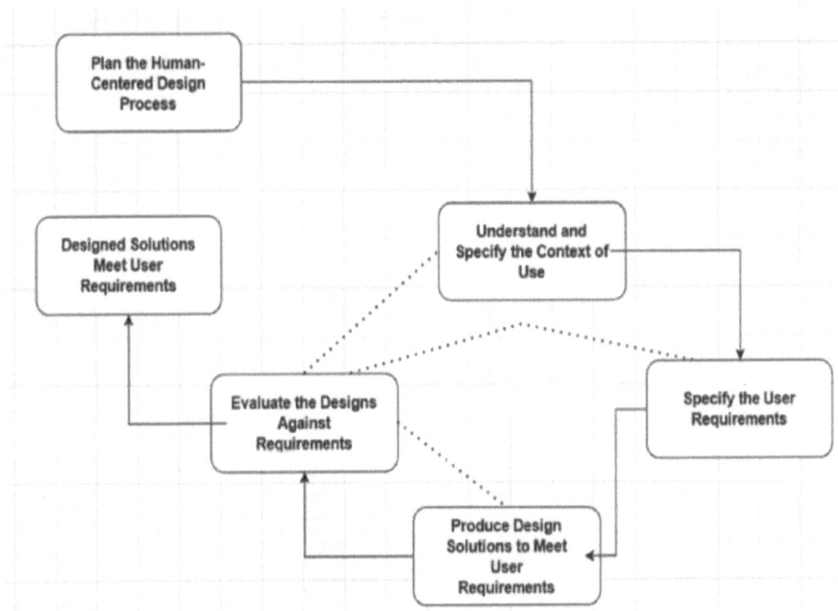

Figure 4.2 Phases of human-centered design.

4.4.1 Importance of human values, emotions, and ethics in AI applications

The integration of human values, emotions, and ethics in AI applications is crucial for ensuring that these technologies align with the needs and values of users. Reinforcement learning is utilized to create a social mechanism that most humans find appealing in the Democratic AI research pipeline (Koster et al. 2022). Democratic AI provides a proof of concept for value-aligned policy innovation by optimizing for human preferences [12].

A reference map of how the intersection between ethics, and governance with AI is in Figure 4.3.

4.4.2 Role of human-centric design in creating learner-centered AI technologies

Human-centric design plays a pivotal role in creating learner-centered AI technologies. It ensures that the design of these technologies is guided by the needs and preferences of learners, leading to more effective and engaging learning experiences. Figure 4.4 proposes a flow of how it should ideally take place

Cirqueira et al. (2020) used a fraud detection context to show the application of scenario-based requirements elicitation for user-centric explainable AI. To identify user or expert requirements for acceptable explanations in

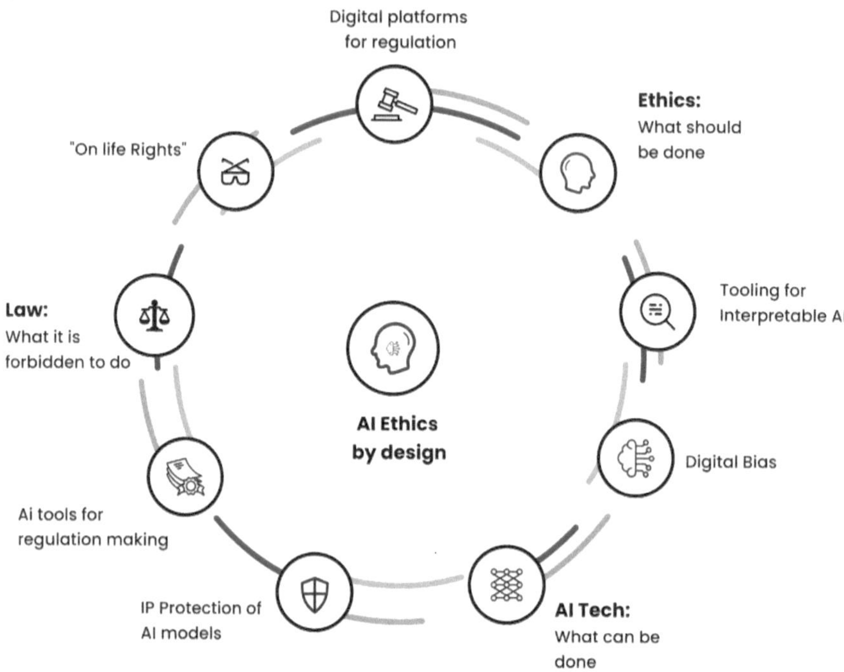

Figure 4.3 The intersection of ethics and governance with AI.

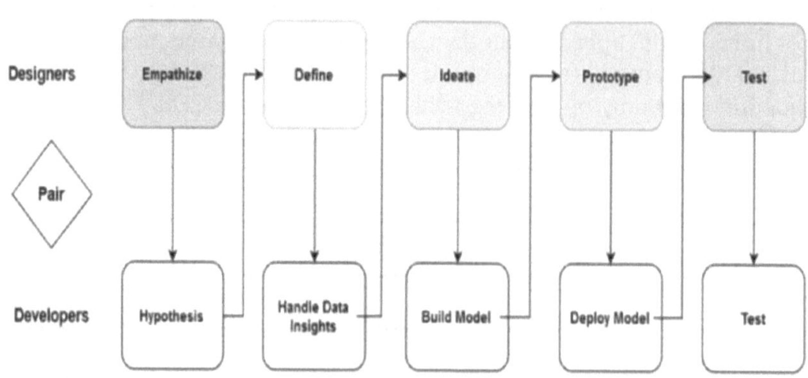

Figure 4.4 Human-centric AI design process.

their daily operations, they constructed scenarios based on banking fraud experts and explained how those scenarios may be adopted [13].

The iterative design and evaluation of Lumilo, a pair of smart glasses that provide teachers with real-time data about their students' learning, metacognition, and behavior in classrooms powered by AI, were presented by Holstein and Aleven in 2021. When teachers and AI tutors collaborate in the classroom, according to the findings of a field research done in K–12 schools, children learn more [11].

These studies collectively highlight the importance of human-centric design in the development of AI applications in education, underscoring the need for a deep understanding of users and their needs, preferences, and limitations.

4.4.3 Integration of human-centric frameworks in AI development for education

The integration of human-centric frameworks in AI development for education is a critical aspect of ensuring that AI technologies are designed and implemented in ways that are beneficial and acceptable to users. These frameworks include user-centered design, participatory design, and value-sensitive design.

4.4.3.1 User-centered design

The user is at the center of the design process according to the user-centered design (UCD) design philosophy. At every level of the design process, it entails being aware of the demands, desires, and constraints of the users. UCD emphasizes the value of including users at every stage of the design process using a range of research and design tools, including observations, usability testing, and interviews. This strategy guarantees that the end product is easy to use and customized to fit the needs of the users. UCD has been used in a variety of domains, including education, where it can aid in the creation of learning-effective and engaging educational technologies [14,15].

4.4.3.2 Participatory design

Another human-centric design strategy is participatory design (PD), which includes end users and other stakeholders in the design process. The premise of PD is that everyone who will be impacted by a design should have a voice in its creation. This could involve involving teachers, students, and administrators in the design of AI-based educational aids in the context of education. This strategy may result in the creation of AI technologies that are more suited to the requirements and environment of the users [16].

4.4.3.3 *Value-sensitive design*

Value-sensitive design (VSD) is a theoretical framework that accounts for human values in the design of technology. VSD involves identifying the values at stake in a particular design context, considering the potential impact of the technology on these values, and incorporating this understanding into the design process. For instance, in the context of AI for education, values such as privacy, fairness, and inclusivity might be particularly relevant. A study by van Wynsberghe and Robbins [17] discusses the application of VSD in the context of care robots, highlighting the importance of considering values such as dignity, privacy, and social connectedness in the design process.

4.5 DEEP LEARNING TECHNIQUES FOR ENHANCED LEARNING

Due to its capacity to analyze massive amounts of data and produce insightful findings, deep learning, a subset of machine learning, is being used more and more in a variety of industries, including education. Partial differential equations (PDEs) are a challenging problem to solve, but deep learning methods, in particular neural networks, have demonstrated amazing effectiveness in a variety of applications. For instance, Lu et al. (2019) demonstrated the potential of deep learning in strengthening computational science and engineering education by introducing physics-informed neural networks (PINNs), which embed a PDE into the loss of the neural network through automatic differentiation [18].

Moreover, deep learning has been used to create more engaging and personalized learning experiences. Abdullah et al. (2021) discussed the use of deep learning for multimodal emotion recognition, which can be applied in educational settings to understand and respond to students' emotional states, thereby creating a more empathetic and effective learning environment [19].

In the context of language learning, Mukhamadiyev et al. (2022) proposed a deep learning-based speech recognition model for the Uzbek language and its dialects. This model demonstrated the potential of deep learning in enhancing language education by providing accurate speech recognition, which can be particularly beneficial for learners of low-resource languages [20].

Furthermore, the integration of deep learning with Internet of Things (IoT) technologies has been explored to enhance physical education. Li et al. (2021) proposed an IoT-assisted physical activity monitoring device that uses deep learning to track and analyze students' physical activities. This approach not only improves the monitoring and assessment of physical education but also encourages students to engage more actively in their physical well-being [21].

These examples illustrate the diverse applications of deep learning in education, from enhancing computational and language learning to facilitating emotional recognition and physical activity monitoring. As deep learning continues to evolve, it is expected to play an increasingly significant role in transforming educational practices and outcomes.

4.5.1 Neural networks in personalized learning

Neural networks can be used for personalized learning as discussed below.

4.5.1.1 Adaptive learning systems

Deep learning techniques are used by adaptive learning systems to offer individualized learning opportunities that are tailored to the requirements of specific learners. One such system is the hybrid intelligent tutoring system (ITS), which was developed by fusing the conversational tutoring system AutoTutor with the adaptive learning system for mathematics called Assessment and Learning in Knowledge Spaces (ALEKS). This method provides a customized learning path based on the student's unique learning style and problem-solving approach by guiding students through step-by-step work examples for algebra issues using self-explanation tutoring dialogs [22].

Another innovative approach to adaptive learning systems is the use of swarm intelligence approaches, such as ant colony optimization (ACO). In the system, an adaptive tutoring system was developed using ACO, where the learning object is personalized based on learning and problem-solving styles. This system generates an optimal learning path that adapts to the learner's style, thereby enhancing the learning experience [23].

4.5.1.2 Intelligent tutoring systems

ITS are increasingly incorporating deep learning techniques to provide personalized feedback and improve student learning outcomes. For instance, Grenander et al. (2021) developed an ITS that uses neural discourse segmentation and classification techniques to decompose students' answers and generate personalized feedback. This approach allows the system to target each student's misconceptions and knowledge gaps, resulting in significantly improved student learning gains [24].

In the realm of physical manipulation tasks, Lima et al. (2021) presented an intelligent control scheme for flexible link manipulators using a sliding mode controller with an embedded artificial neural network. This system adapts itself by learning from experience, demonstrating the potential of neural networks in creating adaptive learning environments for complex physical tasks [25].

These examples illustrate the potential of neural networks in creating adaptive learning systems and ITSs that provide personalized and effective learning experiences.

4.5.1.3 *Attention mechanisms and their role in collaborative learning environments*

Attention mechanisms have been increasingly utilized in collaborative learning environments, particularly in the context of deep learning models. These mechanisms are designed to focus on the most relevant parts of the input data, thereby improving the model's ability to learn complex patterns and relationships. In the context of collaborative learning environments, attention mechanisms can be used to highlight the most relevant interactions or contributions, thereby enhancing the learning process [26].

For instance, in a study by Feng et al. (2020), a Generative Adversarial Network (GAN) based on collaborative learning and attention mechanism was proposed for hyperspectral image classification [26]. The attention mechanism in this model was designed to discard misleading and confounding information, thereby improving the quality of the generated samples. This approach demonstrates how attention mechanisms can be used to enhance the learning process in a collaborative setting.

4.5.1.4 *Transformer models for natural language processing in educational contexts*

Transformer models have revolutionized the field of Natural Language Processing (NLP) and have found significant applications in educational contexts. These models, introduced by Google AI researchers in 2017, use self-attention mechanisms to map input sequences to output sequences, thereby overcoming the limitations of recurrent neural networks (RNNs) [27].

BERT (Bidirectional Encoder Representations from Transformers), one of the most well-liked transformer models, has been extensively employed for a variety of tasks including text classification, question answering, and natural language inference. Transformer models can be employed in educational settings for tasks including automatic grading, content summarization, and individualized learning [28].

For instance, the transformer model was utilized in a study by Vaswani et al. (2017) to increase the accuracy of machine translation by utilizing self-attention and multi-head attention mechanisms. Since then, this approach has been applied to numerous tasks like text summarization, question answering, and natural language inference, proving its usefulness in educational settings.

4.6 LEVERAGING GENERATIVE AI FOR INCLUSIVE LEARNING EXPERIENCES

Generative AI has the potential to revolutionize education by creating inclusive learning experiences that cater to the needs of diverse learners. It can generate personalized content, provide instant feedback, and foster active learning, thereby enhancing the overall learning experience.

4.6.1 Potential benefits of generative AI in education

Generative AI has the potential to revolutionize education by creating inclusive learning experiences that cater to the needs of diverse learners. It can generate personalized content, provide instant feedback, and foster active learning, thereby enhancing the overall learning experience.

The capacity to tailor learning experiences is one of the key advantages of generative AI in education. To create personalized learning materials, generative AI can assess a student's learning style, rate of learning, and competence level. By using a tailored method, students can learn topics more quickly and comfortably, which will improve their learning outcomes [29].

Generative AI can also provide instant feedback to students, which is crucial for their learning process. It can assess students' responses in real time and provide immediate feedback, helping them understand their mistakes and learn from them. This instant feedback mechanism can significantly enhance students' learning experience and boost their confidence [30].

Moreover, generative AI can foster active learning by engaging students in the learning process. It can generate interactive content, such as quizzes, puzzles, and games, that can make learning more engaging and fun. Active learning has been shown to improve students' understanding of the subject matter and enhance their critical thinking skills [31].

In addition, generative AI can help create an inclusive learning environment by catering to the needs of diverse learners. It can generate content in multiple languages and formats, making education accessible to students from different linguistic and cultural backgrounds. It can also cater to the needs of students with disabilities by generating content in accessible formats [32].

Furthermore, generative AI can help educators by automating routine tasks, such as grading and content generation, thereby allowing them to focus more on teaching and less on administrative tasks. It can also provide educators with insights into students' learning progress, helping them tailor their teaching strategies to meet the needs of individual students [33].

4.6.2 Promoting inclusivity through generative AI

Generative AI can play a pivotal role in promoting inclusivity in education by providing customization and adaptability for diverse learners. It can generate

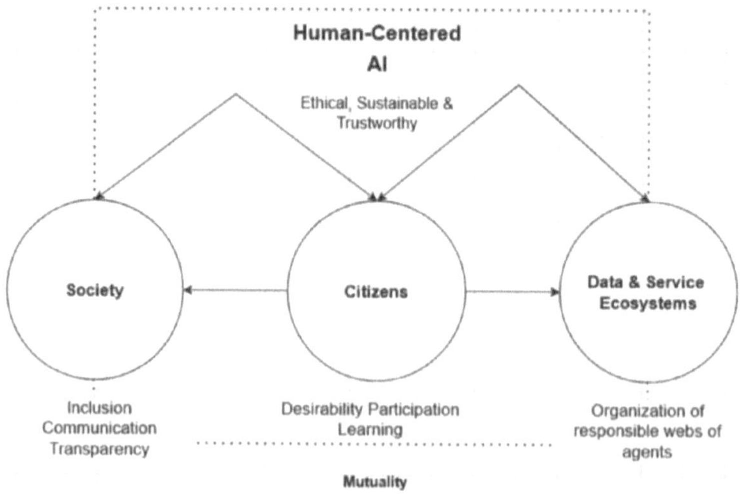

Figure 4.5 An approach to human-centered inclusive AI development and deployment.

personalized learning materials that cater to the unique learning styles, proficiency levels, and interests of individual learners. This level of customization and adaptability can help create an inclusive learning environment where all learners, regardless of their backgrounds or abilities, can thrive [34].

Figure 4.5 depicts how the systematic approach to human-centered AI development promotes inclusive and sustainable social growth based on transparency and communication in decision-making.

4.6.2.1 Customization and adaptability for diverse learners

Generative AI can analyze a learner's interactions with the learning materials and adapt the content based on their learning progress. For instance, if a learner is struggling with a particular concept, the AI can generate additional resources or exercises to help them understand the concept better. This adaptability can help ensure that all learners can keep up with the learning pace and achieve their learning goals [35].

4.6.2.2 Multimodal learning experiences

Furthermore, generative AI can facilitate multimodal learning experiences by generating content in various formats, such as text, audio, images, and videos. This can cater to the diverse learning preferences of learners, as some learners might prefer visual content, while others might prefer auditory or textual content. Multimodal learning experiences can also enhance learners' understanding of the subject matter and improve their retention of the learned content [36].

Moreover, generative AI can generate content in multiple languages, making education accessible to learners from different linguistic backgrounds. It can also generate content in accessible formats, such as braille or sign language, to cater to the needs of learners with disabilities. This can help create a truly inclusive learning environment where all learners, regardless of their language proficiency or physical abilities, can participate and succeed [37].

4.6.3 Enhancing engagement and motivation with generative AI technologies

Enhancing engagement and motivation with generative AI technologies is a burgeoning field of study that has shown promising results in various educational contexts.

4.6.3.1 Gamification and interactive learning

Gamification and interactive learning have been identified as effective strategies for increasing student engagement and motivation. Gamification, the application of game-design elements in non-game contexts, has been shown to enhance learning experiences by making them more interactive and enjoyable. For instance, a study by Chen et al. (2019) demonstrated that gamification in an engineering course led to improved student engagement and learning outcomes [38]. The study utilized a game-based learning platform that allowed students to practice engineering concepts in a fun and interactive way. This approach not only enhanced students' understanding of the course material but also increased their motivation to learn.

Interactive learning, on the other hand, involves the use of technology to facilitate active participation in the learning process. A study by Huang et al. (2019) showed that interactive learning technologies, such as virtual and augmented reality, can significantly enhance students' understanding of complex concepts [39]. These technologies provide immersive learning experiences that allow students to explore and interact with the learning material in a more engaging and meaningful way.

4.6.3.2 Personalized feedback and assessment

Additional applications for generative AI technology include personalized feedback and assessment. Students' learning experiences can be improved by AI by giving them individualized feedback depending on their learning progress. Nguyen and Yang's paper from 2021 covered the usage of AI in education, particularly how it can be used to deliver individualized feedback and assessment [40]. The study focused on how sophisticated algorithms,

NLP, and machine learning are used to build intelligent systems that can give students individualized feedback in real time. By improving their algorithms, these systems may self-adjust to match users' changing needs, resulting in a more customized and productive learning environment.

The study also covered how AI is used to develop adaptive learning systems. These systems can modify themselves to fit the traits, requirements, and preferences of specific learners, resulting in a more tailored learning experience. The use of AI in adaptive learning systems is supported by big data, which allows for the accumulation of diverse individual characteristics and learning styles necessary for intelligent personalization.

However, the study also noted the challenges associated with the use of AI in education, including the need for more profound information retrieval for the assessment of complex educational systems and the need for more research grounded in learning theories to make AI an integral part of learners' lives.

In summary, generative AI technologies hold significant potential for enhancing engagement and motivation in education. Through gamification, interactive learning, personalized feedback, and adaptive learning systems, AI can provide more engaging, interactive, and personalized learning experiences. However, more research is needed to overcome the challenges associated with the use of AI in education and to fully realize its potential.

4.7 PROPOSED WORK

Generating educational content in the form of text, video, and images can have a profound impact on personalized learning. Visual cues can significantly enhance learners' understanding, especially for complex concepts. In this context, Generative AI can create custom educational images tailored to individual learners' needs. Below, we describe the process of using GANs for image content generation in education.

In unsupervised machine learning, a class of AI algorithms known as "Generative Adversarial Networks" (GANs) implements two neural networks competing against one another in a framework similar to a zero-sum game. Candidates are created by one network (the generator), and evaluated by another (the discriminator). Figure 4.6 depicts this in more detail.

4.7.1 Conceptualizing the framework

4.7.1.1 Understanding the problem space

The need for personalized learning solutions arises from the fact that each learner is unique, with different learning styles, abilities, and preferences. Despite this understanding, traditional educational systems often follow a one-size-fits-all approach that does not cater to individual learner needs. The proposed framework will take into account the principles of human-centric

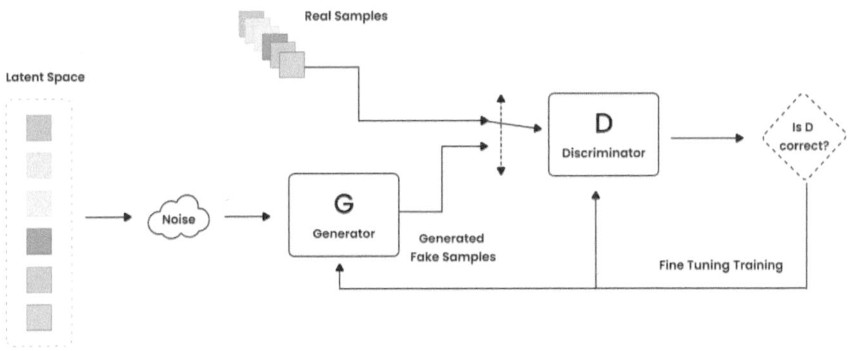

Figure 4.6 Generative adversarial networks (GANs) conceptual flow.

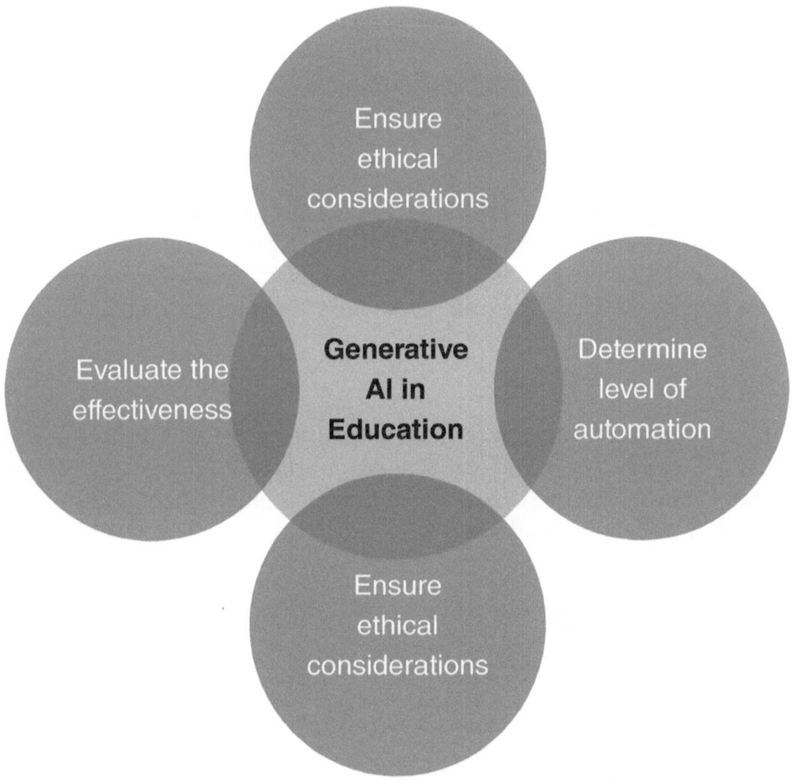

Figure 4.7 Framework overview of using generative AI in education.

design and generative AI to develop a solution that tailors educational content and methods to the individual learner. This personalized approach will aim to enhance learner engagement, improve learning outcomes, and promote inclusivity. An overview of this framework is in Figure 4.7

4.7.1.2 *Proposed solution*

The proposed solution is a human-centric AI system for personalized learning, which will leverage generative AI techniques to create individualized learning materials and experiences. This system will analyze each learner's interactions, learning style, pace, and proficiency level to generate customized learning materials and adaptive learning paths.

4.7.2 Design principles of the proposed framework

Design principles of the proposed framework are as follows.

4.7.2.1 *Human-centric design*

The system will be designed around the needs and preferences of learners. Through a deep understanding of users and their learning contexts, the system will aim to create solutions that are highly usable, accessible, and satisfying to learners. This will involve the integration of UCD, PD, and VSD principles.

4.7.2.2 *Generative AI*

The system will use deep learning techniques to generate educational content that adapts to the learner's needs. It will make use of AI technologies such as neural networks, attention mechanisms, and transformer models to analyze learner data and generate personalized content.

4.7.2.3 *Ethics and privacy*

The framework will adhere to ethical guidelines in AI and respect user privacy. It will follow the principle of data minimization, collecting only the necessary data to deliver personalized learning experiences. Additionally, all data will be anonymized and stored securely.

4.7.3 Development of a generative AI model

For content generation in personalized learning, we use GANs as follows:

i. **Dataset Compilation:** Start by compiling a dataset of educational images relevant to the subject matter. These images can be diagrams, visual illustrations, infographics, etc. This dataset will serve as the training data for the GANs. For personalization, this dataset can include metadata related to students' learning styles and preferences.
ii. **Training the GAN:** A generator and a discriminator are the two models that make up the GAN. The discriminator model is taught to discriminate between real photos from the dataset and fake images created by the generator, while the generator model is trained to create new images.

iii. **Image Generation:** The generator model receives a signal of random noise and outputs an image. The discriminator model assesses this image. The generator modifies its parameters to create more realistic images if the discriminator deems the image to be phony. The generator iterates this process several times, improving its output each time.
iv. **Personalization:** The generated images can be personalized based on student's learning styles and preferences. For instance, if a student prefers diagrammatic representations, the GAN can be trained to generate more diagrams. If a student learns better with real-life images, the GAN can be trained to generate more realistic images. The personalization parameters can be adjusted based on continuous feedback from the learners.
v. **Evaluation:** The generated images are evaluated for their educational value and relevance to the subject matter. Teachers or subject matter experts can review the generated images to ensure they are accurate and helpful. This feedback can be used to fine-tune the GAN model further.

4.7.4 Implementation of attention mechanisms in a collaborative learning environment

Attention mechanisms in AI are powerful tools that enable machine learning models to improve their performance by focusing on specific aspects of their input data. In many different disciplines, particularly in NLP tasks like translation and summarization, these strategies have been applied to great success. Attention processes can be used in the context of education to design more productive and interesting collaborative learning environments.

One potential application of attention mechanisms in collaborative learning is to personalize instruction based on individual student's needs. For example, an AI tutor equipped with attention mechanisms can adapt its teaching strategy according to a student's learning style, proficiency level, and attention span. By focusing on the aspects of the learning material that the student finds challenging or engaging, the AI tutor can provide personalized instruction that caters to the student's unique learning needs. This can help enhance the student's understanding of the subject matter, thereby improving their learning outcomes.

In addition, attention mechanisms can be used to design learning materials that are more engaging and effective. For example, by focusing on the parts of the learning material that students find most engaging, educators can design lessons that capture students' interest and keep them engaged. Similarly, by focusing on the parts of the learning material that students find difficult, educators can design lessons that provide additional support and guidance in those areas.

However, there are a number of difficulties in implementing attention processes in a collaborative learning setting. The necessity for a lot of high-quality data to train the machine learning models is one of the challenges. Keeping student data private and secure is another difficulty. Designing attention mechanisms that can adjust to the dynamic nature of the learning process is a difficulty as well, given that students' learning demands and preferences can alter over time.

Despite these difficulties, adopting attention processes in a setting of collaborative learning has significant potential advantages. These strategies can considerably improve the effectiveness and efficiency of the learning process by providing tailored instruction and fostering efficient student collaboration. To fully utilize the potential of attention mechanisms in transforming education, more study in this field is necessary.

4.7.5 Leveraging transformer models for enhanced NLP in education

Transformer models are emerging as a transformative force in the field of NLP, providing new avenues for enhancing the learning process in educational settings. Transformer models, such as the widely adopted BERT, leverage self-attention mechanisms to understand the contextual relation between words in a text. The self-attention mechanism allows words to interact with each other directly and weigh their importance to each other, which is beneficial for understanding the context and meaning of sentences [28].

In the context of education, transformer models can be used to build ITSs that can understand and respond to students' queries more accurately. For instance, when a student asks a complex question, the system can use transformer models to parse the sentence, understand the context, and generate a more accurate and contextual response [41]. Furthermore, these models can also be used to create automated essay grading systems. These systems can understand the semantics of the essay text using transformer models, making them more reliable and accurate than traditional rule-based grading systems [42].

Implementation of transformer models in education involves training the model on a large corpus of educational data. This data can include textbooks, educational websites, and other resources. Once the model is trained, it can be integrated into the learning platform and used to enhance the NLP capabilities of the platform. This can significantly improve the learning experience for students by providing them with more accurate and personalized responses to their queries [43]. It is important to note, however, that the use of transformer models in education also raises important ethical and privacy concerns that need to be addressed, such as data security and student privacy [44].

In conclusion, transformer models hold great promise in enhancing NLP in education. By leveraging these models, educators can build ITSs that provide more accurate and personalized learning experiences for students. However, the implementation of these models needs to be carried out carefully, taking into account ethical and privacy concerns.

4.8 ETHICAL CONSIDERATIONS AND RISKS IN AI FOR EDUCATION

Ethical considerations play a crucial role in the development and deployment of AI applications in education, particularly in ensuring fairness, transparency, privacy, and accountability. AI systems, such as deep learning and machine learning algorithms, have the potential to significantly influence learners' educational experiences and outcomes. As such, it is critical to consider the ethical implications of these technologies to ensure that they are used responsibly and beneficially [45].

4.8.1 Importance of ethical considerations in AI applications

Fairness is a key ethical consideration in AI applications. As AI technologies become increasingly prevalent in education, there is a risk of bias in AI algorithms, which can lead to unfair treatment of certain groups of learners. For instance, AI algorithms that rely on historical data to make predictions can perpetuate existing inequalities in education if the data is biased [46]. To ensure fairness, it is important to adopt a transparent approach in the development and deployment of AI algorithms, clearly explaining how the algorithms work and making sure that they do not unfairly disadvantage certain groups of learners.

Privacy is another crucial ethical consideration in AI applications. With the increasing use of AI in education, there is a growing concern about the privacy of learners' data. AI systems often require large amounts of data to function effectively, which raises concerns about data privacy and security. Learners' data can include sensitive information, such as their academic performance, personal background, and learning preferences. Ensuring the privacy of this data is paramount to maintaining learners' trust and ensuring the ethical use of AI in education [47].

Accountability is another important ethical factor in AI applications. It speaks to the necessity of holding persons responsible for the design and deployment of AI systems liable for the choices that these systems make. Given the autonomy of many AI systems, it might be challenging to assign blame when these systems take actions that have a big effect on students. To ensure that the advantages of AI in education are achieved and that any negative effects are minimized, it is essential to establish clear lines of accountability [48].

Overall, ethical considerations in AI applications are paramount in ensuring that these technologies are used responsibly and beneficially in education. By considering fairness, transparency, privacy, and accountability, we can help ensure that AI technologies enhance learners' educational experiences and outcomes, rather than exacerbating existing inequalities or infringing on learners' rights [49].

4.8.2 Ethical challenges specific to AI in education

Ethical considerations in AI for education extend beyond the universal principles of AI ethics and require a close examination of issues specific to the education sector. These include data privacy and security, bias and fairness in AI algorithms, and social and psychological impacts on learners.

4.8.2.1 Data privacy and security

The use of AI in education inevitably involves the collection and processing of vast amounts of personal data about students, including their academic performance, learning preferences, and behavioral patterns. While this data can be instrumental in providing personalized education, it also raises serious concerns about data privacy and security.

As Klimova and Simonova (2019) highlighted, AI systems in education can potentially expose sensitive student data to unauthorized parties, leading to privacy violations and identity theft. These risks are exacerbated by the fact that many educational institutions lack robust data protection mechanisms.

Furthermore, the sharing of student data with third-party AI service providers can potentially result in the misuse of this data for purposes unrelated to education. As such, it is essential to establish strong data governance policies and practices to ensure that student data is used responsibly and securely.

4.8.2.2 Bias and fairness in AI algorithms

Bias and fairness in AI algorithms are another significant ethical challenge in the use of AI in education. AI systems are trained on existing data, which can often reflect historical biases and inequalities. If not addressed properly, these biases can be perpetuated and amplified by the AI systems, leading to unfair and discriminatory outcomes.

For instance, a study by Eirinaki et al. (2018) found that AI-based recommendation systems used in e-learning platforms can inadvertently favor certain types of content or learners, thereby perpetuating existing disparities in learning opportunities. Similarly, AI systems used for student assessment can potentially be biased against certain student groups if the assessment criteria are not properly calibrated.

Therefore, it is crucial to incorporate fairness and bias mitigation strategies into the design and deployment of AI systems in education. This includes the use of diverse training data, rigorous algorithmic auditing, and continuous monitoring and adjustment of the AI systems.

4.8.2.3 Social and psychological impact on learners

The use of AI in education can also have profound social and psychological impacts on learners. As Buckingham et al. (2019) argued, AI systems can potentially create a sense of surveillance and control in the learning environment, leading to increased stress and anxiety among students.

Moreover, the reliance on AI for learning support can potentially reduce human interaction in the learning process, which can negatively impact students' social skills and emotional development. As Bostrom and Yudkowsky (2014) note, the potential dehumanizing effect of AI is a serious ethical concern that needs to be addressed in the design and deployment of AI systems in education.

Additionally, the use of AI for personalized learning can potentially create a 'filter bubble' effect, where students are only exposed to content that aligns with their existing knowledge and preferences. This can limit their exposure to diverse perspectives and hinder their critical thinking skills.

Addressing these social and psychological impacts requires a human-centric approach to AI in education, where the well-being of learners is prioritized over the efficiency and convenience of AI systems. This includes ensuring that AI systems are designed and used in a way that respects students' privacy, promotes fairness, and supports their social and emotional development.

4.8.3 Responsible implementation strategies and guidelines

A responsible approach toward the implementation of human-centric AI solutions for personalized learning requires strategic planning, stakeholder engagement, and ethical considerations. The key strategies and guidelines include:

4.8.3.1 Ethical considerations

To avoid problems like data misuse, discrimination, and privacy infringement, it is essential to ensure the ethical usage of AI in education. Aiming for openness, inclusion, and responsibility in AI systems, the OECD Principles on AI and the IEEE Ethically Aligned Design establish ethical criteria that implementers should follow [50,51].

4.8.3.2 Stakeholder involvement

The successful implementation of AI solutions necessitates the involvement of all stakeholders, including educators, learners, parents, and policymakers. A PD process [17] can help ensure the technology meets the needs of its users and is effectively integrated into the learning process.

4.8.3.3 Professional development and training

To ensure the effective use of the AI solution, ongoing professional development, and training for educators should be provided. This can help them understand how to best use the technology to support personalized learning and address any technical challenges that may arise [52].

4.8.3.4 Regular evaluation and improvement

After implementation, regular evaluation of the AI system should be conducted to measure its effectiveness, identify any issues, and make necessary improvements. This could involve collecting feedback from users, monitoring system performance, and conducting impact assessments [53].

4.8.3.5 Data protection and privacy

With the use of AI systems, data protection, and privacy must be prioritized. Measures must be in place to secure student data, and clear policies should outline how data is collected, stored, and used [54].

4.8.3.6 Equal access and inclusivity

Ensuring equal access to the AI solution is paramount to avoid exacerbating existing inequalities. This involves considering factors such as the digital divide and ensuring the system is designed to be inclusive and accessible for all learners [55].

4.9 INSIGHTS AND RECOMMENDATIONS FOR EDUCATORS, RESEARCHERS, AND DEVELOPERS

Insights and Recommendations for Educators, Researchers, and Developers are discussed in this section.

4.9.1 Best practices for integrating AI technologies in educational settings

AI technologies can provide a great deal of support in educational settings, but they must be carefully integrated to ensure their efficacy. Effective strategies include the gradual introduction of AI tools, starting with simple applications and progressing to more advanced ones, allowing educators and students to acclimate to the technology [45]. Also, both teachers and students should receive training on how to use these tools effectively, and the technology should be used to augment the teacher's work, not replace it [49].

Lastly, evaluation mechanisms should be established to assess the impact and effectiveness of AI technologies on teaching and learning [53].

4.9.2 Collaboration between educators and AI developers

Effective AI integration in educational settings requires robust collaboration between educators and AI developers. Educators provide context-specific insights and ensure that the developed tools meet the educational needs [17]. Conversely, AI developers can offer technical guidance to ensure that AI tools function optimally in educational settings [48]. This mutual exchange of knowledge can lead to the creation of AI tools that are both efficient and contextually relevant.

4.9.3 Ethical guidelines and policies for AI in education

There are many ethical issues with using AI in teaching. These include confidentiality, equity, and openness. To be fair and transparent in their functioning, AI technologies must be built and used in a manner that respects students' privacy [46,47]. Guidelines for ethical AI development and use can be found in documents like the OECD Principles on AI and the IEEE's Ethically Aligned Design [50] (OECD, 2019; IEEE, 2019). When using AI tools, it is equally essential to protect student data privacy and digital literacy [54].

4.9.4 Future directions and emerging trends in AI for education

AI in education is a rapidly developing field. Emerging trends include personalized learning, where AI tailors educational content to individual student needs, and the use of AI for efficient administrative tasks [45]. Further, AI can provide data-driven insights to help improve teaching strategies and outcomes [49]. Educators, researchers, and developers need to stay informed about these trends and adapt their practices accordingly. However, continued research and ethical guidelines are necessary to navigate the challenges and potentials of these developments [48].

4.10 CONCLUSION

The related work explored a range of topics, such as the integration of AI technologies in educational settings, collaboration between educators and AI developers, ethical considerations, and future trends. AI's potential to augment teaching and learning processes was widely recognized, but the importance of careful implementation, ongoing training, and regular assessment was also highlighted [45,49]. Collaboration between educators and AI developers was identified as essential for creating contextually relevant and technically robust AI tools [17] Significant ethical concerns included privacy, fairness, and transparency, with several policy frameworks proposed to guide ethical AI use [46,47].

The findings have significant implications for research and practice in AI-based education. For practice, the importance of collaboration between developers and educators, careful integration of AI technologies, and ongoing training is evident. These findings call for a pedagogy-driven approach to AI integration in education. For research, the findings highlight the need for more studies on the ethical implications of AI, the efficacy of AI tools in various learning contexts, and the development of guidelines for fair and transparent AI use.

Despite the considerable progress made, several open challenges remain. These include ensuring the equity of AI tools, addressing privacy concerns, and managing the risks of reliance on AI in education. Future research could explore adaptive learning systems that modify their behavior based on student performance, AI tools that assist teachers with administrative tasks, and the development of best practices for implementing AI in various educational contexts. In addition, as AI continues to evolve, there will be a need for ongoing exploration of ethical guidelines and policies to ensure that AI benefits all learners without compromising their rights or well-being.

REFERENCES

[1] Geirhos, R., Jacobsen, J., Michaelis, C., Zemel, R., Brendel, W., Bethge, M., & Wichmann, F. (2020). Shortcut learning in deep neural networks. *Nature Machine Intelligence*, 2, 665–673.

[2] Saha, S., Gan, Z., Cheng, L., Gao, J., Kafka, O. L., Xie, X., Li, H., Tajdari, M., Kim, H. A., & Liu, W. K. (2021). Hierarchical deep learning neural network (HiDeNN): an artificial intelligence (AI) framework for computational science and engineering. *Computer Methods in Applied Mechanics and Engineering*, 373, 113452.

[3] Saravi, B., Hassel, F., Ülkümen, S., Zink, A., Shavlokhova, V., Couillard-Després, S., Boeker, M., Obid, P., & Lang, G. (2022). Artificial intelligence-driven prediction modeling and decision making in spine surgery using hybrid machine learning models. *Journal of Personalized Medicine*, 12(4), 509.

[4] Raissi, M., Perdikaris, P., & Karniadakis, G. E. (2019). Physics-informed neural networks: a deep learning framework for solving forward and inverse problems involving nonlinear partial differential equations. *Journal of Computational Physics*, 378, 686–707.

[5] Atan, O., Jordon, J., & van der Schaar, M. (2018). Deep-treat: learning optimal personalized treatments from observational data using neural networks. *Proceedings of the AAAI Conference on Artificial Intelligence*, 32(1), 11841.

[6] Schultz, R. B., & DeMers, M. N. (2020). Transitioning from emergency remote learning to deep online learning experiences in geography education. *Journal of Geography*, 119(6), 214–225.

[7] Gillaspy, M. L., & Vasilica, C. (2021). Developing the digital self-determined learner through heutagogical design. *International Journal of Educational Technology in Higher Education*, 18(1), 1–19.

[8] Li, M. (2021). Learner-centered learning tasks in higher education: a study on perception among students. *Education Sciences*, 11(5), 230.

[9] Fernando, K. R. M., & Tsokos, C. P. (2021). Dynamically weighted balanced loss: class imbalanced learning and confidence calibration of deep neural networks. *IEEE Transactions on Neural Networks and Learning Systems*, 32(6), 2424–2437.

[10] Lee, K. S., Kim, J. Y., Jeon, E. T., Choi, W. S., Kim, N. H., & Lee, K. Y. (2020). Evaluation of scalability and degree of fine-tuning of deep convolutional neural networks for COVID-19 screening on chest X-ray images using explainable deep-learning algorithm. *Journal of Personalized Medicine*, 10(4), 213.

[11] Holstein, K., & Aleven, V. (2021). Designing for human-AI complementarity in K-12 education. *Proceedings of the AAAI Conference on Artificial Intelligence*, 35(18), 15795–15803.

[12] Koster, R., Balaguer, J., Tacchetti, A., Weinstein, A., Zhu, T., Hauser, O., Williams, D., Campbell-Gillingham, L., Thacker, P., Botvinick, M., & Summerfield, C. (2022). *Human-Centered Mechanism Design with Democratic AI. Nature Human Behaviour.* Springer Nature, Berlin.

[13] Cirqueira, D., Nedbal, D., Helfert, M., & Bezbradica, M. (2020). Scenario-based requirements elicitation for user-centric explainable AI – a case in fraud detection. In A. Holzinger, P. Kieseberg, A. Tjoa, & E. Weippl (Eds.) *Machine Learning and Knowledge Extraction* (pp. 288–305). University of Groningen, Groningen, Netherlands.

[14] Norman, D. A., & Draper, S. W. (1986). *User Centered System Design; New Perspectives on Human-Computer Interaction.* Lawrence Erlbaum Associates, Inc, Mahwah, NJ.

[15] Abras, C., Maloney-Krichmar, D., & Preece, J. (2004). User-centered design. In W. Bainbridge (Eds.) *Encyclopedia of Human-Computer Interaction* (,pp 445-456). Sage Publications, Thousand Oaks, CA.

[16] Simonsen, J., & Robertson, T. (Eds.). (2012). *Routledge International Handbook of Participatory Design.* Routledge, London.

[17] van Wynsberghe, A., & Robbins, S. (2019). Ethicist as designer: a pragmatic approach to ethics in the lab. *Science and Engineering Ethics*, 25(1), 291–314.

[18] Lu, L., Meng, X., Mao, Z., & Karniadakis, G. (2019). DeepXDE: a deep learning library for solving differential equations. *SIAM Review*, 61(4), 808–830.

[19] Abdullah, S., Ameen, S., Sadeeq, M. A., & Zeebaree, S. R. M. (2021). Multimodal emotion recognition using deep learning. *Journal of Advanced Science and Technology Trends*, 2(2), 1–14.

[20] Mukhamadiyev, A., Khujayarov, I., Djuraev, O., & Cho, J. (2022). Automatic speech recognition method based on deep learning approaches for Uzbek language. *Sensors*, 22(10), 3683.

[21] Li, Q., Kumar, P., & Alazab, M. (2021). IoT-assisted physical education training network virtualization and resource management using a deep reinforcement learning system. *Journal of Ambient Intelligence and Humanized Computing*, 8, 12.

[22] Nye, B. D., Pavlik, P., Windsor, A., Olney, A., Hajeer, M. H., & Hu, X. (2018). SKOPE-IT (Shareable Knowledge Objects as Portable Intelligent Tutors): overlaying natural language tutoring on an adaptive learning system for mathematics. *International Journal of STEM Education*, 5(1), 1–23.
[23] Rastegarmoghadam, M., & Ziarati, K. (2017). Improved modeling of intelligent tutoring systems using ant colony optimization. *Education and Information Technologies*, 22(6), 2885–2909.
[24] Grenander, M., Belfer, R., Kochmar, E., Serban, I., St-Hilaire, F., & Cheung, J. (2021). Deep discourse analysis for generating personalized feedback in intelligent tutor systems. *Proceedings of the AAAI Conference on Artificial Intelligence*, 35(17), 15015–15023.
[25] Lima, G. S., Porto, D. R., Oliveira, A., & Bessa, W. M. (2021). Intelligent control of a single-link flexible manipulator using sliding modes and artificial neural networks. *Electronics Letters*, 57(18), 798–804
[26] Feng, J., Feng, X., Chen, J., Cao, X., Zhang, X., Jiao, L., & Yu, T. (2020). Generative adversarial networks based on collaborative learning and attention mechanism for hyperspectral image classification. *Remote Sensing*, 12(7), 1149.
[27] Vaswani, A., Shazeer, N., Parmar, N., Uszkoreit, J., Jones, L., Gomez, A. N., Kaiser, L., & Polosukhin, I. (2017). Attention is all you need. In I. Guyon & U. Von Luxburg (Eds.) *Advances in Neural Information Processing Systems* (pp. 5998–6008). NIPS, Long Beach, CA.
[28] Jeyaprakash, A. (2022, December 26). NLP TRANSFORMERS. LinkedIn. https://www.linkedin.com/pulse/nlp-transformers-akshaya-jeyaprakash
[29] ISTE. (n.d.). *Personalized Learning: A Guide for Engaging Students with Technology*. ISTE, New York.
[30] Shute, V. J. (2008). Focus on formative feedback. *Review of Educational Research*, 78(1), 153–189.
[31] Freeman, S., Eddy, S. L., McDonough, M., Smith, M. K., Okoroafor, N., Jordt, H., & Wenderoth, M. P. (2014). Active learning increases student performance in science, engineering, and mathematics. *Proceedings of the National Academy of Sciences*, 111(23), 8410–8415.
[32] Smith, T. E., & Borkoski, C. (2019). Artificial intelligence and inclusive education: speculations on the future. In T. E. Smith & C. Borkoski (Eds.) *Artificial Intelligence and Inclusive Education: Speculations on the Future* (pp. 1–7). Springer, Cham.
[33] UNESCO. (2019). *Artificial Intelligence in Education: Challenges and Opportunities for Sustainable Development*. UNESCO, Paris, France.
[34] Zhou, J., & Salvendy, G. (Eds.). (2019). *Human Aspects of IT for the Aged Population. Design for the Elderly and Technology Acceptance*. ITAP 2019. Lecture Notes in Computer Science, vol. 11593. Springer, Cham.
[35] UNESCO. (2019). Artificial intelligence for education: Challenges and opportunities for sustainable development. UNESCO Digital Library. https://unesdoc.unesco.org/ark:/48223/pf0000366994
[36] Axon Park. (2023, June 13). Will AI Change Education? https://www.axonpark.com/will-ai-change-education/
[37] American University School of International Service. (2023, May 23). Four Questions on AI Global Governance Following the G7 Hiroshima Summit. https://www.american.edu/sis/news/20230523-four-questions-on-ai-global-governance-following-the-g7-hiroshima-summit.cfm

[38] Chen, X., Zou, D., Cheng, G., & Xie, H. (2019). Detecting latent topics and trends in educational technologies over four decades using structural topic modeling: a retrospective of all volumes of computer & education. *Computers & Education*, 130, 37–53.

[39] Huang, Y. M., Liang, T. H., Su, Y. N., & Chen, N. S. (2019). Empowering personalized learning with an interactive e-book learning system for elementary school students. *Educational Technology Research and Development*, 60(4), 703–722.

[40] Nguyen, T., & Yang, Y. (2021). Artificial intelligence in education: a review. *Complexity*, 8, 75264–75278.

[41] Yang, Z., Dai, Z., Yang, Y., Carbonell, J., Salakhutdinov, R. R., & Le, Q. V. (2019). XLNet: generalized autoregressive pre-training for language understanding. *Advances in Neural Information Processing Systems*, 32, 5754–5764.

[42] Shermis, M. D., & Burstein, J. (2013). *Handbook of Automated Essay Evaluation: Current Applications and New Directions*. Routledge, London.

[43] Hutchison, D., Kanade, T., Kittler, J., Kleinberg, J. M., Mattern, F., Mitchell, J. C., & Steffen, B. (2020). *Artificial Intelligence in Education*. Springer Nature, Berlin.

[44] Zeide, E. (2017). The structural consequences of big data-driven education. *Big Data & Society*, 4(1), 2053951716661691.

[45] Belfield, R. E. (2019). The ethics of AI in education. *British Journal of Educational Technology*, 50(6), 2755–2765.

[46] Barocas, S., Hardt, M., & Narayanan, A. (2019). Fairness and Machine Learning. Retrieved from FairMLBook.org.

[47] Kieseberg, S., Leitner, P., Mulazzani, M., Munroe, L., Schrittwieser, S., Sinha, M., & Weippl, E. (2017). Privacy aware machine learning and the right to be forgotten. Paper presented at the European Symposium on Research in Computer Security (ESORICS 2017), Oslo, Norway, pp. 198–212.

[48] Dignum, V. (2019). *Responsible Artificial Intelligence: How to Develop and Use AI in a Responsible Way*. Springer Nature, Berlin.

[49] Russell, S., Dewey, D., & Tegmark, M. (2015). Research priorities for robust and beneficial artificial intelligence. *AI Magazine*, 36(4), 105–114.

[50] Organization for Economic Co-operation and Development. (2019). *OECD Principles on AI*. OECD, Paris, France.

[51] IEEE. (2019). *Ethically Aligned Design: A Vision for Prioritizing Human Well-being with Autonomous and Intelligent Systems*. IEEE, New York.

[52] Karsenti, T., & Collin, S. (2013). Why are new teachers leaving the profession? Results of a Canada-wide survey. *Education*, 3(3), 141–149.

[53] Danielson, C. (2013). *The Framework for Teaching Evaluation Instrument*. The Danielson Group, Chicago, IL.

[54] Sweeney, D. (2018). *Student Data Privacy, Equity, and Digital Literacy*. Future Ready Schools, Arkansas, CA.

[55] Dabner, N. (2012). 'Breaking Ground' in the use of social media: a case study of a university earthquake response to inform educational design with Facebook. *The Internet and Higher Education*, 15(1), 69–78.

Chapter 5

Machine learning/deep learning use cases in education

Radha Sridharan, N. S. Sukanya, B. Sivasakthi, and Vinithra Sankaran

5.1 INTRODUCTION

At its core, AI involves utilizing computers or machines to imitate the problem-solving and decision-making abilities of the human mind. Within the realm of AI, machine learning focuses on employing various self-learning algorithms to extract knowledge from data and make predictions. Deep learning, a subset of machine learning (ML), can be seen as a scalable form of machine learning that automates much of the feature extraction process and reduces the need for human intervention, enabling the handling of large datasets. Machine learning and deep learning are instrumental in the field of education, as they have the potential to bring about revolutionary changes in multiple aspects of the learning process. Presented below are several compelling justifications highlighting their significance in the educational realm.

Machine learning can be adopted by teachers in the field of education to save time on tasks outside the classroom. For instance, teachers can utilize remote virtual assistants to assist their students, thereby enhancing the learning experience and promoting academic advancement. By employing machine learning, personalized learning becomes achievable, allowing educators to comprehend their students' progress better. Consequently, teachers can tailor the curriculum to cater to the specific requirements of the learners. AI advancements enable intelligence moderation in education, facilitating the analysis of data by human tutors and moderators [1].

The goals and objectives of this topic revolve around optimizing learning outcomes, promoting personalized instruction, and deriving insights to improve educational practices. Some of the use cases of machine learning/deep learning in education include predicting academic performance, personalized learning and adaptive education, natural language processing in education, intelligent learning management systems, computer vision applications in education, and recommender systems in education.

 DOI: 10.1201/9781003534761-5

5.2 USE CASES OF ML/DL IN EDUCATION

The integration of machine learning in education revolutionizes the learning process, providing educational institutions with novel tools to monitor and enhance student performance and engagement. The personalized approach enabled by machine learning fosters inclusivity, accessibility, and engagement, thereby making education more dynamic and engaging for all. We will delve into the possibilities unleashed by machine learning in the field of education, examining real-world examples and applications. Subsequently, we will examine the potential challenges associated with implementing machine learning in education and explore how experts in the field can overcome them.

5.2.1 Predicting academic performance using machine learning

Modern learning institutions operate in a highly competitive and complex environment. Thus, analyzing performance, providing high-quality education, formulating strategies for evaluating the students' performance, and identifying future needs are some challenges faced by most universities today [2]. Student intervention plans are implemented in universities to overcome students' problems during their studies. Student performance prediction at the entry level and during the subsequent periods helps the universities effectively develop and evolve the intervention plans, where both the management and educators are the beneficiaries of the students' performance prediction plans. Figure 5.1 shows the use of machine learning in predicting students' academic performance. By leveraging demographic data, it is possible to identify at-risk students. Employing data mining algorithms on datasets can prove advantageous for all stakeholders in educational institutions [3].

By training machine learning models on historical data, they can learn to make accurate predictions about future academic outcomes. Here's how the process typically works:

1. **Data Collection:** Data collection involves gathering student demographics, previous academic records, socioeconomic backgrounds, attendance records, study habits, and other pertinent variables.
2. **Preprocessing the Data:** In this stage, the collected data is processed to handle missing values, normalize features, and rectify any inconsistencies or outliers.
3. **Train and Evaluate the Model:** Preprocessed data is used to train machine learning algorithms such as linear regression, decision trees, random forests, support vector machines, or neural networks. These algorithms learn patterns and relationships between the features and

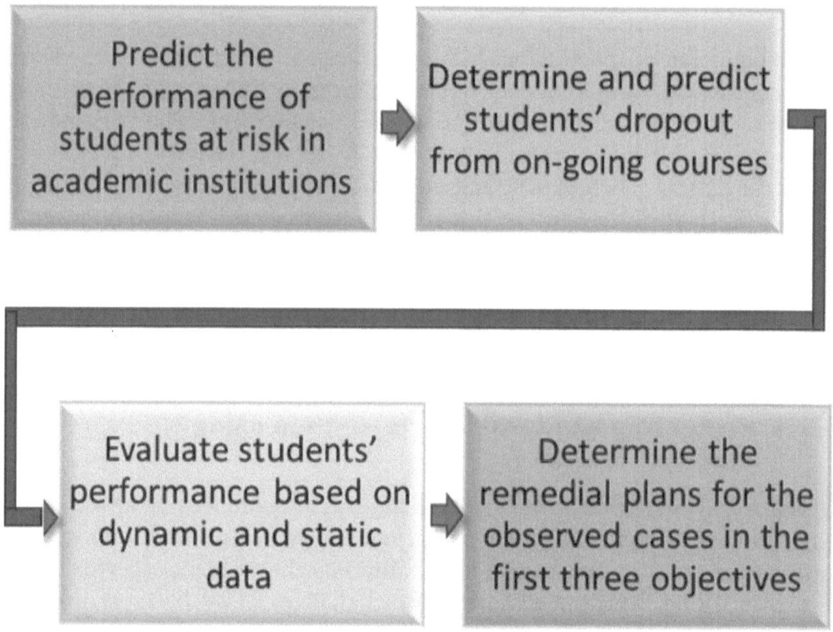

Figure 5.1 Prediction of students' academic performance using machine learning.

the target variable, which usually represents academic performance metrics such as SGPA, exam scores, or pass status. This model is then evaluated using appropriate evaluation metrics such as mean squared error (MSE), accuracy, precision, recall, or *F*1 score.

4. **Prediction and Analysis of the Performance:** After the model is trained and evaluated, it can be utilized to make predictions on new, unseen data. These predictions offer valuable insights into individual student performance and early detection of at-risk students,

5.2.2 Personalized learning and adaptive education

Artificial intelligence (AI) techniques have been harnessed in personalized adaptive educational systems to surpass the constraints posed by rigidly predefined learning styles. These methods leverage machine learning algorithms to address the task of tailoring online learning experiences. This involves automatically and fluidly associating students' behavioral characteristics with specific learning styles, thus enhancing the effectiveness of individualized learning journeys. The rise of online education has been driven by technological advancements and the impact of the COVID-19 pandemic. E-learning has made it possible for learners around the world to access information, resulting in an increase in data flows and the use of big data technology. As a result, e-learning platforms must accommodate the growing number of users and the substantial amount of data being generated. Additionally, e-learning has introduced physical distance and reduced personal engagement between teachers and learners. This highlights the importance of developing a more personalized approach to e-learning in order to meet the diverse needs of individual learners within the growing student population [4]. Considering an efficient adaptive e-learning system, it is crucial to develop a student profile and model that accurately represents the learner's characteristics, including their learning styles (LSs). This ensures the formation of an effective framework that can effectively cater to the individual needs of the learner [5].

AI approaches have garnered considerable attention and are being utilized in various applications to customize and adapt e-learning experiences. Figure 5.2 shows various platforms that stimulated research on ML-based personalized adaptive learning to identify LSs of students. These approaches aim to classify students based on their LSs, ultimately enhancing their e-learning journey. In a study conducted by H. Anantharaman et al. [6], a personalized e-learning system was introduced, which utilized deep learning (DL) and machine learning (ML) algorithms. This system was designed to adapt to the learner's specific learning style and level, offering a tailored e-learning experience. Table 5.1 shows the frequency of ML algorithms used in the detection of LSs in various studies [6].

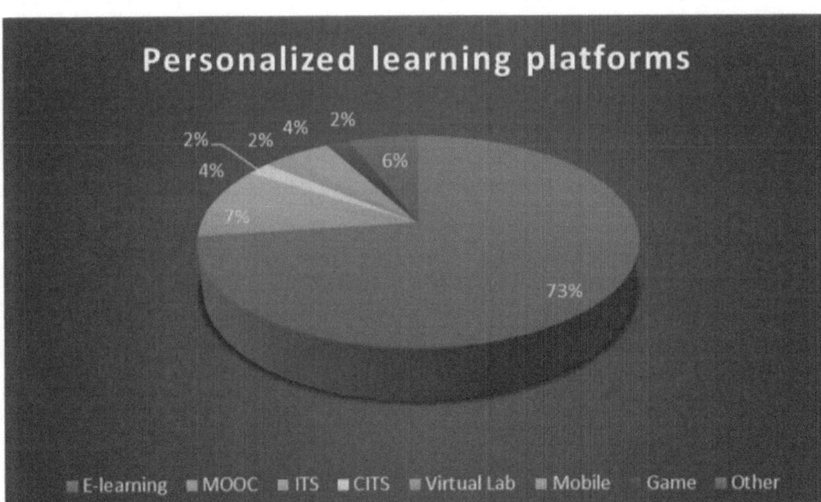

Figure 5.2 Platforms that stimulated research on personalized adaptive learning based on ML to identify learning styles.

Table 5.1 Frequency of learning style classification (ML) algorithms

Sl. No.	*Machine learning algorithms*	*#Articles*
1	Bayesian Network	13
2	Neural Network	14
3	Decision Trees	19
4	Fuzzy Cognitive Maps	6
5	K-nearest Neighbor	8
6	Support Vector Machines	7
7	K-means	10
8	Genetic Algorithms	2
9	Ant Colony System	3
10	Particle Swarm Optimization	2
11	Incremental Dynamic Case based Reasoning	3
12	Convolution Neural Network	1
13	Recurrent Neural Network	1
14	Logistic Regression	3
15	Other	6

5.2.3 Natural language processing in education

Natural Language Processing (NLP) is an AI technique that enables communication between humans and computers using natural language. NLP systems take speech or text as input and produce output in the form of natural language. NLP can be broadly divided into two parts: Natural Language Understanding (NLU) and Natural Language Generation (NLG). NLU involves converting an input into a meaningful representation and analyzing various aspects of the language. NLG, on the other hand, focuses on generating coherent and meaningful sentence structures in natural language based on internal representations [7]. NLP holds immense potential for educational purposes, making it crucial to develop innovative software systems and advanced techniques within educational settings. The primary objective of incorporating NLP in the educational domain is to enhance the educational system by implementing efficient and effective policies. These policies enable the utilization of advanced technologies to drive improvements in the educational landscape [8]. Some of the applications of NLP in education are given in the following sections.

5.2.3.1 Automated essay scoring

In the realm of education, Automated Essay Scoring (AES) employs NLP to automatically assess and rate essays written by students. By utilizing NLP techniques, AES streamlines the essay grading procedure, delivering prompt and consistent feedback to learners. Through the analysis of linguistic features and patterns within the essays, NLP algorithms can evaluate aspects such as grammar, vocabulary, coherence, and argumentation. This facilitates objective and efficient evaluation of essays, thereby enhancing the essay grading process in educational contexts.

5.2.3.2 Intelligent tutoring systems

Intelligent tutoring systems (ITSs) are software programs designed to deliver personalized instruction based on individual student requirements. These systems perform various tutorial functions, such as presenting essential information, assigning tasks or asking questions, and providing feedback, with the aim of inducing cognitive and motivational transformations in students. To achieve this objective, ITSs harness AI and ML techniques to define content models, determining the subject matter to be taught, as well as the tutoring strategies to be employed for each student. In essence, ITSs outline both the "what" and "how" of teaching, utilizing AI and ML methodologies [9].

5.2.3.3 Language learning and translation

NLP plays a vital role in translation tasks and language learning by facilitating automated translation and improving translation tools. Powered by NLP, machine translation systems can automatically convert text from one language to another, providing learners with easier access to information across different languages. Moreover, NLP techniques enhance computer-assisted translation tools by supporting human translators with tasks such as extracting terminology, aligning parallel texts, and editing machine-generated translations.

In educational settings, NLP technology significantly contributes to enhancing the efficiency, accuracy, and accessibility of language-related activities, benefiting both language learners and translators alike.

5.2.3.4 Sentiment analysis in education

Within the field of education, sentiment analysis proves to be a valuable application of ML, offering numerous practical applications. Through the utilization of ML algorithms, sentiment analysis can effectively examine and decipher the emotions, opinions, and attitudes conveyed within textual data pertaining to educational settings. By employing sentiment analysis in ML applications, educational institutions gain access to valuable insights regarding the emotional and attitudinal dimensions of education. This, in turn, empowers them to make well-informed decisions, improve the overall learning experience, and provide effective support for student success [10].

5.2.4 Computer vision applications in education

Computer vision systems have indeed made significant progress in recent years, thanks to advancements in DL and other ML techniques. The three basic steps you mentioned—acquiring video frames, processing the images, and understanding the content—are fundamental to computer vision systems [11]

5.2.4.1 Object detection and recognition for educational materials

The Internet has brought the world closer together, facilitating easy access to information and enabling individuals to explore any topic of interest with just a few clicks. This has greatly expanded the opportunities for learning and skill development. With a plethora of online courses and skill-oriented programs available, people now have the freedom to pursue their interests and acquire knowledge on a wide range of subjects.

The growing demand for online learning has led to the rise of Massive Open Online Course (MOOC) platforms, which offer a diverse selection of courses, including those from renowned institutions. In addition to traditional college courses, individuals can now enroll in online certification programs that provide audio and video lectures, as well as digital materials like e-books, to facilitate their learning experience. These online courses aim to create an immersive classroom-like environment, providing students with a comprehensive learning experience.

One of the main privileges of online learning is its flexibility. Flexibility is a significant benefit of online learning, granting students the freedom to acquire knowledge and skills at their own rhythm and convenience, aligning with their personal schedules. This is particularly beneficial for individuals who are juggling multiple commitments, such as full-time employment or concurrent enrollment in other courses. Online learning allows them to pursue additional skills and knowledge without the constraints of traditional classroom settings.

Overall, online learning has revolutionized the education landscape by offering a wide array of courses and programs that cater to diverse interests and needs. It provides individuals with the opportunity to expand their skill sets, acquire new knowledge, and enhance their professional prospects. With its convenience and flexibility, online learning has become an attractive option for those seeking continuous learning and personal growth in today's fast-paced world [12].

5.2.4.2 Facial recognition for attendance and security

In traditional classrooms, computer vision can be employed to track student engagement, offering insights into their postures, behaviors, and eye movements. Figure 5.3 depicts the steps in facial recognition for

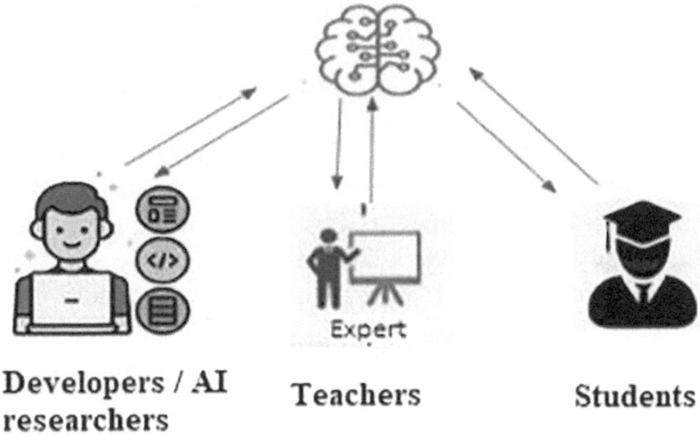

Why	Verification Improvement	Learning Adoption	Legislative Compliance
What	Global Model Data representation	Local explanation Why and Why not	Why not
How	Intrinsic	Visualization Natural language	Plain language
Eval	Completeness tests Performance	Test of comprehension Survey of trust	Satisfaction questionnaire

Figure 5.3 Steps in facial recognition for attendance and security.

attendance and security. This analysis helps monitor their level of interest and attentiveness. By examining this data, teachers can gain a better understanding of each student's behavior and identify fluctuations in their interest levels over time. With this information, educators can adapt their teaching methods to capture the attention of the majority of students and enhance their interest in the subject matter. This fosters open interaction between teachers and students, allowing teachers to assess and comprehend student reactions to their teaching techniques. Genuine feedback from students can be compared with the data collected through computer vision, enabling teachers to make improvements in their teaching methods and offer customized courses and materials tailored to students' individual learning abilities.

It is important to acknowledge that the implementation of computer vision in education has raised ethical concerns and sparked debates. The continuous surveillance of students has been a topic of discussion, questioning the privacy implications. Furthermore, there are arguments that the technology itself may not be fully matured for widespread implementation in educational settings.

It is crucial to recognize that the use of computer vision in education has its limitations and challenges [13].

5.2.4.3 Handwriting recognition and digitization

Handwriting recognition, also known as handwritten text recognition (HWR), is a technology that involves the recognition and interpretation of handwritten data, converting it into machine-readable output.

Recent developments in DL, particularly the introduction of transformer architectures, have significantly accelerated progress in the field of handwritten text recognition. This process is often referred to as Intelligent Character Recognition (ICR) because the algorithms required for ICR demand a higher level of intelligence compared to conventional Optical Character Recognition (OCR) used for printed text.

Incorporating handwriting recognition into the digitization process not only preserves history but also opens up new avenues for global access and engagement [14].

5.2.4.4 Augmented reality and virtual reality in education

The potential of augmented reality (AR) powered by DL to revolutionize education is immense. It possesses the capability to transition from conventional teaching approaches to experiential learning by providing immersive experiences. Applications of AR in education are diverse, ranging from virtual laboratories to machine learning-driven robots. However, there is still room for improvement in the implementation of AR technology in education, which is the driving force behind ongoing research and development efforts.

The objective of this endeavor is to establish a model that enhances the functions and services of AR using DL techniques. By leveraging the power of DL, it is possible to develop user-friendly, modern, and personalized intelligent AR applications. These applications can provide a seamless integration of digital elements into the real world, resulting in an enhanced and interactive learning experience.

It is important to differentiate between AR and virtual reality (VR). While VR transports users to entirely different digital environments, AR enriches the existing real-world environment by overlaying digital elements onto it. VR immerses users in a simulated reality, whereas AR enhances the reality they are already in. It's akin to comparing a journey to a new place with a guided tour of your hometown. AR enhances the learning experience by blending virtual content with the real-world context, offering new possibilities for education and knowledge acquisition.

By combining the capabilities of DL and AR, education can be transformed into a more engaging, interactive, and immersive experience. The potential for learning through AR is vast, and ongoing advancements in DL will continue to enhance the effectiveness and impact of AR in education [15].

AR in the education sector offers a plethora of benefits that make learning more interesting, interactive, and practical.Following are the benefits of AR in Education:

5.2.4.4.1 Encouraging Student Involvement

With its ability to make abstract concepts more concrete and understandable, AR can improve student engagement and motivation. In addition, rich visuals and immersive experiences in AR allow students to truly engage with the subject matter, making learning more exciting and fun.

5.2.4.4.2 Encouraging Collaboration and Communication

AR offers opportunities for collaborative learning by enabling students to interact with virtual objects and scenarios in a shared space. Through AR applications, students can collaborate on solving complex problems, conducting virtual experiments, or exploring virtual environments. This collaborative approach promotes active engagement, stimulates discussions, and encourages students to exchange ideas and perspectives.

5.2.4.4.3 Enhancing Retention of Information

By incorporating AR as a supplement to traditional teaching methods, students can gain a deeper understanding and visualization of complex concepts, leading to improved long-term retention of information.

5.2.4.4.4 Cutting Costs

Integrating AR into the curriculum takes advantage of the widespread availability of smartphones among students, making it a convenient and cost-effective solution. By leveraging students' existing devices, the need for expensive textbooks and other physical learning materials can be reduced, resulting in significant cost savings for educational institutions.

5.2.4.4.5 AR Classroom Activities and Lesson Plans

To maximize the potential of AR in the classroom, educators should integrate technology into their lesson plans. By incorporating AR apps, they can create interactive visual aids, virtual field trips, and hands-on learning experiences that enhance student engagement and understanding.

An excellent example of incorporating AR into the classroom is using an AR app like Merge Cube hologram to explore the solar system. With this app, students can hold the Merge Cube, which serves as a marker, and view a virtual representation of the planets and celestial bodies in 3D. They can rotate and interact with each planet, zoom in for a closer look, and access additional information about their features, composition, and other relevant details. This immersive experience allows students to visualize and understand the characteristics of different planets in a hands-on and engaging way.

Indeed, the possibilities of using AR in the classroom are virtually limitless. By integrating AR into the learning environment, educators can create dynamic and immersive experiences that captivate students' attention and foster a genuine interest in the subject matter.

5.2.4.4.6 Implementing AR in the Classroom

Absolutely! When incorporating AR into instructional methods, it's important to align the technology with specific learning objectives and choose appropriate AR apps to support those goals. By leveraging AR apps, educators can enhance traditional instructional methods and create immersive and interactive learning experiences for students.

By using AR triggers strategically placed around the classroom, science teachers can create interactive and engaging lab simulations that provide safety warnings and guidance to students as they navigate through the virtual lab environment [16].

5.2.5 Intelligent learning management systems

5.2.5.1 Automated content generation and adaptation

AI tools that focus on automated content generation utilize NLG techniques NLG systems use algorithms and language models to analyze data, extract relevant information, and generate text that is fluent, contextually appropriate, and tailored to the specific task or purpose.

These systems can generate a wide range of content, including reports, summaries, product descriptions, news articles, and more.

The process of NLG involves several steps, such as data preprocessing, data analysis, content structuring, language generation, and post-processing. During language generation, the NLG system utilizes predefined rules, templates, or ML techniques to generate meaningful and coherent sentences based on the input data.

NLG technology has advanced significantly in recent years, with the emergence of sophisticated DL models like GPT-3. These models can generate highly accurate and contextually relevant content that closely resembles human-written text. NLG-powered automated content generation systems are widely used in various industries, including journalism, e-commerce, finance, and customer service, to produce high-quality content efficiently and at scale [17].

5.2.5.1.1 Automated Content Creation

Indeed, AI has revolutionized content creation in various ways. Content creators and businesses are increasingly leveraging AI algorithms and NLG systems to automate the process of generating written content. This approach offers several benefits and opportunities for both efficiency and creativity.

Automated Content Generation: AI-powered content creation tools can produce articles, blog posts, product descriptions, social media posts, and other forms of written content automatically. These systems use advanced algorithms to analyze data, understand context, and generate coherent and relevant text.

5.2.5.1.2 Enhanced User Experience with AI Chatbots

Chatbots are computer programs that utilize AI techniques such as NLP and ML to simulate human-like conversations with users. They have become increasingly popular in various industries, including customer support.

One prominent example is Facebook Messenger, which incorporates chatbots to interact with users and address their inquiries or issues in real time. Instead of requiring users to fill out forms or navigate complex menus, chatbots allow for more interactive and conversational interactions. Users can simply type in their questions or concerns, and the chatbot responds with relevant information or assistance.

Absolutely! AI chatbots can be versatile tools that extend beyond customer support and play a significant role in content marketing as well. They can engage users in meaningful conversations and gather valuable information in the process.

5.2.5.2 Learning analytics and data-driven decision-making

Learning analytics is the process of collecting, analyzing, and interpreting data from educational activities to gain insights and make data-driven decisions. It involves the use of data and analytics techniques to understand student learning behaviors, improve educational outcomes, and enhance the overall learning experience. By leveraging the power of data, learning analytics enables educational institutions to make informed decisions at both the individual student level and the institutional level.

Here's how learning analytics and data-driven decision-making can benefit education.

5.2.5.2.1 *Student Performance Monitoring*

Learning analytics allows educators to track and analyze individual student performance. By collecting data on factors such as test scores, assignments, attendance, and engagement, educators can identify students who are struggling or at risk of falling behind. This enables timely interventions and personalized support to improve student outcomes.

5.2.5.2.2 *Identifying Learning Patterns and Trends*

By analyzing large datasets, learning analytics can identify patterns and trends in student learning behavior. This includes understanding how students interact with learning materials, identifying the most effective teaching strategies, and uncovering areas where students commonly face challenges. These insights can inform instructional design and delivery, leading to more effective teaching practices.

5.2.5.2.3 *Personalized Learning*

Learning analytics facilitates personalized learning experiences by providing insights into individual student needs and preferences. Through the examination of data related to student performance, interests, and learning preferences, educators can customize instructional content and approaches to cater to the unique requirements of individual students. This enhances engagement, motivation, and overall learning outcomes.

5.2.5.2.4 *Early Warning Systems*

Learning analytics can help build early warning systems that identify students who may be at risk of dropping out or experiencing academic difficulties. By analyzing various data points, such as attendance, assessment scores, and engagement levels, educators can intervene early and provide targeted support to struggling students, increasing their chances of success.

5.2.5.2.5 *Curriculum and Instructional Design*

Data-driven decision-making in education involves using learning analytics to inform curriculum development and instructional design. By analyzing data on student performance and engagement, educators can identify areas where the curriculum needs improvement or where additional support is required.

5.2.5.3 *Automated assessment and feedback systems*

Electronic scoring (ES), the task of machine-grading essays or constructed-response items, can refer to the use of technology to evaluate and score various types of constructed-response items, including but not limited to essays.

Automated Assessment and Feedback Systems in DL involve the utilization of AI and ML techniques to autonomously evaluate and deliver feedback on student work or assessments.

Handcrafted features refer to manually designed and engineered characteristics or attributes of the essays that are deemed relevant for scoring. These features can include aspects such as vocabulary usage, sentence structure, coherence, or the presence of specific keywords or phrases. Researchers have developed algorithms and models that utilize these handcrafted features to assess the quality and proficiency of essays. Due to the abundant availability of large volumes of trainable collection extracted from online sources and the progress made in developing models for word representation, significant advancements have been achieved. DL approaches in NLP enable accurately classifying and assigning scores to various types of text, such as essays, reviews, or sentiment analysis tasks.

Here are some key aspects of automated grading and feedback systems in DL:

Data Training: These systems require large amounts of annotated data to train the DL models. Initially, human experts grade and provide feedback on a substantial number of sample submissions to create a labeled dataset for training the models.

Feature Extraction: DL models are designed to extract relevant features from the student submissions. For example, in essay grading, features like grammar, vocabulary, coherence, and argument strength might be considered. In code evaluation, factors like correctness, efficiency, and adherence to coding conventions are important.

Model Development: DL models, such as convolutional neural networks (CNNs) or recurrent neural networks (RNNs), are commonly employed for automated grading and feedback tasks. These models learn from the labeled dataset to understand patterns and make predictions or assessments.

5.2.6 Recommender systems in education

5.2.6.1 *Personalized course recommendations*

Absolutely, recommender systems have become crucial in e-learning platforms to assist users in selecting the most suitable courses based on their preferences and needs. By leveraging user data and algorithms, these systems can automate and optimize the course selection process, leading to a more personalized and tailored learning experience.

The use of machine intelligence (MI) techniques, such as collaborative filtering, K-nearest neighbor (KNN), Singular Value Decomposition (SVD), and neural network-based collaborative filtering (NCF) models, showcases the application of AI algorithms in generating course recommendations.

Based on the preferences and historical data of users, the recommender system can analyze and understand their preferences to suggest courses that align with their interests. By utilizing a large dataset of Coursera's course reviews from Kaggle, this work aims to provide learners with a personalized and tailored course selection experience.

The implementation of this recommender system using Python language indicates the practical application of programming to develop AI-based solutions. By leveraging the proposed system, learners can benefit from accurate and relevant course recommendations, enabling them to select e-learning courses that suit their preferences and learning goals.

It's important to note that the specific implementation details, performance, and effectiveness of the proposed recommender system would require further analysis and evaluation. Nonetheless, the utilization of AI techniques in e-learning platforms holds significant potential for enhancing the learning experience and improving learner satisfaction.

It's interesting to note that the performance of the models in the recommender system for e-learning course recommendation is evaluated using various performance metrics such as hit rate (HR), average reciprocal hit ranking (ARHR), and mean absolute error (MAE). These metrics provide insights into the accuracy and effectiveness of the recommendation models.

Here are some key aspects of personalized course recommendations in education using DL:

Learner Profiling: Personalized course recommendation systems start by building learner profiles. These profiles include information about the learner's academic background, interests, learning preferences, goals, and possibly other relevant data points. The system may gather this information through self-reported surveys, user interactions, or historical data.

Data Collection: The system collects data related to learners' activities, such as course enrollments, completion rates, grades, quizzes, and assessments. This data provides insights into learners' performance, progress, and engagement levels.

DL Models: DL models, such as neural networks, are trained using the collected data to identify patterns and relationships between learners' characteristics and course outcomes. The models can learn complex representations of learner profiles and course attributes to make accurate recommendations.

Recommendation Generation: Based on the trained DL models, the personalized course recommendation system generates tailored suggestions for each learner. The recommendations can include specific courses, learning materials, supplementary resources, or even personalized learning paths that align with the learner's goals and preferences [18].

5.2.6.2 Adaptive learning pathways

The research mentioned aims to explore teaching strategies for ceramic art in the context of contemporary students, taking advantage of the rapid expansion of Internet technology. The focus is on leveraging DL techniques to establish an automatic question answering (QA) system. This system aims to analyze new teaching strategies and combine them with the capabilities of the Internet to assist students in resolving challenges they encounter during the learning process.

Firstly, the related theories of DL and personalized learning are analyzed. DL techniques such as Back Propagation Neural Network (BPNN), Convolutional Neural Network (CNN), Long Short-Term Memory (LSTM), and Gated Recurrent Unit (GRU) are compared to implement both a single model and a mixed model.

Next, student questions are gathered, chosen, and analyzed to conduct comparative experiments using different models. Experimental parameters are set to evaluate the performance of these models. The experiments reveal that traditional retrieval methods have limitations, with average accuracy and Mean Reciprocal Rank (MRR) reaching only about 0.5. On the other hand, in the basic neural network models, specifically the LSTM and GRU structural models, the average accuracy improves significantly to around 0.81. The findings suggest that utilizing neural network models, such as LSTM and GRU, can improve the accuracy and effectiveness of answering student questions in the context of personalized learning in ceramic art education. By leveraging the capabilities of these models, students can receive more accurate and relevant responses to their queries, enhancing their learning experience and understanding of the subject matter.

Finally, the hybrid model achieves an accuracy of about 0.82, indicating an improvement compared to the basic neural network models. The hybrid model combines multiple approaches or components to leverage their respective strengths and enhance the performance of the QA system. This model outperforms the other models, including the traditional retrieval methods and basic neural network models like LSTM and GRU, and provides the best results in answering student questions accurately.

The high accuracy of 0.87 suggests that the BiGRU-Attention model provides correct answers to a significant portion of the student questions. Additionally, the MRR of 0.89 indicates that, on average, the model's top-ranked answer is closer to the desired response compared to the other models [19].

Through a diverse range of techniques, flexible learning technology scans identify areas of strengths and weaknesses, provide targeted feedback, and adjust the difficulty and content of subsequent learning activities to maximize learning outcomes It is characterized by:

1. Intelligent adjustment of content, questions and answer choices based on performance
2. Self-paced study
3. Optimal, individualized learning pathways
4. Immediate feedback and remediation
5. A focus on attaining expertise and proficiency.

5.3 ETHICAL CONSIDERATIONS AND CHALLENGES IN ML/DL IN EDUCATION

5.3.1 Privacy and security concerns

Data privacy is a critical ethical consideration in the implementation of ML in education [20,21]. Access to a large volume of data that comprises personal information of students' academic records, and behavioral patterns is needed for ML algorithms toward making right predictions and recommendations. Clear understanding of these issues and devising mechanisms for addressing them that involve stakeholders, along with civil society, to ensure that the benefits of these technologies surpass their disadvantages is much necessary [22]. Some of the key considerations for maintaining data privacy in ML for education are given below:

i. **Student Consent:** Educational Institutions must obtain consent from students or their legal guardians before collecting and using their data for ML purposes. It is required to mention clearly in the consent, the types of data being collected and details of its usage. If any third-party involvement is present for data processing, those details also need to be specified in the consent.
ii. **Anonymity:** To protect the privacy of the individual, information that discloses personal identity should be removed from the data used for ML. Encryption and decryption techniques can be implemented to assist in re-identification.
iii. **Limited Data Collection:** It is very vital to design ML models to gather and maintain only the data that serves the purpose. Surplus data collection may lead to privacy breaches and may result in the privacy of students being compromised.
iv. **Policies and Regulations:** Institutions must carefully evaluate their data privacy policies when collaborating with third-party ML service providers. Responsibilities of third-party related to data privacy and

security, restrictions on usage of data and sharing must be clearly outlined in contracts and agreements.

v. **Transparent Medium of Communication:** It is the primary responsibility of the educational institution to make the process of data collection transparent. Parents, students and all stakeholders must be informed prior about the purpose of using ML in the education sector and its merits.

5.3.2 Bias and fairness in educational data and algorithms

Biases are set at various levels in ML models in the current AI platforms [23]. Data, design, and algorithms for decision-making can be inherently biased. Historical data might encompass biases based on race, gender, socioeconomic status, or other factors. When algorithms learn from this biased data, the predictions they make carry these biases, leading to unfair outcomes for certain groups of students. Gender bias is one of the apparent problems encountered by students when they learn language courses as part of their curriculum. It is seen when they use AI for translating a text from gender-neutral language to gender-specific language. For instance, Google Translate rendered the translation of "She/he is a nurse" in Turkish into the feminine form as well as "She/he is a doctor" in Turkish into the masculine form [24].

Handling bias and thereby ensuring fairness in educational algorithms is a very challenging task. Biases become undetected when they get embedded in complex algorithms.

5.3.2.1 Approaches for promoting fairness

To mitigate bias and promote fairness in educational algorithms, a few approaches that can be adopted are discussed below:

i. **Diversity in Training Data:** Ensuring that training data is diverse and exempts biases is not easy. Training data must comprise observations of all possible groups including minority student groups.
ii. **Bias Mitigation Techniques:** Techniques like data preprocessing to remove bias, post-processing to fine-tune predictions toward achieving fairness can be employed.
iii. **Guidelines and Policies:** Creating guidelines and policies that cater to algorithmic fairness in education facilitates the development, implementation, and evaluation of educational algorithms.

With continuous evaluation and improvement of educational algorithms, we can aim at achieving equitable outcomes for all student groups.

5.3.3 Transparency and explainability

Making the educational algorithms transparent and their decision-making processes explainable is vital. Transparency in educational algorithms refers to the data they use, how decisions are made and the logic behind their predictions. Explainability moves a step forward and allows users to understand and validate the recommendations made by the algorithm. Both these characteristics enhance trust in educational algorithms and provide valuable insights into the learning process for educators and students.

Educators can customize and improve their instructional strategies by understanding how algorithms assess performance and provide recommendations. Students can also obtain insights into their progress, get right directions, and make informed decisions about their path of learning. Detailed documentation that describes the algorithms, their inputs, and their decision-making processes can enhance transparency. This documentation should be easily accessible and understandable by all types of users, including technical and non-technical stakeholders.

Presenting explanations in the context of the user facilitates explainability more than when we try to develop explainable systems that would only satisfy the needs of a general explanation. If various explanations are particularly designed for different users, each one can be designed with a specific purpose, content, and presented in a distinct manner. Research in explainability suggests three big families of explanations [25]:

The system gives explanations to various types of users, taking into account their goals, and renders appropriate and tailored information that meets their needs and expectations. Evaluation methods are also customized for each explanation.

Figure 5.4 represents the three aspects/components on which explanations are created: goals of an explanation, explanation content, and types of explanations. The last aspect is how explanations can be assessed, which is very important for advancements in explainable AI.

5.3.4 Teacher-student relationship and automation

The teacher-student relationship plays an indispensable role in education, encompassing guidance, mentoring, and personalized support. As technology and automation are more common in education, it becomes essential to understand its impact on this relationship. Some of the key factors to be considered in this scenario are as follows:

i. **Role of Automation:** Automation and technology are valuable tools to support teaching and improvise student learning experiences. They can be used to automate administrative tasks done by teachers, deliver personalized learning resources, and provide insights into student progress. By reducing the time spent on repetitive tasks, teachers can focus more on student-centric and tailored instructions.

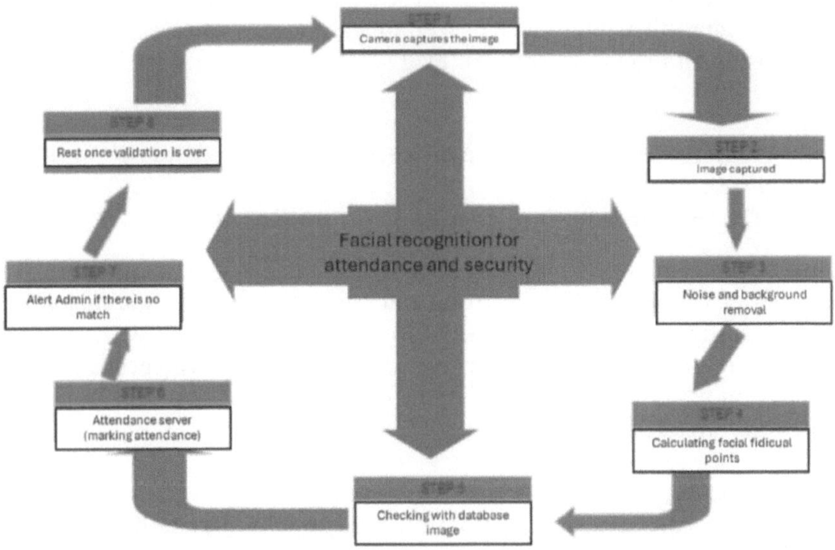

Figure 5.4 Types of users and aspects of explanations.

ii. **Improved Personalization:** Automation can drive toward adaptive learning systems that tailor content delivered based on individual student needs and preferences. This personalization can help students to progress at their own pace and style. Teachers can get a deep understanding of students' learning patterns and their limitations and thereby offer guidance and support accordingly.
iii. **Individualized Feedback:** With automation, timely and valuable feedback can be given to students. It can help them understand their strengths and areas to be focused for improvement. Automated grading systems and feedback tools can assist teachers in providing in-depth feedback and support based on the particular needs of each student.
iv. **Emotional and Social Support:** Automation can enhance certain aspects of teaching, but it is necessary to realize that technology cannot replace the human touch. Emotional and social support can be given only by teachers and not by technology. A strong teacher-student relationship is essential for having an encouraging learning environment that takes care of the socio-emotional needs of students.

Thus automation and technology can be used for efficient administrative processes and for providing personalized support and individualized feedback. However, it is important to recognize the unique characteristics and values that teachers possess in terms of emotional support and mentoring cannot be replaced by technology. Striking a balance between automation and the human element is much needed to maintain a strong teacher-student relationship and thereby provide effective learning experiences.

5.4 FUTURE DIRECTIONS AND EMERGING TRENDS

5.4.1 Deep reinforcement learning in education

Deep reinforcement learning (DRL) is a computational framework for the automation of goal directed learning and sequential decision-making. DRL focuses on training intelligent agents from direct interaction with its environment. It is applicable for scenarios involving an agent who needs to learn a policy on what decision to take in different situations [26,27].

Few key areas of application of DRL are as follows:

i. Curriculum Personalization

   DRL can be used to train an instructional policy to help students with a personalized curriculum. Given a particular student, instructional policy maps the history of responses of students with the next task toward maximizing learning opportunities.

ii. Educational Content Generation

   Another notable research direction is to apply DRL for generating educational content like quizzes, videos, and other exercises. Teachers can assign new practice tasks as homework or as a quiz to check students' knowledge. Students can automatically receive new similar tasks after they fail to solve a given task.

iii. Automation of Feedback and Assessment

   DRL agents can be trained to evaluate and grade student responses. This helps in the design of more effective and personalized feedback. Students receive immediate responses to their queries.

Though DRL in education has marked its significance, there are several challenges that need to be addressed. These include data efficiency, interpretability, and scalability of AI in educational settings.

5.4.2 Lifelong learning and continuous education

Lifelong learning and continuous education represent the acquisition of knowledge and skills on a continuous basis beyond formal education. Many benefits obtained from this are:

i. Professional Development

   Continuous education plays an essential role in professional development. It enables individuals to upgrade their skills and stay in pace with advancements in their fields. Professional development workshops, seminars, online courses, and certifications are common platforms for acquiring new capabilities.

ii. Personal Development

This helps to develop new interests, become familiar with various topics, associate with new people and expand the network and enhance self-esteem [28].

iii. Societal Benefits: Lifelong learning offers numerous benefits for individuals and society as a whole. For individuals, it results in high employability, advancement in career, and personal fulfillment. From a societal perspective, lifelong learning contributes to a skillful and dynamic workforce, innovation, social relationship, and economic development.

5.5 CASE STUDIES AND SUCCESS STORIES

Knewton is a platform for adaptive learning that delivers personalized content. It analyzes student performance data and identifies the strengths and weaknesses of each student. It then makes customized recommendations based on students' proficiency. Carnegie Learning is an education technology company that provides personalized learning platforms by incorporating ML algorithms. This platform analyzes student performance and engagement and generates tailored recommendations. The ML algorithms adapt to individual student needs, allowing for personalized instruction and improved learning outcomes. Currently, universities rely much on algorithm-based solutions to support learning success and curricula. One of such successful projects is based in Australia at Deakin University. It integrates IBM's supercomputer, Watson, that provides 365 days of feedback to students. Austin Peay State University has built software, called Degree Compass. It fetches the past grades of several students in various courses and based on them it analyzes the student's academic record. It predicts how well a specific student is likely to perform in a particular course long before the first day of class. Smart Sparrow is another adaptive learning platform that creates interactive and adaptive course content thereby enabling students to learn at their own pace. Gradescope is an assessment platform that incorporates ML techniques to automate grading and provide feedback on assignments. Feedback can reshape the learning experience of a student by encouraging them to learn better, and remain motivated [29]. Online instruction delivery necessitates quite more frequent feedback on performance for both faculty and students. By analyzing performance data, teachers can understand data patterns and alter their delivery of instruction accordingly, thereby improving their performance.

5.6 CONCLUSION

Integration of ML and DL in education has high potential and prospects to transform the conventional teaching and learning process. It facilitates empowering educators and rendering learners a personalized, very engaging, and effectual learning experience, thereby improving educational outcomes.

While ML/DL in education provides innumerable opportunities, it also becomes essential to address concerns about data privacy, algorithm bias, and ethical considerations. Continuous research, collaboration among technology and education sectors, and application of effective strategies are vital to increase the benefits of ML/DL and reduce the potential risks.

REFERENCES

[1] Nafea, I. (2018). *Machine Learning in Educational Technology*. InTech, Houston, TX. https://doi.org/10.5772/intechopen.72906

[2] Tair, M. M. A., & El-Halees, A. M. (2012). Mining educational data to improve students' performance: a case study. *International Journal of Information*, 2(2). https://www.researchgate.net/publication/258501044_Mining_Educational_Data_to_Improve_Students'_Performance_A_Case_Study

[3] Rufai, A., Suru, D. U., & Afrifa, J. (2021). The role of machine learning and data mining techniques in predicting students' academic performance. *International Journal of Computer Applications Technology and Research*, 10, 198–207.

[4] Essa, S. G., Celik, T., & Human-Hendricks, N. E. (2023). Personalized adaptive learning technologies based on machine learning techniques to identify learning styles: A systematic literature review. *IEEE Access*, 11, 48392–48409. https://doi.org/10.1109/ACCESS.2023.3276439

[5] Aissaoui, O. E., Madani, Y. E. A. El, Audit, L., & Allioui, Y. E. (2019). Combining supervised and unsupervised machine learning algorithms to predict the learners' learning styles. *Procedia Computer Science*, 148, 87–96.

[6] Anantharaman, H., Mubarak, A., & Shobana, B. T. (2018). Modelling an adaptive e-learning system using LSTM and random forest classification. *Proc. IEEE Conferences on e-Learning e-Management e-Services (IC3e)*, Langkawi, Malaysia, pp. 29–34.

[7] Sharma, H. (2021). Improving natural language processing tasks by using machine learning techniques. *2021 5th International Conference on Information Systems and Computer Networks (ISCON)*, Mathura, India, pp. 1–5. https://doi.org/10.1109/ISCON52037.2021.9702447

[8] Khaled, D. (2014). Natural language processing and its use in education. *International Journal of Advanced Computer Science and Applications*, 5, 12. https://doi.org/10.14569/IJACSA.2014.051210

[9] Sleeman, D., & Brown, J. S. (1982). *Intelligent Tutoring Systems*. Academic Press, New York.
[10] Zhou, J., & Ye, J.-M. (2023) Sentiment analysis in education research: A review of journal publications. *Interactive Learning Environments*, 31(3), 1252–1264. https://doi.org/10.1080/10494820.2020.1826985
[11] Klingler, N. (2023). https://viso.ai/applications/computer-vision-in-education
[12] Kuppusamy, P., & Suresh, J. K. (2021) (2020). A Deep Learning Model to Smart Education System, e-Conference on Artificial Intelligence and Machine Learning, Mumbai, India.
[13] Tang, J., Zhou, X., & Zheng, J. (2019). Design of intelligent classroom facial recognition based on deep learning. *Journal of Physics: Conference Series*, 1168, 022043. https://doi.org/10.1088/1742-6596/1168/2/022043
[14] Preetha, S., Afrid, I. M., Karthik Hebbar, P., & Nishchay, S. K. (2020). Machine learning for handwriting recognition. *International Journal of Computer*, 38, 1.
[15] Caudell, T. P., & Mizell, D. W. (1992). Augmented reality: An application of heads-up display technology to manual manufacturing processes, *Proceedings of the TwentyFifth Hawaii International Conference on System Sciences*, Kauai, HI, USA, vol. 2, pp. 659–669.
[16] Ravi Varma, R. B., Umesh, I. M., & Upendra, R. S. (2021). Augmented reality and deep learning in e-learning - a new approach. *International Journal of Applied Engineering Research*, 16(9), 749–751.
[17] Chaitali, D., Srinath, S., Gandharv, S., Saksham, A., & Prasad, R. (2023). AI-based learning content generation and learning pathway augmentation to increase learner engagement. *Computers and Education: Artificial Intelligence*, 4, 100110.
[18] Qinglong, L., & Jaekyeong, K. (2021). A deep learning-based course recommender system for sustainable development in education. *Journals Applied Sciences*, 11(19), 8993. https://doi.org/10.3390/app11198993
[19] Raj, N. S., & Renumol, V. G. (2022). An improved adaptive learning path recommendation model driven by real-time learning analytics. *Journal of Computers in Education*, 11(1), 121–148. https://doi.org/10.1007/s40692-022-00250-y
[20] Regan, P. M., & Jesse, J. (2019). Ethical challenges of edtech, big data and personalized learning: Twenty-first century student sorting and tracking. *Ethics and Information Technology*, 21, 167–179. https://doi.org/10.1007/s10676-018-9492-2
[21] Remian, D. (2019). Augmenting education: Ethical considerations for incorporating artificial intelligence in education (Unpublished master's thesis). University of Massachusetts, Boston, MA.
[22] Stahl, B. C., & Wright, D. (2018). Ethics and privacy in AI and big data: Implementing responsible research and innovation. *IEEE Security & Privacy*, 16, 3.
[23] Krutka, D., Manca, S., Galvin, S., Greenhow, C., Koehler, M., & Askari, E. (2019). Teaching "against" social media: Confronting problems of profit in the curriculum. *Teachers College Record*, 121(14), 1–42.

[24] Johnson, M. (2020). A scalable approach to reducing gender bias in Google translate. https://ai.googleblog.com/2020/04/a-scalable-approach-to-reducing-gender.html

[25] Ribera, M. & Lapedriza, A. (2019). *Can We Do Better Explanations? A Proposal of User-Centered Explainable AI.* IUI Workshops, Marina del Ray, CA.

[26] Sutton, R. S., & Barto, A. G. (2018). *Reinforcement Learning: An Introduction.* MIT Press, Cambridge, MA.

[27] Singla, A., Rafferty, A. N., Radanovic, G., & Heffernan, N. T. (2021). Reinforcement Learning for Education: Opportunities and Challenges. *arXiv preprint arXiv:2107.08828.* https://rl4ed.org/edm2021/

[28] Laal, M., Laal, A., & Aliramaei, A. (2014). Continuing education, lifelong learning. *Procedia - Social and Behavioral Sciences*, 116, 4052–4056.

[29] Amos, J. R., Angra, S., Castleberry, C., & Stadie, O. (2021, March). Using Gradescope to Facilitate Tag-Enhanced Student Feedback. In *Proceedings of the 52nd ACM Technical Symposium on Computer Science Education* (pp. 1354–1354). https://doi.org/10.1145/3408877.3432498

Chapter 6

Digital dark arts

Combating trolls with AI-based solutions

Susan Mathew K, Richa Sharma, and Arti Arya

6.1 INTRODUCTION

As of January 2023, the estimated number of social media users is 4.72 billion [1]. It would be a probable assumption that the reader of this chapter uses at least one form of social media. The users of interest in this chapter are not organic users but a deviant type of social media users known as trolls. Normal users of social media seem to interpret trolling subjectively. Some classify those who comment abusively under a popular personality's post as trolls. Seasoned users on platforms like Reddit label those who post irrelevant, outrageous, and difficult-to-believe content as trolls.

The pivotal question persists: who is a troll?

Before we try to answer that question, we must elucidate why it is vital that we reach a consensus on definitive markers to identify trolling online. Social media facilitates the easy exchange of ideas and information on a diverse number of topics. Similar to real life, you will encounter those who hold differing views from your own. Since social media eliminates the distance factor from conversations, you can eventually reach a point where you are consuming more information than you can process in a healthy manner leading to more volatile outbursts online. If the platform is one that uses moderators to keep the online community healthy, the moderators should not be able to suppress or stifle unpopular stances by labelling them as trolling.

Additionally, not being exposed to differing viewpoints is counterproductive to having organic conversations. This in turn can reduce the quality of discussions on the platform. Ethically, this can also foster a power imbalance in the platform with decision-making concentrated in the hands of a few or with the platform itself. By establishing some clear criteria, we can ensure a more effective and consistent approach to addressing the issue of trolling across various online platforms. Lastly, social media provides the privilege of being anonymous. This worsens trolling since users can conduct themselves as they please in online spaces while simultaneously avoiding accountability and consequences.

DOI: 10.1201/9781003534761-6

Engaging with trolls can lead to detrimental effects on both psychological and physiological well-being, including trauma, suicidal ideation, and the manipulation of opinions on a global scale via state-sponsored trolls. It can result in targeted harassment of prominent users at the behest of reputation management agencies and the proliferation of propaganda-driven ideologies and misinformation.

With respect to trolling, an incident that sparked global interest in trolling occurred during the 2016 US Presidential Elections. The US alleged the influence of Russia in engaging in social media (Twitter) manipulation and disseminating misinformation by launching a coordination campaign. This was coordinated by the Internet Research Agency (IRA). A lot of work that has happened regarding state-sponsored trolls was based on the dataset that was made public following this revelation.

Considering the global reach of social media and some of the consequences listed above, it is imperative to understand how we can leverage AI-based systems to identify trolls. This work will highlight some problems faced in this domain of work, psychology of trolls, related characteristics and features proposed for troll identification, some notable systems and their results, problems faced by researchers, and lastly, countermeasures platforms can adopt against trolling.

6.2 THE DEFINITION PROBLEM

The origin of the word troll has roots in Scandinavian mythology. Trolls are creatures who are short, unpleasant to look at, and interact with. They are found underneath bridges. They wait for travellers who attempt to cross the bridge and refuse to let them do so until they correctly answer a riddle. At a very high level, there are several parallels these mythical creatures have with online trolls, but the definition problem is about the details. Subjectivity and not being able to discern intent with certainty are problems that follow at every level. Another usage refers to a method of baiting fish with lures. This is more in line with the layperson's high-level perception of what an online troll must be like.

Our area of interest is in academic definitions of who a troll is and what constitutes trolling.

Hardaker provides the following definition in [2]:

> "A troller is a CMC[1] user who **constructs the identity** of sincerely wishing to be part of the group in question, including **professing, or conveying pseudo-sincere intentions**, but whose real intention(s) is/are to cause **disruption** and/or to **trigger or exacerbate conflict** for the **purposes of their own amusement.**"

In her work, Hardaker also carefully differentiates between disruption and aggression and notes that posting aggressive messages can cause disruption. The troll has failed if the users identify the true intent of the troll and refuse to take the bait.

The use of an identity-based deception is captured by this definition along with deception by explicitly conveying a pseudo-stance. The troll's primary motivation here is their own amusement. How this trolling might occur is varied but the expected outcome is disruption or worsening conflict. Clearly, deception is a key marker in defining a troll.

Dynel expands on deception in [3]. Deception can occur about anything that cannot be objectively verified as false. Therefore things like exaggeration for humour cannot be categorised as deception. Dynel expands on the different ways that deception might manifest:

i. Identity-Based Deception
ii. **Message-Based Deception:** In an attempt to establish credibility, the troll can end up constructing a pseudo-identity for themselves.
iii. **Response Deception:** Deceiving the target about their believing the assertions to be true.
iv. **Enterprise Deception or Pseudo-Stance:** Deceive the target by presenting his/her assertion as something he/she believes to be true and is true.
v. Deception by Omission
vi. **Subtle Art of Verbal Deception:** exploring non-assertion-based deceptive tactics employed by trolls where the deceptive intent of the troll disregards the recipient's perception despite creating the illusion of genuine interest.

The identification of deception is complicated because it is not possible to determine another's intent with absolute surety. Since users can usually create multiple accounts on the same platform (using different email addresses), identity-based participation can be difficult to track and monitor. Dynel also stated that deception and humour are necessities for trolling but aggression is not.

On humour as a marker, in [4], the author Bishop J states the difference between classical and anonymous trolling in the following definition(s). This presents trolling as a means of entertainment and the difference is noted when you analyse who is/are the benefitting parties:

Classical trolling can be seen to be done for the **community's consensual entertainment** in order to **build bonds between users.**

Anonymous trolling on the other hand can be seen to be done **at the expense of someone outside of a particular community** for that person's **own sick enjoyment** or to share with others who are part of a **'clubhouse' that encourages trolling.** Anonymous trolling is therefore driven by self-amusement. Even in the case where someone engages in trolling for the 'clubhouse', the troll gains pleasure from the validation he/she might receive from those who are a part of this group.

Additionally, in this work, different troll characters and trolling types have been illustrated and noted with examples. Here, the distinction is drawn around whether the intent behind trolling is positive or negative. Classical trolling is a benevolent and enriching activity but anonymous trolling is negative and carried out for the troll's pleasure (direct and indirect). More recent works and popular definitions around trolling tend to generalise it to mean anonymous trolling where the intentions and effects are negative. There is a lot of diversity in the approaches trolls might take on and they also differ in their interactions. Dynel rightly described this as an "internally diversified phenomenon" in [3].

The existence of troll farms and the rise of state-sponsored trolls bring motives into question. Troll farms refer to multiple users working in a coordinated manner towards a common agenda. Both troll farms and state-sponsored trolls (state-sponsored trolls usually belong to troll farms) have the potential to cause opinion manipulation on global levels.

It is widely agreed that trolls are motivated by self-amusement. In fact, trolling has been linked to sadism. But this begs the question, **is self-entertainment the only primary motive?** Bringing in the possibility of **monetary benefits** challenges the known definitions further.

In [5], the authors write about ideologically driven trolls who "seek to draw attention to social problems or to support socially or politically marginalized groups". This opens the possibility of further classification that can be done on the basis of the content of the troll's post.

For a troll (anonymous trolling, non-benevolent) to be successful, one noted dimension captured by academic and colloquial definitions seems to be some frustration on the target's end. Below is a non-academic definition:

The art of **deliberately**, cleverly, and **secretly making people angry**, usually via the Internet **using dialogue**. Trolling does not mean just making rude remarks: Shouting swear words at someone doesn't count as trolling; it's just flaming and isn't funny [6].

Even in the quoted academic definitions, provoking or inciting some volatile or emotional response from the victim seems to be the end goal. But when you bring in more sinister end goals for trolling like public opinion manipulation, the expected outcome from the victim gets trickier to capture in definitions. We have no choice but to expand beyond the understanding of trolling just for lulz or lolz [3].

A more well-rounded definition is the one coined by [7] where the authors describe trolling as an **intentionally disruptive behaviour** that occurs (a) in the context of Internet discourse, and (b) among **users having no existing relationship in real life.** Trolls may **act alone or with others;** either **indiscriminately, or selectively toward certain individuals, social groups, political parties, corporate entities, and so on.** This addresses disruption and the singular or group-existence of trolls while capturing the possibility of being oriented towards a particular cause or goal.

There is an insufficiency of definitions to provide an all-encompassing, objective standard of identification. The existence of varied manifestations of trolling coupled with how quickly trolling is evolving makes this

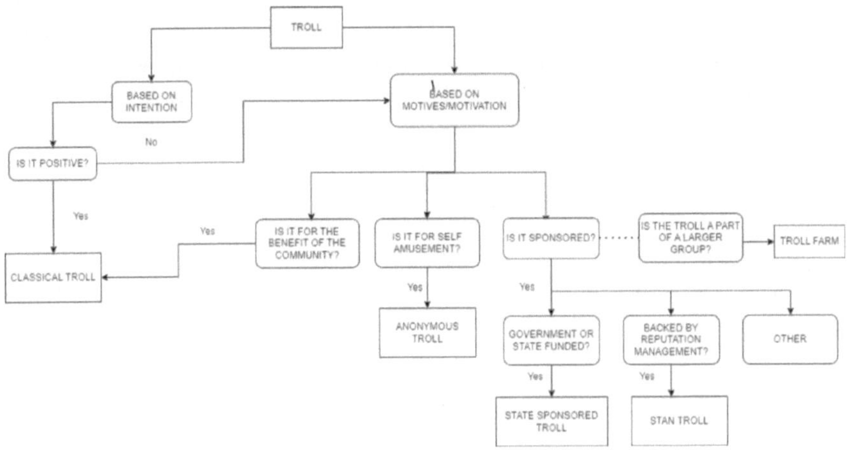

Figure 6.1 Typology of trolls.

a rigorous challenge. Figure 6.1 presents a simple typology that helps differentiate between different types of trolls based on some of the discussions in this section. Classification can be done based on various other parameters like the content, the style adopted by the troll, and so on. It is best if researchers identified what specific type(s) of troll(s) their work would be focussing on, built off of whatever existing characteristics have been observed in past works, and reported expansions or deviations observed in behaviour in their works to bring in more clarity and cohesion into the different schools of thought.

6.3 THE PSYCHOLOGY OF TROLLS

Following are the psychology of trolls.

6.3.1 Antisocial behaviour

A number of works have chosen to define trolls as online users who display antisocial behaviour. In [8], the authors view antisocial behaviour online as an extension of similar behaviour offline and this includes acts of aggression, harassment, and bullying [9,10]. We cannot classify all manifestations of antisocial behaviour online as trolling. For instance, cyberbullying is not the same as trolling [11]. It is important for researchers to use a definition in their work that separates trolling from other manifestations of antisocial behaviour.

"People's understanding of what constitutes antisocial behaviour (ASB) is determined by a series of factors including context, location, community tolerance and quality of life expectations" [12,13]. Therefore, antisocial behaviour has no objective standard of measure. There can be platform-specific markers for what can be deemed as antisocial behaviour. In [8], the authors have used a definition of trolling that includes flaming, grieving, swearing, or personal attacks, including behaviour outside the acceptable bounds defined by several community guidelines for discussion forums. Those following this line of thought for defining trolling on the basis of antisocial behaviour must couple it with markers identified in Section 6.2 to distinguish trolls from other malicious users online.

6.3.2 Dark Tetrad

It is worth looking at some traits associated with trolling. Sadism is one that naturally follows from the previous section since trolls do what they do for self-amusement. Here, we refer to trolling in a strictly negative sense. Work has been undertaken to study trolling with respect to the Dark Tetrad.

The Dark Tetrad is a collection of subclinical personality traits that include Sadism, Psychopathy, Narcissism, and Machiavellianism. Sadism refers to deriving pleasure from inflicting or perceiving pain on another [14]. Narcissism is defined as excessive self-love or egocentrism [15]. A Machiavellian is someone who regards other people as tools to be used, if necessary, for purposeful deception, in order to further their own interests [16]. Psychopathy is a synonym for antisocial personality disorder [17], which is described as the existence of a chronic and pervasive inclination to disregard and violate the rights of others. Repeated transgressions of the law, exploitation of others, deceitfulness, impulsivity, aggressiveness, reckless disregard for the safety of self and others, and irresponsibility are manifestations, which are accompanied by a lack of guilt, regret, and empathy [18].

In [19], the authors found that sadism and psychopathy were good predictors of trolling but narcissism and Machiavellianism were not. They explained that since trolling is more concerned with disrupting someone else's activity, trait narcissism would find little incentive to engage in this activity. Machiavellians are calculating and display impulse control. For that reason, they would not find the quick response time trolling demands to be optimal. In [20], the authors conclude that sadism and Machiavellianism have strong associations with trolling while narcissism was negatively related to trolling enjoyment which seems consistent with previous findings. There is research that states that the traits Machiavellianism and sadism are the strongest predictors of trolling [21].

Regarding psychopathy itself, in [22], the authors provide additional insight into why trolling would appear attractive to those with this trait.

"As trait psychopathy is characterised by impulsivity and thrill-seeking behaviour [23], it is possible that the thrill-seeking aspect of creating mayhem online is a central motivator to troll."

6.3.3 On empathy - cognitive and affective

Cognitive empathy is the ability to discern the feelings and emotions of another person while affective empathy is the ability to feel the emotions of another.

Research has shown that those with lower affective empathy are more likely to participate in trolling. In [22], affective empathy is negatively correlated with trolling. If trolls could "feel" what their victims felt, it would act as a deterrent to trolling. On cognitive empathy, they state that "if trait psychopathy is high, then individuals with higher levels of cognitive empathy are more likely to engage in controlling behaviours".

This simply implies that trolls with high cognitive empathy can understand what the victim is feeling. While affective empathy can work as a deterrent, cognitive empathy alone is insufficient to discourage trolling. Instead, it can worsen the damage the troll can cause.

6.3.4 Why is this important?

Another important point to note is the findings of [8] in which the authors state that users can be incited to troll based on mood and the context of discussion. While all the traits discussed above are not an exhaustive discussion on this topic, they matter because the development of subsequent countermeasures depends on them. The problem does not stop at just identifying trolls. It extends to what can be done after identification. Some proposed countermeasures include total bans, shadow bans, temporary suspensions, unmasking the troll (revealing their real identity), etc. What is the best course of action to tackle a troll? Researchers believe and propose that mitigation strategies should understand the troll [24] and the people who are likely to respond to trolls [25] to build a truly efficient solution.

Various attempts have been made to understand the online patterns of trolls through data analysis. From text-based measures like readability and sentiment analysis to variations in posting frequency, different systems and approaches have been proposed that are based on rich and diverse features. Some of these are explained in the next section.

6.4 CHARACTERISTICS OF A TROLL AND FEATURES

After a cursory understanding of the psyche of the troll, let's explore some characteristics of their online behaviour as recorded in Table 6.1.

Table 6.1 Characteristics and features of trolls

Sl. No.	*Characteristics*	*Proposed or implemented features*	*Reference*
1	Trolls write less similarly to other users	Cosine similarity	[27]
2	Posts by trolls are more difficult to understand	Readability indexes: ARI, dale, smog, flesch, fog, etc.	[27,28]
3	Trolls use less positive words	Sentiment Analysis, Lexical Analysis	[28–30]
4	Trolls write more negative sentences	AffectiveSpace, Hourglass of Emotions, Parrott's socio-psychological mode, lexicon use	[28–30]
5	Trolls use more profanity	Profanity filters, Bad word count	[31]
6	Trolls post more frequently	Daily and hour-wise user activity, Date of publication - weekday or weekend, Time of publication - day time or night time	[28,32]
7	Trolls use less tentative or conciliatory language ("could", "perhaps", "consider")	Can be computed from user posts	[27]
8	The increase in the rate of post deletions over time is higher for trolls than non-trolls	Can be computed longitudinally	[27]
9	Text quality decreases over time as compared to normal users	Longitudinal analysis of readability, Higher post frequency can indicate low-quality posts	[27]
10	Trolls write shorter comments compared to non-trolls	Number of syllables, number of words, average number of sentences, word count	[27,28,33]
11	Degree of controversy	Hostility - higher number of negative words	
12	Trolls tend to be active users	Ratio of total number of active actions to total number of passive actions performed by user, Regular user to one-time user ratio	[31]
13	Trolls engage in multiple conversations	Higher number of replies	[27,28]
14	Trolls have fewer followers than non-trolls	Follower count, User Up/Down Ratio	[28,31]
15	Trolls link more external media	Number of URLs	[28]
16	Trolls use more hashtags to gain more visibility and they talk about a more diverse range of topic	Hashtag count	[28]

(Continued)

Table 6.1 (Continued) Characteristics and features of trolls

Sl. No.	*Characteristics*	*Proposed or implemented features*	*Reference*
17	Measure of influence	Modified h-index	[28]
18	Trolls write more aggressive, rude or hostile sentences	Number of negative words in a sentence, Usage of capital letters – a post written entirely in capital letters is considered to be rude, Abusiveness, Aggression	[28,29,31,33]
19	Trolls tend to be less factual	Count of numbers used in posts	[33]
20	Structure of sentences for trolls	Number of verbs and auxiliary verbs, Number of conjunctions, pronouns, prepositions, articles, subordinations, Number of hapaxes, Number of rare words, Lexical Analysis	[28,29]

Several works have analysed the data they have collected and their findings give us insights into what to look for. In general, while studying works where the definition of a troll seems insufficient, a good practice can be to check if a particular characteristic can be extrapolated to trolling. Each characteristic must be critically evaluated with respect to the definition provided and the expected markers in Section 6.2. Additionally, we will also check if the finding is corroborated by other works. Lastly, the platform has to be kept in mind while understanding interaction-specific patterns like those that relate to comments. A recommendation to gain a broad overview of the features and approaches explored for troll identification is [26].

6.5 SYSTEMS FOR TROLL DETECTION

In this section, several approaches that explore the use of Artificial Intelligence for the identification of trolls have been described. The inclusion of a work in this section was decided after evaluating it for a variety of parameters like how creative the approach is, what type of troll is identified, and the results of the approach.

The definition of a troll used in each work is also mentioned as a direct quotation. The reader must critically evaluate the definition against the expected markers described in Section 6.2.

6.5.1 Do not feel the trolls [30]

Year: 2015

Authors: Cambria E, Chandra P, Sharma A et al.

Definition of a Troll Used: In Internet slang, a troll is someone who posts inflammatory, extraneous, or off topic messages in an online community, such as an online discussion forum, chat room, or blog, with the primary intent of provoking other users into a desired emotional response or of otherwise disrupting normal on-topic discussion [34].

In social web context, emotional attacks on a person or a group through malicious and vulgar comments in order to provoke response are referred to as 'trolling' and the generator is called 'a troll'.

The first thing that caught our eye was the clever wordplay in the title. The phrase "Do Not Feed the Trolls" is one that is frequently echoed in online communities. This phrase asks community members to not interact with trolls or take the bait. The modification of the word "Feed" to "Feel" tells the readers a lot about their proposed approach. In this work, they propose the use of Sentic Computing. In [35], it is defined as "multidisciplinary approach ... that aims to bridge the gap between statistical NLP and many other disciplines that are necessary for understanding human language, such as linguistics, commonsense reasoning, semiotics, and affective computing."

With respect to trolling, we can observe the focus placed on the emotional nature of the posts of trolls apart from the phrasing and overall meaning. Logically then, it makes sense to combine semantics with affective knowledge to identify trolls from non-trolls.

The authors have defined Sentic Computing as a combination of common sense knowledge and affective knowledge. To realise this, two Sentic Computing tools were used. Additionally, the authors have proposed a novel emotion categorisation model which is based on Plutchik's model for emotion classification and Minsky's ideas about emotions.

1. **AffectiveSpace:** It is defined as a language visualisation tool that converts text into a multidimensional space. Let us first understand what makes up this multidimensional space. AffectiveSpace combines a semantic network for common sense knowledge with a lexical representation for emotions. This combination yields AffectiveNet, a matrix of the order 14,301 × 117,365. Concepts are represented row-wise and columns stand for related common sense attributes or emotions. Therefore, the values in every cell would be the truth value of the resultant assertion made.

   The sign of these values indicates whether the assertion has a positive (+) or negative valence[2] (–). The similarity between the two concepts would be the dot product of their vector representations in the AffectiveNet. To deal with the high dimensionality problem, Principal Component Analysis and Truncated Singular Value Decomposition are used for dimensionality reduction. The first 100 principal components are used to build the resultant AffectiveSpace. Similar concepts are likely to appear near each other in the vector space.

2. **The Hourglass of Emotions:** four independent affective dimensions are used: Pleasantness, Attention, Sensitivity, and Aptitude. These four dimensions are defined with respect to Human–Computer Interaction (HCI). Each dimension is characterised by six levels to identify the intensity of the emotion. This can be a float in the range [–3,3]. They are broadly used to represent 24 basic emotions. From the dimensional form, we can obtain the sentic vectors such that each sentic vector has four dimensions which stand for Pleasantness, Attention, Sensitivity, and Aptitude.

The system is called Troll Detector which aims to calculate the probability of a post being "edited" by a troll and to later flag troll users.

Apart from identifying troll posts, they also propose conditions under which a user can be flagged as a troll and how to handle subsequent posts of a flagged troll user.

First, a set of concepts associated with trolls is constructed using CF-IOF weighting. For a post, the concepts in the post are individually evaluated to see if they belong to the known set of concepts associated with trolls. This is done through spectral association. After this, the sentic vector is obtained. Post manual experimentation on a set of 1,000 offensive texts obtained from WordNik, the authors concluded that a high absolute value of Sensitivity and a very low value for polarity demonstrated a greater likelihood for a post to be edited by a troll. The final formula used to compute the "trollness" (t) for a user is:

$$t = \frac{5}{9}\sum_{i=1}^{N} \frac{9\ s_i(c_i) + 10\,|\text{Sensitivity}(c_i)| - \text{Pleasantness}(c_i) - |\text{Attention}(c_i)| - \text{Aptitude}(c_i)}{N} \tag{6.1}$$

where N refers to the total number of cncepts in a post, c_i refers to individual concepts, s_i is the similarity of c_i with the known troll concepts and 5/9 is the normalisation factor.

When the Troll Detector receives an input, it first goes through an NLP Module. This module recognises emoticons, whole uppercase words, onomatopoeic repetitions, exclamation points, degree adverbs, special punctuation, and other textual cues that indicate the text's affective valence. It then performs lemmatisation which groups together inflected forms of a word as a single term. This then goes to the Semantic Parser. The Semantic Parser extracts concepts from the text and for each concept, returns its occurrence within the text, the associated valence, and the corresponding intensity. The AffectiveSpace Module projects the concepts onto the AffectiveSpace after clustering with the Hourglass Model to get the affective valence as the sentic vector. This vector coupled with semantic characteristics is used to calculate the trollness.

Table 6.2 Evaluation metrics

Metric	*Score*
Precision	82%
Recall	75%
F1	78%

There is an interaction database that stores trollness, post, post type, sender ID, and receiver ID. Posts with a high trollness score (threshold set as 60% after trial and error) are flagged as troll posts. If one user has more than 2 troll posts that share a common receiver or a common community, that user is flagged as a troll. After someone is flagged as a troll, their posts are deleted from the website but not completely deleted from the database. Further posts from a user labelled a troll will be displayed on the website only if their posts to a previous victim (receiver) have a low trollness score.

To evaluate their approach, the authors first constructed a dataset of 500 tweets from WordNik. These posts were then manually annotated to be a troll post or a non-troll post. A point to note would be that here, the authors have considered true positives to mean posts with both a positive troll-flag and a trollness $\in [0.6, 1]$ and those with both a negative troll-flag and a trollness $\in [0, 0.6)$. The results are tabulated in Table 6.2.

Compared to related works in troll identification previous to the publication of this work, the authors have stated that this approach has demonstrated improvements in results. As a part of their future work, the authors discuss the possibility of extending their approach beyond single post analysis to evaluating the user's holistic behaviour within the same social network. We will explore an approach that identifies trolls based on the holistic user activity next.

6.5.2 A holistic system for troll detection on Twitter [28]

Year: 2018

Authors: Fornacciari P, Mordonini M, Poggi A et al.

Definition of a Troll Used: A troll is an individual adopting an antisocial behaviour that provokes and irritates other users of an online social platform, often using aggressive or offensive language, and whose effect is to derail the normal evolution of an online discussion and possibly to stop it. Thus, "trolling" refers to "a specific type of malicious online behaviour, intended to disrupt interactions, aggravate interactional partners and lure them into fruitless argumentation" [36].

A user has to be considered a troll if his/her activities are driven by antisocial behaviour.

In this work, the authors aimed to build a system to identify trolls by taking a holistic view of the user and their online activity. This goal was achieved with an accuracy of 95.5%. The platform of focus was Twitter. This work is also notable for the rich features used. The features used are based on findings and approaches adopted in previous works and to the best of our knowledge, this work is the first that attempts to use them all together to gain a holistic understanding of the user. The features have been grouped together to create six feature groups.

The proposed system is called TrollPacifier. For its implementation, the authors used ActoDES [37]. ActoDES is defined as a software framework for the development of large concurrent and distributed systems. It is an actor-based model where the two primary actors are a scheduler and a service provider. Twitter services and additional actors are leveraged along with this system.

The authors have constructed their own dataset for this work. They first obtained a dataset that was supposed to be made up of 3,000 trolls and 3,000 non-trolls. Users mentioned by others in tweets containing words like "troll", "ban", "harass", "block", "stalk", etc. were considered to be trolls. This was based on [32], where the authors considered a user to be a troll if N separate users call them so. More general tweets with common words were used to identify non-troll accounts. Upon a manual review, the authors determined that at least a quarter of this dataset labelled trolls do not exhibit antisocial behaviour. The final dataset used had 500 trolls and 500 non-trolls (1,000 users in total). This dataset is a subset of the originally extracted dataset after manual inspection. The user's recent timeline and their role and attitude in long discussions formed the basis of the judgement.

Feature Groups: The features used in this work are grouped together into six feature groups utilised by the TrollPacifier. The six feature groups are - Sentiment (SENT), Time and frequency of actions (TIME), Text content and style (TEXT), User behaviours (BEHA), Interactions with the community (COMM) and Advertisement of external content (ADVE). The features within the groups have been tabulated below in Table 6.3.

The classification algorithms for the detection of trolls from non-trolls are Naive Bayes (NB), Sequential Minimum Optimisation (SMO), and Random Forest (RF). For evaluating the approach, the authors have reported accuracy, F-measure, Kappa, AUC, and Recall.

While RF works best when individual feature groups are used by itself for the classification tasks, when all the feature groups are utilised for the classification task, SMO performs better. Limiting the results to just accuracy, the metrics have been recorded below in Table 6.4.

Table 6.3 Tabulation of feature groups used

Sl. no	*Feature group*	*Total number of features*	*Features*
1	Sentiment	26	Sentiment Based: minimum, maximum, and average of positive, negative, and neutral sentiment of tweets. Number of objective and subjective tweets; Number of positive and negative tweets, Number of tweets expressing one of the six emotions for subjective tweets. SenticNet Based: Sensitivity, Attention, Pleasantness, Attitude, Polarity, Trollness Abusiveness[a]
2	Time and frequency of actions	57	Activity represented in four-hour intervals[b], Distinguished time intervals, Frequency of actions in the recent timeline and throughout the history of user's activity on platform
3	Text content and style	31	Readability Indices: Kincaid, ARI, Colemaniau, FleschReadingEase, GunningFogIndex, LIX, SMOGIndex, RIX Text-Based: Average word and sentence length, by number of characters, syllables, or words; number of long words, number of complex words Grammar-Based: Number of verbs, number of specific auxiliary verbs, number of conjunctions, number of pronouns, number of prepositions, number of articles, number of subordinations in the middle or at the beginning of a phrase, number of hapaxes, number of rare words
4	User behaviours	38	Total number of tweets, retweets, replies, favourites, citations, and quotes, proportion between active and passive actions, count of different actions associated with a single item, maximum repetitions of a single action
5	Interactions with the community	34	Number of followers and following count, ratio of followers and following, ratio of tweets per follower, number of posts retweeted or favourited by other users, counts of given and received mentions, number of users mentioned, counts of different actions relating to a single user or to a single tweet, H-index based on retweets, likes, and their sum

(*Continued*)

Table 6.3 (Continued) Tabulation of feature groups used

Sl. no	*Feature group*	*Total number of features*	*Features*
6	Advertisement of external content	38	Number and frequency of URLs in posts and comments and in profile information, number and frequency of published or advertised media, number and frequency of hashtags

[a] The abusiveness of a post is measured using an ad hoc classifier trained on **two** classes of inputs - one belonging to trolls and other to non-trolls. The initial dataset of 3,000 trolls and 3,000 non-trolls were used for this task. Although there were cases of mislabelling in this dataset, we can see why they were considered for this task since the authors define abusiveness as "improper or wrongful use, provoking others to finally report the author as a troll". Something about their content online caused others to report them as a troll. Abusiveness as a feature, targets to quantify this. Before training, the text data from these 6,000 users underwent preprocessing with the number of tweets being balanced between both classes. The bag-of-words model is used. Information Gain (IG) is used to identify the most discriminant features since the resulting corpus is quite large. The features are ranked based on the computed IG score and the top k ones are selected. The value of k is determined to be around 30,000 through Grid Search. This classifier is then used to compute the abusiveness for each user.

[b] The division into 4-hour intervals was decided specifically after comparisons between different values using automatic classifiers based on KNN, Naive Bayes, and SMO (Sequential Minimum Optimisation).

Table 6.4 Tabulation of accuracy (%)

Feature group(s)	*Sequential minimum optimisation*	*Naive Bayes*	*Random forest*
Sentiment	78.80	72.31	78.97
Time and frequency of actions	67.84	61.28	75.65
Text content and style	67.56	64.97	68.47
User behaviours	75.88	58.62	79.59
Community interactions	80.45	74.96	83.16
Advertisement of external content	78.07	71.70	85.01
All feature groups	95.52	80.25	88.28

On why RF does not perform well when the entire dataset is considered, the authors explain that RF does not work well with highly dimensional data. NB performs the worst among the three. The authors explain that since NB assumes independence of features which does not hold true for this dataset, it does not perform well at this task. The authors experimented with model ensembling by stacking these models together and feeding the individual prediction confidence levels to a neural network that acts as the meta-learner. The neural network was optimised using grid search. This method gives an accuracy of 93.6%.

The highest obtained accuracy was when all feature groups were taken with SMO (95.52%). Provided a well-rounded definition of a troll is used while labelling the dataset, a holistic dataset such as the one the authors here used can help with the subjectivity issue to an extent since multiple dimensions of the user's online behaviour are captured. The authors also found that the COMM or community feature group has better predictors for trolling. We will focus on another work that takes the user's community into account in the next section.

6.5.3 Predicting the role of political trolls in social media [38]

Year: 2019

Authors: Atanasov A, De G, Morales F et al.

Definition of a Troll Used: Internet "trolls" are users of an online community who quarrel and upset people, seeking to sow discord by posting inflammatory content.

In this work, the authors have specifically focused on state-sponsored trolls (trolls with a political agenda, funded by state) and related troll farms.

In this work, the authors propose two methods - one is a supervised learning task and the other is a distant supervision learning task for the identification of trolls. The dataset used in this work is the IRA Political Trolls Dataset and the platform used in this work is Twitter. Both the learning tasks use embeddings. The direct supervision task or T1 takes a straightforward approach of prediction since the labels provided. The distant supervision task or T2 is designed for a scenario where user labels are not available. Labels are assigned based on the news media mentioned by a particular user. News media outlets tend to display a bias towards some political party. This, along with the low churn rate of news media accounts is leveraged to assign expected labels.

The authors aim to identify state-sponsored trolls in a more collective sense. The authors use word embeddings and graph embeddings in this work. Graph embeddings are latent vector representations of a graph. They store useful information of the graph like topology and contextual relationships between nodes of the graph. The authors used node2vec to train the graph embeddings. Two primary graph embeddings are used in this work. Before diving into them, it is important to understand the labelling used and what they refer to.

The IRA Political Trolls Dataset has three types of political trolls: Right Troll, Left Troll, or News Feed Troll [39].

i. Right Trolls spread right leaning messages. With respect to the 2016 US Presidential Elections, these trolls assume the role of a supporter of Donald Trump.
ii. Left Trolls engage in discourse about identity: gender, sexual, religious, and racial. They criticise mainstream democrats and appear to support Bernie Sanders before the election.
iii. News Feed Trolls are those that present themselves as US local news aggregators.

Graph Embeddings:

i. **User-To-Hashtag (U2H):** A bipartite, undirected graph with the nodes as users and hashtags. Where U refers to the set of users and H refers to the hashtags, an edge e connecting node u and node h means the user u mentioned the hashtag h in their tweet. The graph created has 108,410 nodes and 443,121 edges. The output embedding for each node has 128 dimensions. This embedding captures troll similarity based on hashtag use.
ii. **User-To-Mention (U2M):** An undirected graph designed to capture troll similarity based on who they mention (@username) in their tweets. There exists an edge connecting node u to node v if user u has mentioned (@v) user v in their tweets. The resulting graph has 591,793 nodes and 832,844 edges. The generated embedding for a node has 128 dimensions.

Word Embeddings:

BERT: Based on transformer architecture, the pretrained BERT model is used on the troll's tweets to generate an embedding with 768 dimensions. The final embedding for a user is the average of all the embeddings from the user tweets.

These three embeddings are given as inputs. These embeddings can either be fed to the model by concatenating them together to form one long feature vector or the embeddings can be used individually to make predictions that are combined together (ensembling). Both these variations have been explored by the authors for both tasks.

T1: Supervised Approach

The set of trolls who have labels readily available are used here. The embeddings are generated and fed to an L2 Regularised Logistic Regression Classifier. L2 Regularisation or Ridge Regression is implemented to prevent overfitting.

Both concatenation and ensembling are used. The evaluation metrics are accuracy and macro F1 reported with fivefold cross-validated scores. Individual embeddings work well on their own for the prediction. U2M outperforms other individual embeddings. Combining embeddings together improves the accuracy. Concatenating embeddings together seems to give a better result than ensembling. The detailed metrics are tabulated in Table 6.5.

T2- Distant Supervision

As summarised briefly previously, this is a learning task for when the labels are not available for the trolls. This situation is more reflective of a real world scenario. The labels are derived from the news media mentioned by the user. News media accounts are relatively stable for this task. Media Bias/Fact Check (MBFC) is used to label news media sites. They provide seven classes based on the bias displayed by the news media in question: Extreme Left, Left, Centre Left, Centre, Centre Right, Right, and Extreme Right. The authors chose to reduce these seven classes to three. This is recorded in Table 6.6.

Table 6.5 Evaluation metrics for T1

Method	*Accuracy (%)*	*Macro-F1*
BERT	86.9	83.1
User2Hashtag	87.1	83.2
User2Mention	88.1	83.9
U2Hashtag (concat) User2Mention	88.3	84.1
U2Hashtag (ensemble) User2Mention	88.7	84.4
U2Hashtag (concat) User2Mention (concat) BERT	89.0	84.4
U2Hashtag (ensemble) User2Mention (ensemble) BERT	89.2	84.7

Table 6.6 Derived labels

Media BIAS/fact check label	*Label assigned in this work*	*Count*	*Equivalent troll class*
Extreme Left	Left	341	Left
Left			
Centre Left	Centre	372	News Feed
Centre			
Centre Right			
Right	Right	619	Right
Extreme Right			

How are labels propagated? The authors have explored two possible approaches – one via a proxy model and another using label propagation for semi-supervised learning.

i. **Proxy Model:** A proxy model is trained for the media which is later applied on the users. For U and M where U refers to the set of user and M refers to the set of media, a subset of users who link a specific medium m is generated ($C_m \subseteq U$). From this, a representation of medium m is created by aggregating the embeddings of users that mention m.

$$R(m) = \frac{1}{|C_m|} \sum_{u \in C_m} R(u) \tag{6.2}$$

where $R(m)$ is the representation for medium m, C_m is the set of users who have linked m and $R(u)$ is the representation for the user u.

A logistic regression model is trained which uses $R(m)$ as features and the label for the medium $l(m)$ which is used to generate labels for users.

ii. **Label Propagation:** Where a small set of labelled items are available, label propagation generates labels for the unlabelled items. The generated labels in each iteration of the algorithm depend on the most frequent label held by its neighbouring nodes. Convergence is achieved when the label held by a node matches that held by the majority of its neighbours. Based on two versions of the similarity graph, two versions for label propagation are given:
   a. **LP by Direct Mention (LP1):** There is an edge in the similarity graph $(u, v) \in E$ for two users $u, v \in U$ if u and v share a common hashtag or a common user mention signified by logical OR. This expression is extended by adding links also based on the media M (if user u belongs to C_m). The final expression is depicted below:

$$(u, h) \in U2H \wedge (v, h) \in U2H \vee (u, w) \in U2M \wedge (v, w) \in U2M \tag{6.3}$$

   b. **LP by Similarity Graph (LP2):** A similarity graph is constructed between media and users. For each pair $x,y \in U \cup M$ there is an edge in the similarity graph$(x,y) \in E$ if

$$sim\left(R(x), R(y)\right) > \tau \tag{6.4}$$

   where sim is a similarity function between vectors and τ is used to specify the sparseness of the resulting graph. Here the value of τ is taken to be 0.55.

The conclusions show promising results. Since both the evaluation metrics (accuracy and macro averaged F1) provide consistent rankings for the models, the authors have chosen accuracy as the primary metric while discussing their findings.

For both T1 (fully supervised task) and T2 (distant supervision task), among BERT, U2H, and U2M embeddings, U2M embeddings performed better.

The authors provide three possible reasons why this may be so – the U2M graph is larger, the social circle of a troll is a better predictor than hashtags used and text content is noisy and therefore BERT embeddings alone yield a comparatively lower accuracy.

Concatenating the embeddings together gave an accuracy of 89.2% in T1 and 78.2% in T2 which is marginally higher than what is obtained after ensembling separate predictions of each embedding together (89% and 78%, respectively).

With label propagation we see some more improvement with both LP1 and LP2, with LP2 (based on similarity graph) giving slightly better results over L1 with an 89.6% accuracy for T1 and 78.5% for T2.

6.5.4 How "troll" are you? measuring and detecting troll behaviour in online social networks [40]

Year: 2022

Authors: Ezzeddine F, Luceri L, Ayoub O et al.

Definition of a Troll Used: None provided

The authors have stated that they are focussing on state-sponsored trolls acting in misinformation operations.

In this work, the authors first use the trajectories of users to classify if they are trolls or not. From this, they propose the troll score. Troll score is a metric that aims to quantify the extent to which a user acts like a troll. The social media platform used in this work is Twitter.

State-sponsored trolls are different from anonymous trolls. This difference is noted when we observe what drives their behaviour. While anonymous trolls are primarily motivated by their own amusement, state-sponsored trolls are motivated by monetary benefits. This is important to note as one assumption the authors have made in their work is that trolls will execute their agenda irrespective of the feedback they receive from the community. Their posting habits remain unchanged despite community feedback. This assumption cannot be reasonably supported in the case of anonymous trolls.

The proposed approach builds sequences of state action pairs of user activity and feedback received.

There are four actions that can be performed by an account:

1. **Original Tweet (tw)**: generates original content.
2. **Retweet (rt)**: retweet content generated by others.

3. **Interact with Others (in)**: replies or mentions.
4. **No Action (no)**: the user does not perform any action.

There are three states that refer to the possible feedback an account can receive:

1. **Retweeted (RT)**: an original tweet by the account is retweeted.
2. **Interacted with (IN)**: replies to or mentions the account's tweets.
3. **No Interaction (NO)**: no feedback is provided.

There are 11 combinations of state action pairs that can be considered. The authors do not consider (no, NO) to be a valid combination. The derived state action pairs are ordered according to time to get the sequence of activities associated with a specific user account. From these ordered sequences of activities, trajectories are created of length *L*. The number of trajectories depends on the value of *L*.

A sliding window of length *L* was used to create trajectories. The sliding window was incremented by one step in every iteration until all pairs were consumed in trajectories. This also means that two consecutive trajectories will have *L*-1 elements in common. Deciding an optimal value for *L* is important from a computational perspective and from the point of view of model performance. The authors experimented with five different values of *L* 50, 65, 80, 100, 150, and 200.

For the classification task, the model of choice was LSTM (long short-term memory networks) which works well with time series data. The hyperparameters were optimised using random search. The LSTM model is reported to have four LSTM layers, four dropout layers, and one dense layer. The dropout layers occur after each LSTM layer to prevent overfitting. The activation function used is the sigmoid function. The model training minimises the binary cross entropy using Adam optimisation. The evaluation metrics reported used tenfold cross-validation.

This work used the dataset associated with the 2016 US Presidential Elections. Upon evaluating their approach, the authors reported an AUC of close to 99% using state action pairs with an accuracy of 97%. Curiously the authors found that the value of *L* had little effect on the accuracy. There is a slight increase in metrics as the value of *L* increases.

Additionally, the authors studied the contributions of states to this classification task. They classified trajectories using actions and contrasted them against those that used state action pairs. Using Actions alone, the AUC fluctuated around 82%.

Troll score captures the "trollness" of an account. It is the ratio of the number of trajectories classified as troll to the total number of trajectories for that account. The output is in the range [0, 1] where a value closer to 0 is indicative of regular behaviour and a value close to one is indicative of more troll-like behaviour. The threshold value to separate trolls from ordinary users was determined experimentally.

$$t_s = \frac{Nc_t}{N} \quad (6.5)$$

where t_s is the troll score, Nc_t is the number of trajectories classified as troll trajectories and N is the total number of trajectories for a user.

An inspection of the troll score distribution revealed that while most of the trolls obtained a high troll score (~0.8), there were some trolls who obtained low troll scores as they exhibited user-like behaviour. The authors performed behavioural clustering to understand why this occurred.

Finally, the authors wanted to extend their approach beyond the identification of trolls to the identification of misbehaving accounts in general and additionally, they sought to test if their approach worked with conversation around a different topic and not just a political one. They constructed a dataset based on accounts involved in conversations about Covid-19. They filtered out bots, the suspended accounts were considered to be misbehaving accounts and the others were broadly labelled generic users. The approach scored an AUC of 79.9% with an $F1$ score of 75.24%.

Time-ordered sequences of user's action-feedback pairs can indeed prove beneficial in differentiating state-sponsored trolls from non-trolls. Since the approach was evaluated on a highly specific subset of the Political Trolls dataset, the authors have agreed that they do not expect similarly high results in a more challenging scenario. In the present cyberspace, there are multiple types of malicious users. Trolls are just one subset with further subdivisions based on the type and nature of trolling. There is still room to improve the robustness of the approach with respect to limiting it to the identification of state-sponsored or sponsored trolls alone irrespective of the context. It is also worthwhile to study the feasibility of this approach on other social networking platforms.

6.6 CHALLENGES

1. **API/Data Restrictions:** On February 9, 2023, Twitter stopped providing free access to its previously free API. From July 1, 2023, Reddit is expected to follow suit. Both Twitter and Reddit were instrumental in providing data needed to combat trolling. It is a well-accepted fact that the success of AI-based models and intelligent systems lies in choosing the right features and in the quality of the final dataset aggregated. This is where synthetic datasets fall short. Synthetic datasets fail to be representative of the real world online environment. Post the alleged Russian influence in the 2016 US elections, the dataset made publicly available as a result of this investigation has acted as the foundation on which many researchers have developed new frameworks and approaches to identify state-sponsored trolls. The data itself has yielded crucial

insights into patterns of such sponsored campaigns. Lastly, trolling as we know it is evolving along with people's response to it. This is noted in communities like Reddit where trolling to some extent seems to be an accepted part of the platform [31]. It is the need of the hour for researchers and curious minds alike to have access to data to not only work towards finding frameworks for identifying trolls but so that we may also develop effective strategies to combat trolling.

2. **The Data Problem:** A lot of aspects mentioned in definitions of who a troll is and what trolling is cannot be known for certain. For example, deception in general cannot be identified definitively. For technical works, this poses a challenge especially during dataset construction. While constructing a dataset, it is imperative that all the annotators know what to look for with respect to who a troll is on their platform of focus. Since different works seek to identify specific types of trolls, there are further variations in definitions. This is another important distinction to be maintained. For example, a state-sponsored troll is driven by different motives when compared to an anonymous troll and this difference is present in their online behaviour. It will be easier to bridge the gap if researchers release the datasets built for their work. This not only helps in solving the data problem, it also helps the community understand each other's work with greater clarity. Finally, with respect to the dataset itself, it is important for researchers to account for the fact that the present online cyberspace is not only dominated by trolls. There are bots, cyborgs, cyberbullies, spammers, and other malicious actors online. The dataset should be designed to prevent conflating trolls from other actors. A system designed to account for these will be more effective at the identification of trolls in a real world scenario since there undoubtedly exists some shared characteristics among them.
3. **Learning Loops for A Working System:** As stated previously, considering how trolling is evolving and how community reception towards trolling is changing, it is important for future works to incorporate feedback loops into their proposed design or architecture. There is a need for fresh data especially for methods that rely on text-based data for the identification of tolls. Incorporating correctly classified examples can be beneficial in achieving this. This has been proposed in [28].
4. **Next Step Problem:** The aftereffects of a troll interaction could be something short-lived like a brief period of anger or it could lead to something more damaging like opinion manipulation or lasting trauma. When analysed at multiple levels, the consequences or aftereffects of trolling differ in severity, magnitude, and reach. Therefore, it would be right to conclude that not all trolling is equal and not all trolling deserves the same consequence. In [27] the authors describe a situation that is concerning. For two users with similar or the same text post quality, if one of them is unfairly censored by an undeserved

post deletion, that user would tend to write worse. Misuse of power by the platform can further worsen the problem as opposed to solving it. It was observed in [31] that platforms like Reddit benefit from trolling since it has contributed to the culture of the platform.

So this begs the question "what next?" What should the platform do after identifying a troll? The troll who has been recognised to have influenced the elections of another country is not the same as the troll in the common section trying to get a rise for the lulz.

It is important to mention the responsibility of the normal social media user at this point. There is power in the hands of an organic user of the platform to turn the environment of the community back to a state of normalcy and to diffuse the situation in some capacity. Opinions spoken with a sense of absolute conviction that leaves no room for doubt in their phrasing are more likely to be accepted as true. It is up to the normal user to critically examine every such statement they come across and not blindly accept any information that comes their way. Snap judgements are encouraged on social media platforms because of the quick rate at which opinions are consumed online. We would advise against that and ask users to exercise caution before accepting and propagating another's opinion or assertion as valid.

6.7 CONCLUSION

In this chapter, we delve into the problem of troll identification on social media. We analysed the problem with definitions that academic works run into while trying to define a troll. We then establish and highlight some distinguishing markers that can differentiate trolls from other malicious online users. Technical works highlight observable, verifiable characteristics of trolls. In the next section, we elaborate on the psychology of a troll. This is important to understand so that effective policies and countermeasures against trolling can be developed by platforms. We compile some characteristics of trolls with features that can encapsulate them. While this is not an exhaustive list, it is a good starting point for those who seek to understand how trolls differ in their online activity from non-trolls and what features can be used to represent the same. We have explored some systems based on a variety of AI-based approaches. The definitions used in each work have been presented. The approaches have been explained with brief technical explanations wherever necessary and the results are also recorded. Given how trolling can hurt users at an individual level, degrade trust within a group at a community level, and the weaponisation of trolling at a global level, it calls for policies and regulations to be enforced. Trolling is not always a negative thing, but it is undoubtedly a multifaceted problem that calls for a lot of nuance. The last section lays out both challenges and possible recommendations.

NOTES

1 CMC stands for computer mediated communication.
2 Valence refers to the pleasantness or unpleasantness of an emotional stimulus. https://www.ncbi.nlm.nih.gov/pmc/articles/PMC6676801/

REFERENCES

[1] Chaffey, D. (2023). *Global Social Media Statistics Research Summary 2023 [June 2023]*. Smart Insights. Retrieved June 10, 2023, from https://www.smartinsights.com/social-media-marketing/social-media-strategy/new-global-social-media-research/

[2] Hardaker, C. (2010). Trolling in asynchronous computer-mediated communication: From user discussions to academic definitions. *Journal of Politeness Research* (Vol. 6, Issue 2, pp. 215–242). https://doi.org/10.1515/JPLR.2010.011

[3] Dynel, M. (2016). Trolling is not stupid: Internet trolling as the art of deception serving entertainment. *Intercultural Pragmatics* (Vol. 13, Issue 3, pp. 353–381). https://doi.org/10.1515/ip-2016-0015

[4] Bishop, J. (2014). Representations of 'trolls' in mass media communication: A review of media-texts and moral panics relating to 'internet trolling'. *International Journal of Web Based Communities* (Vol. 10, pp. 7–24).

[5] Sanfilippo, M., Yang, S., & Fichman, P. (2017). Managing online trolling: From deviant to social and political trolls. *Hawaii International Conference on System Sciences* (Vol. 2017, pp. 219). https://doi.org/10.24251/HICSS.2017.219

[6] Zerotrousers. (2009). *Trolling*. Urban Dictionary. Retrieved June 14, 2023 from https://www.urbandictionary.com/define.php?term=Trolling

[7] Steinhauer, M., Fichman, P., & Sanfilippo, M. R. (2017) Online trolling and its perpetrators: Under the cyberbridge. *International Journal of Communication* (Vol. 11, pp. 201). https://ijoc.org

[8] Cheng, J., Bernstein, M., Danescu-Niculescu-mizil, C., & Leskovec, J. (2017). Anyone can become a troll: Causes of trolling behavior in online discussions. *Proceedings of the ACM Conference on Computer Supported Cooperative Work, CSCW*, pp 1217–1230, Portland, OR, USA. https://doi.org/10.1145/2998181.2998213

[9] Akbulut, Y., Şahin, Y. L., & Erişti, B. (2010). Cyberbullying victimization among Turkish online social utility members. *Journal of Educational Technology & Society* (Vol. 13, pp. 192–201).

[10] Kayany, J. M. (1998). Contexts of uninhibited online behavior: Flaming in social newsgroups on Usenet. *Journal of the American Society for Information Science* (Vol. 49, Issue 12, pp. 1135–1141).

[11] Moor, L., & Anderson, J. R. (2019). A systematic literature review of the relationship between dark personality traits and antisocial online behaviours. *Personality and Individual Differences* (Vol. 144, pp. 40–55).

[12] Nixon, J., Blandy, S., Hunter, C., Jones, A., & Reeve, K. (2003). *Developing Good Practice in Tackling Anti-Social Behaviour in Mixed Tenure Areas*. Sheffield: Sheffield Hallam University.

[13] Home Office Research Directorate, Great Britain. (2004). *Defining and Measuring Anti-Social Behaviour.* Great Britain: Home Office.
[14] American Psychological Association. (n.d.). *Sadism.* APA Dictionary of Psychology. American Psychological Association. Retrieved June 15, 2023 from https://dictionary.apa.org/sadism
[15] American Psychological Association. (n.d.). *Narcissism.* APA Dictionary of Psychology. American Psychological Association. Retrieved June 15, 2023 from https://dictionary.apa.org/narcissism
[16] American Psychological Association. (n.d.). *Machiavellianism.* APA Dictionary of Psychology. American Psychological Association. Retrieved June 15, 2023 from https://dictionary.apa.org/machiavellianism
[17] American Psychological Association. (n.d.). *Psychopathy.* APA Dictionary of Psychology. American Psychological Association. Retrieved June 15, 2023 from https://dictionary.apa.org/psychopathy
[18] American Psychological Association. (n.d.). *Antisocial Personality Disorder.* APA Dictionary of Psychology. American Psychological Association. Retrieved June 15, 2023 from https://dictionary.apa.org/antisocial-personality-disorder
[19] Craker, N., & March, E. (2016). The dark side of facebook(r): The dark tetrad, negative social potency, and trolling behaviours. *Personality and Individual Differences* (Vol. 102, pp. 79–84). https://doi.org/10.1016/j.paid.2016.06.043
[20] Buckels, E. E., Trapnell, P. D., & Paulhus, D. L. (2014). Trolls just want to have fun. *Personality and Individual Differences* (Vol. 67, pp. 97–102). https://doi.org/10.1016/j.paid.2014.01.016
[21] Karlsson, A. M. C., & Kajonius, P. J. (2020). Not only trolls are trolling the internet: A study on dark personality traits, online environment, and commentary styles. *International Journal of Personality Psychology* (Vol. 6, pp. 12–23). https://doi.org/10.21827/ijpp.6.37214
[22] Sest, N., & March, E. (2017). Constructing the cyber-troll: Psychopathy, sadism, and empathy. *Personality and Individual Differences* (Vol. 119, pp. 69–72). https://doi.org/10.1016/j.paid.2017.06.038
[23] Paulhus, D. L., & Williams, K. M. (2002). The dark triad of personality: Narcissism, machiavellianism, and psychopathy. *Journal of Research in Personality* (Vol. 36, Issue 6, pp. 556–563).
[24] Bentley, L. A., & Cowan, D. G. (2021). The socially dominant troll: Acceptance attitudes towards trolling. *Personality and Individual Differences* (Vol. 173, pp. 110628).
[25] Sun, Q., & Shen, C. (2021). Who would respond to a troll? A social network analysis of reactions to trolls in online communities. *Computers in Human Behavior* (Vol. 121, pp. 106786).
[26] Tomaiuolo, M., Lombardo, G., Mordonini, M., Cagnoni, S., & Poggi, A. (2020). A survey on troll detection. *Future Internet* (Vol. 12, Issue 2, pp. 31). https://doi.org/10.3390/fi12020031
[27] Cheng, J., Danescu-Niculescu-Mizil, C., & Leskovec, J. (2021). Antisocial behavior in online discussion communities. *Proceedings of the International AAAI Conference on Web and Social Media* (Vol. 9, Issue 1, pp. 61–70). https://doi.org/10.1609/icwsm.v9i1.14583
[28] Fornacciari, P., Mordonini, M., Poggi, A., Sani, L., & Tomaiuolo, M. (2018). A holistic system for troll detection on Twitter. *Computers in Human Behavior* (Vol. 89, pp. 258–268). https://doi.org/10.1016/j.chb.2018.08.008

[29] Capistrano, J. L. C., Suarez, J. J. P., & Naval, P. C. (2019). SALSA: Detection of cybertrolls using sentiment, aggression, lexical and syntactic analysis of tweets. *ACM International Conference Proceeding Series* (Vol. 10, pp. 1–6). https://doi.org/10.1145/3326467.3326471

[30] Cambria, E., Chandra, P., Sharma, A., & Hussain, A. (2010). Do not feel the trolls. *CEUR Workshop Proceedings* (Vol. 664, pp. 1–12).

[31] Dollberg, S., Langner, T., & Seidel Roger Wattenhofer, J. (2015). The meta-data troll detector. *ETH, Zürich* (Vol. 2015, pp. 1–14).

[32] Mihaylov, T., Georgiev, G., & Nakov, P. (2015). Finding opinion manipulation trolls in news community forums. *Proceedings of the Nineteenth Conference on Computational Natural Language Learning,* 310–314. Toronto, Canada: Association for Computational Linguistics.

[33] MacHova, K., Porezany, M., & Hreskova, M. (2021). Algorithms of machine learning in recognition of trolls in online space. *SAMI 2021 - IEEE 19th World Symposium on Applied Machine Intelligence and Informatics, Proceedings,* 349–353, Slovakia. https://doi.org/10.1109/SAMI50585.2021.9378699

[34] Troll https://en.wikipedia.org/wiki/Troll - Wikipedia (Accessed on 6th August 2024)

[35] *Sentic Computing.* SenticNet. (n.d.). Retrieved June 20, 2023 from https://sentic.net/computing/

[36] Coles, B. A., & West, M. (2016). Trolling the trolls: Online forum users constructions of the nature and properties of trolling. *Computers in Human Behavior* (Vol. 60, pp. 233–244).

[37] Bergenti, F., Iotti, E., Poggi, A., & Tomaiuolo, M. (2016). Concurrent and distributed applications with actodes. *MATEC Web of Conferences* (Vol. 76, pp. 04043).

[38] Atanasov, A., De Francisci Morales, G., & Nakov, P. (2019). Predicting the role of political trolls in social media. *Proceedings of the 23rd Conference on Computational Natural Language Learning (CoNLL),* 1023–1034. Toronto, Canada: Association for Computational Linguistics.

[39] Boatwright, B. C., Linvill, D. L., & Warren, P. L. (2018). Troll factories: The internet research agency and state-sponsored agenda building. *Resource Centre on Media Freedom in Europe* (Vol. 29, pp. 192).

[40] Ezzeddine, F., Luceri, L., Ayoub, O., Sbeity, I., Nogara, G., Ferrara, E., & Giordano, S. (2022). *How "Troll" are You? Measuring and Detecting Troll Behavior in Online Social Networks.* https://arxiv.org/abs/2210.08786

Chapter 7

Artificial intelligence for smart manufacturing using digital twin technology

Manju Sadasivan, M T Vasumathi, and Jayanthi M

7.1 INTRODUCTION

Digital twin technology, in integration with artificial intelligence (AI), plays a pivotal role in revolutionizing smart manufacturing. By combining these two innovative technologies, businesses can unlock numerous benefits and optimize their production processes. Digital twin technology encompasses the design of a simulated or virtual duplicate of a tangible product, procedure, or system. This virtual representation mirrors the real-world counterpart in real time, enabling manufacturers to gather valuable data and insights. By integrating AI into this digital twin framework, manufacturers can harness the capability of advanced analytics and machine learning algorithms to enhance decision-making and operational efficiency [1].

The potential of advanced analytics and machine learning within the realm of intelligent manufacturing powered by digital twins offers several advantages. They enable live supervision and examination of production processes, providing valuable insights into performance, quality, and efficiency. By leveraging AI algorithms, manufacturers can identify patterns, anomalies, and potential bottlenecks, enabling proactive decision-making and optimization.

Furthermore, AI-driven digital twins facilitate predictive maintenance, ensuring the timely detection of equipment failures or maintenance needs. By examining data from sensors, machines, and other sources, AI algorithms can predict when a component is likely to fail, allowing manufacturers to plan maintenance actions in advance. This proactive approach reduces the amount of time when a system or process is not operational. It also reduces costs, and extends the lifespan of equipment [2].

The merging of AI and digital twin technology also enables intelligent scheduling and resource optimization. By simulating various production scenarios and considering elements such as machine availability, material availability, and production demand, AI algorithms can generate optimized schedules that maximize efficiency and minimize waste.

The fusion of artificial intelligence and digital twin technology presents significant opportunities for smart manufacturing. By creating virtual

 DOI: 10.1201/9781003534761-7

replicas of physical systems and leveraging AI algorithms, manufacturers can gain valuable insights, optimize processes, enable predictive maintenance, and enhance scheduling and resource allocation. This combination empowers businesses to achieve higher productivity, improved quality, and increased competitiveness in the dynamic landscape of modern manufacturing [3].

7.1.1 Smart manufacturing

Smart manufacturing, commonly referred to as Digital Manufacturing or Intelligent Manufacturing, encompasses the use of communication and computing technologies to establish digital connectivity and data analytics-driven operations among all stakeholders involved in the product value chain, including the interconnection among supply chain, corporate operations, and the production area tiers. This connectivity facilitates smart coordination to efficiently align demand and supply, shorten time-to-market, enhance customization abilities, and gain cost benefits [4].

In the past, manufacturing was traditionally characterized by a linear sequence of processes aimed at converting raw materials (RM) into finished goods (FG). However, the introduction of Smart manufacturing (SM) has brought about a paradigm shift through the utilization of innovative technological methods and remedies like cyber-physical production systems (CPPS), the Internet of Things (IoT), automation/robotics, extensive data analysis, cloud computing, and profound learning. The objectives of SM encompass multiple aspects. A few of them are briefed below.

1. **Live Monitoring, Control, and Optimization of the Production Process:** SM enables the comprehensive collection of data across every level within the supply network, encompassing the production area, factory, and supply chain. By harnessing this data and transforming it into valuable knowledge, SM facilitates instantaneous tracking, control, and fine-tuning of the production process. This results in enhanced efficiency and productivity.
2. **Enhancing Industry Efficiency, Profitability, and Sustainability:** The primary goal of SM is to boost the efficiency, profitability, and long-term viability of the industry. By leveraging state-of-the-art technologies and data-driven insights, SM optimizes resource utilization, minimizes waste, and enhances overall operational performance. This leads to increased profitability and improved sustainability for manufacturing enterprises [5].

7.1.2 Digital twin technologies

The digital twin (DT) concept is gaining substantial attention owing to the swift growth of cyber-physical systems and intelligent manufacturing. A DT denotes a virtual depiction of an actual object, system, or procedure.

It is formed by gathering real-time information from sensors, devices, and other origins to imitate and replicate the actions, traits, and functioning of its real-world equivalent. DTs are used to monitor, control, and optimize physical assets throughout their lifecycle. They enable better understanding, analysis, and decision-making by providing a digital replica that can be interacted with, tested, and simulated in a risk-free environment. The synergy between AI and digital twin technologies enables organizations to leverage real-time data, predictive analytics, and advanced simulations to improve efficiency, reduce costs, and optimize capability across a diverse array of applications and industries [6–8].

7.1.3 Industry 4.0- an overview

Today's industry requires the computerization and intelligentization of the manufacturing methodology. The shift from mass production to customized production is currently taking place in manufacturing industries. The advancements in production technologies and practices and their applications contribute to increased productivity. The term "Industry 4.0" refers to the fourth industrial revolution, characterized by enhanced organization and control throughout the entire product lifecycle to meet individual customer requirements. Industry 4.0 comprises notions like the IoT, Industrial Internet, Intelligent Manufacturing, and Manufacturing based on Cloud technology. It emphasizes the consolidation of humans in the production process to drive constant improvement, focus on value-added activities, and reduce waste [9]. The key technologies that comprise Industry 4.0 are shown in Figure 7.1.

7.1.3.1 Artificial intelligence

Industry 4.0 is witnessing significant advancements with the incorporation of AI. The focus of industries lies in enhancing product consistency, productivity, and cost-effectiveness through the collaborative integration of robotics and human expertise. In intelligent manufacturing environments, interconnected processes heavily rely on AI-driven automation systems that efficiently collect and analyze diverse data types. By harnessing smart automation platforms, the landscape of modern production undergoes transformative changes. AI facilitates informed decision-making by providing relevant insights and also alerts individuals about potential malfunctions [10].

7.1.3.2 Internet of Things

The objective of IoT is to extensively infiltrate our daily surroundings and the items within it, establishing a bridge between the physical and virtual domains. This connection enables individuals and objects, commonly referred to as "things," to be seamlessly interconnected at all times and

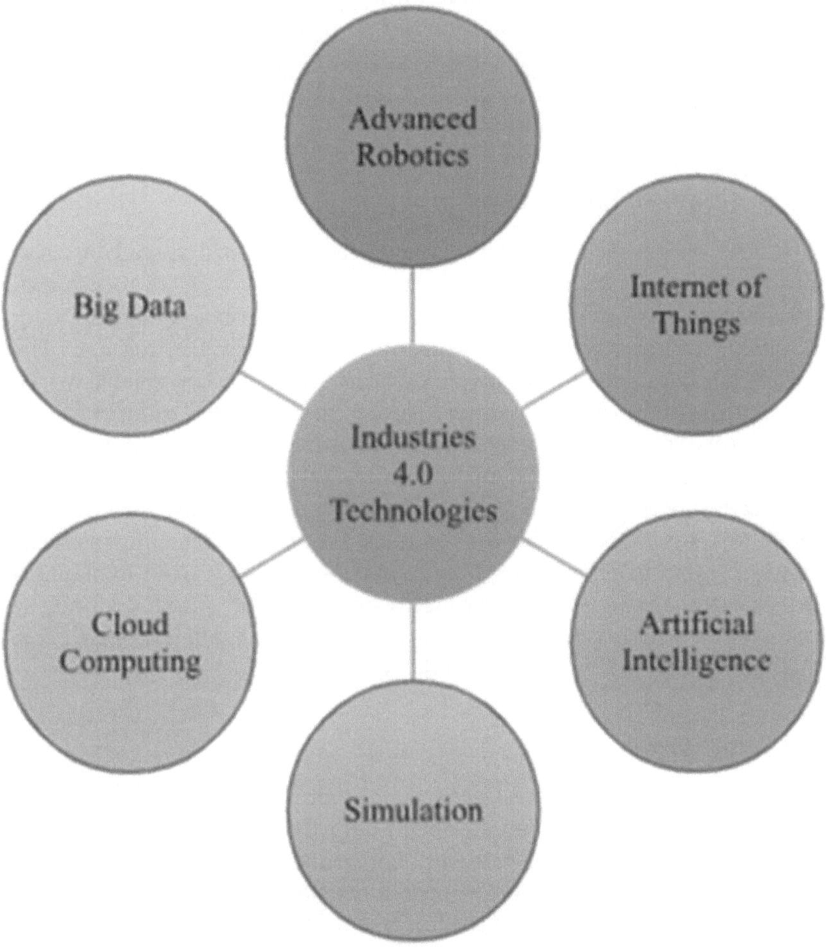

Figure 7.1 The key technologies that comprise Industry 4.0.

in any location, with the flexibility to connect with any network or service. IoT is recognized as a dynamic and worldwide network consisting of interlinked objects that have unique addresses. These objects operate on standardized and interoperable communication protocols and possess the ability to configure themselves automatically [11].

7.1.3.3 Cloud computing

Cloud computing is revolutionizing the field of enterprise IT architecture, with companies of all sizes embracing Software-as-a-Service (SaaS) solutions to streamline their internal IT operations and accelerate the adoption of enterprise systems. One possible initial step in adopting cloud computing is migrating one or more enterprise systems to the cloud. This includes Customer Relationship Management (CRM), Enterprise Resource Planning (ERP), Product Lifecycle Management (PLM), Advanced Planning System, and Manufacturing Execution Systems [12].

7.1.3.4 Advanced robotics

As the use of robotic systems becomes increasingly widespread in various industries and everyday life, the importance of having a DT of robots grows in practical scenarios. These scenarios include multi-robot coordination, collaboration, and interactions with humans, particularly in situations where ensuring human safety is a top priority. By having a DT, a sustainable working environment can be created, considering the safety and well-being of humans.

Traditionally, many have attempted to implement robot digital twins using simulation or cloud frameworks. In recent times, new concepts and cases have emerged that leverage AI for semi- and fully self-governing robotic systems. These include techniques like transfer learning, as well as imitation learning referred to as acquiring knowledge from explanation. These methods have proven to be highly beneficial for complex robotic systems where constructing highly accurate dynamics models through traditional methods is not feasible [13,14].

7.1.3.5 Big data

Smart Manufacturing systems generate a wide range of data, which comprises a fusion of diverse and compiled data from interconnected systems spanning various levels and domains. The primary aim of big data analysis in advanced manufacturing is to unearth novel correlations, influential elements, and patterns embedded within this data. Furthermore, the analysis aims to observe these findings in real time through the observation of big data streams [15].

The analysis of this data has several potential benefits:

- It allows for the generation of knowledge from vast and continuously changing data streams. This knowledge can act as a foundation for decision support.
- It facilitates the examination of value flows. By leveraging the identified associations and patterns. This can be useful for simple monitoring purposes or for detecting anomalies.
- It enhances the clarity of smart manufacturing systems, which is valuable for optimizing models and predicting future events.
- It facilitates new possibilities for diagnosis within the systems.
- It supports the optimization of key performance indicators (KPIs).

7.1.3.6 Simulation

Manufacturing simulation involves employing computer modeling to virtually assess and evaluate manufacturing methods and procedures. This encompasses a range of operations such as manufacturing, assembly, inventory management, and transportation. By utilizing simulations, the need for physical testing of a manufacturing system is significantly reduced, resulting in considerable time and cost savings.

In the realm of industrial manufacturing, there are numerous critical components, including manufacturing monitoring systems, managerial systems, and extensions for enhancing CAD/CAM designs. Historically, a considerable number of researchers have engaged in the creation of simulations for robots with the aim of achieving accurate and flexible systems to simulate robot actions. Nonetheless, there are discernible differences referring to the process of generating robotic models and incorporating them into digital surroundings. Some simulators are limited to simulation or approximation of the robot's dynamics and real-world actions [16].

7.2 FUNCTIONAL COMPONENTS OF DT

The existence of DTs is enabled by the convergence of the IoT, machine learning, AI, and large-scale data technologies. It is worth mentioning that a digital twin does not always have to precisely replicate the physical object in three-dimensional form [6,8,14,17,18], although having a 3D model is highly advantageous. Anyway, the function model has a vital contribution in defining the purpose and functional needs of the DT. It involves assessing whether the DT will enhance, safeguard, oversee, or forecast real-world processes. Additionally, it requires analyzing which specific functions or activities of the physical process will be incorporated into the DT [17,19]. This involves developing a process plan that considers both the physical process and the functionalities of the DT. The plan includes determining the equipment to be included, the aspects of modeling to be incorporated (such as mechanics, 3D space, electronics, etc.), and establishing the correlation between the physical and digital components in terms of information exchange. Furthermore, it is imperative to recognize the non-functional requirements of the DT.

7.3 IMPORTANCE OF DTS IN SMART MANUFACTURING

Smart manufacturing has remarkably changed the manufacturing process by optimizing productivity, reducing costs, and minimizing waste. What was once a conventional manufacturing process involving human labor, tools, and equipment has now transformed into a smart process through digitalization. DT technology has a significant contribution in driving innovation in smart manufacturing. By merging digital twins and smart manufacturing services, significant changes can manifest in areas such as product design and process optimization [20]. Through continuous analysis of data streams via DT technology, trends in the performance of manufacturing processes can be identified. Comparative analysis based on these insights can optimize real-world manufacturing processes. Digital twins create virtual models of physical objects, enabling simulations to understand behavior, assess states, and analyze dynamic changes. This iterative optimization in the virtual environment during the design phase ultimately leads to error-free production, avoiding costly mistakes later in manufacturing [21].

The convergence of digital twins and smart manufacturing benefits end-users by providing optimized products tailored to market needs at competitive prices. At the same time, companies achieve cost savings by producing error-free goods.

The construction industry faces challenges related to low productivity and limited technology adoption. Researchers have recognized the significant potential of DT technology in addressing these challenges and transforming the construction sector. DTs are also gaining traction in smart cities, enhancing connectivity and enabling IoT applications that benefit communities. Moreover, digital twins have garnered widespread acclaim in various sectors, driven by industry demand and their ability to facilitate data flow between digital and physical entities [22].

Smart buildings, with their costly operation and maintenance requirements, can find value from digital twin technology. By creating digital twins from intelligent assembling management systems, operational efficiency can be improved, leading to reduced maintenance costs. Digital twins and smart manufacturing have also found applications in prognostics and health management systems. These systems leverage interconnectivity and extensive data gathering to detect faults and improve predictions and deployment across different manufacturing operations. Intelligent health diagnosis and maintenance methods enabled by digital twins empower maintenance personnel to identify critical influencing factors without prior knowledge of equipment health degradation [18,23].

Although smart manufacturing is successfully implemented in various industries, research on the potential benefits of integrating the new 5G connectivity remains limited. Furthermore, studies in the shipbuilding and automotive industries have demonstrated the positive impact of digital twins and smart manufacturing. These technologies have generated superior results relative to conventional methods, improving productivity, quality, and the ability to predict production possibilities.

7.4 INTEGRATION OF AI AND DT

AI aids in crafting novel intelligent machines capable of responding in ways equivalent to human intelligence. The combination of digital twins and AI will lead to incredible transformations in every aspect of life. AI can help DTs by providing a sophisticated analytical tool capable of automatically analyzing gathered data and delivering valuable insights, such as predictions about outcomes, offering recommendations for eliminating future difficulties, etc.

7.4.1 How AI enhances digital twin technology in smart manufacturing

Artificial intelligence in DTs is an abstract engineering system that has numerous utilizations, including product design, equipment manufacture, medical analysis, aerospace, and others. Intelligent manufacturing is now gaining the most attention in the field of research [24].

The efficient and prompt response of the utilization state of the machinery in the manufacture, as well as preemptive alert of faults or breakdowns, is a major concern in today's intelligent manufacturing sector. Autonomy and active self-optimization are the essential features of intelligent manufacturing. In the sphere of intelligent manufacturing, DT technology can result in the creation of virtual simulations for products, manufacturing processes, and even entire factories. This, in turn, contributes to the enhancement of production efficiency within product development and manufacturing companies. Additionally, it possesses the capability to conceive designs within a three-dimensional virtual environment. It is possible to simplify product geometry verification, assembly feasibility verification, and process implementation by altering the components and products of varied sizes and assembly relationships. Therefore, by using AI technology, the production times, time, and cost of physical prototype making, etc. can be minimized.

7.4.2 Application of AI in DTs' technology in intelligent manufacturing

Digital twin technology can be applied in various domains to leverage AI technologies and achieve significant benefits. Here are some applications of digital twins in AI technologies.

- **Manufacturing and Industrial Processes:** Intelligent manufacturing has been facilitated by the development of modern and advanced information technology, including big data, AI, IoT, edge computing, etc. Digital twins combined with AI can optimize manufacturing processes by monitoring and analyzing real-time data from sensors, machines, and production lines. AI algorithms can identify patterns, predict equipment failures, optimize production schedules, and suggest process improvements. This helps in reducing downtime, increasing productivity, and ensuring efficient resource allocation.

- **Smart Cities and Infrastructure:** Digital twins combined with AI can be used to monitor and manage complex urban systems. AI algorithms can analyze data from various sources such as traffic sensors, weather forecasts, and utility grids to optimize energy consumption, traffic flow, waste management, and other city services. This can lead to improved sustainability, resource efficiency, and quality of life for citizens.
- **Healthcare and Medical Devices:** Digital twins integrated with AI can enable personalized healthcare and remote patient monitoring. By analyzing patient data, medical records, and sensor information, AI algorithms can provide real-time insights, identify irregularities, and predict health conditions. Digital twins can also simulate the effects of treatments or interventions, enabling healthcare professionals to undergo well-informed choices and optimize patient care.
- **Transportation and Logistics:** Digital twins combined with AI can optimize transportation systems and logistics operations. AI algorithms can analyze data from vehicles, routes, weather conditions, and customer demand to optimize routing, reduce fuel consumption, improve delivery schedules, and enhance supply chain management. This leads to improved efficiency, reduced costs, and better customer satisfaction.
- **Energy Management and Utilities:** Digital twins integrated with AI can optimize energy generation, distribution, and consumption. By analyzing data from power grids, weather forecasts, and energy usage patterns, AI algorithms can optimize energy flow, predict demand, and recommend energy-saving strategies. This helps in reducing energy waste, improving grid stability, and enabling more sustainable energy management.
- **Building and Facility Management:** Digital twins combined with AI can optimize the operation and maintenance of buildings and facilities. AI algorithms can analyze data from sensors, HVAC systems, occupancy patterns, and energy consumption to optimize energy efficiency, improve comfort levels, and detect equipment faults or maintenance needs. This results in reduced energy costs, improved occupant satisfaction, and proactive maintenance [25].

7.4.3 Benefits of AI innovations to the manufacturing industry

a. **Predictive Maintenance:** The IoT sensors of the Twin system produce large amounts of data in real time. This data can be actively examined by businesses to detect issues within the system. This capability allows companies to better plan ahead for predictive maintenance, leading to enhanced efficiency in production lines and reduced maintenance expenditures [26].

b. **Predictive Production Defect:** During the manufacturing process, it is feasible to address potential important issues and faults without halting the production line and to intervene quickly to resolve problems that could cause considerable cost and time harm. AI algorithms can detect any flaw in machines. The data stored in the cloud analyses or monitors equipment disruptions.
c. **Quality Control:** Taking special care to ensure the quality of the productions. AI offers more innovative manufacturing processes and higher product quality. It also decreases process downtime and boosts productivity. AI systems can alert workers to production flaws that could lead to quality difficulties in the future. Manufacturers can maintain quality levels by addressing these challenges.
d. **Efficaciousness:** AI enables total visibility across warehouses. This provides users with information that can be utilized to detect quality concerns, decrease scrap, and make other production improvements. This enables producers to boost production throughput while also improving quality.
e. **Productivity:** Engineers can quickly identify the material type that should be utilized for a certain product, and manufacturers can use reports to forecast orders.
f. **Profitability:** Manufacturers can use AI to assess the state of their equipment and anticipate when maintenance should be conducted. It is possible to reduce unforeseen machinery downtime and minimizing maintenance costs can be achieved through the application of machine learning techniques for predictive maintenance [27,28].

Despite the difficulties, the advantages of AI advancements for the manufacturing sector are evident. Ranging from anticipatory upkeep to assuring quality, AI technologies are enhancing effectiveness, output, and financial gain in global manufacturing plants. As AI technologies progress, we anticipate witnessing further advantages in the upcoming years.

7.4.4 Design of smart manufacturing system using digital twin

Manufacturing System Design (MSD) conventionally encompasses the traditional practices of modeling, analyzing, and optimizing various aspects such as system configuration, output capability, manufacturing process flow, material handling system, manufacturing methods, system flexibility, and operation strategies [29]. The DT is utilized at different stages of design

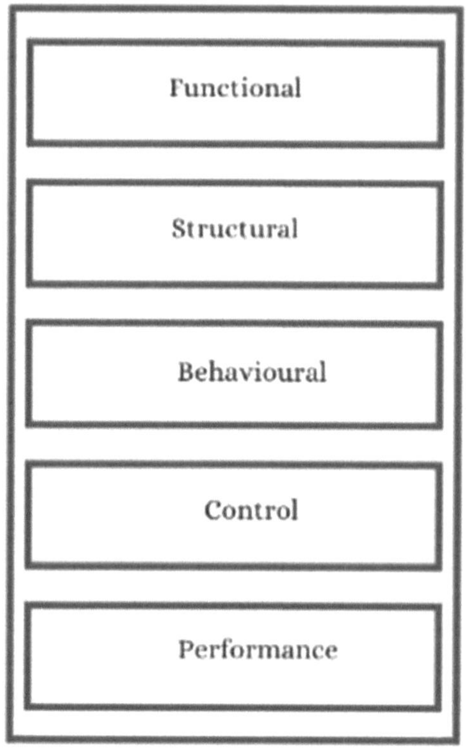

Figure 7.2 Various levels in which the DT is employed in the smart manufacturing design.

throughout the manufacturing process. Figure 7.2 illustrates the specific levels at which the DT is employed in smart manufacturing design.

- **Functional Level:** The functional requirements of smart manufacturing vary depending on the specific product and process involved. These requirements encompass several key aspects. Firstly, there is a need for a process that efficiently converts raw materials into the final product. Additionally, simulation and monitoring of the manufacturing process are crucial to ensure that it is executed correctly from the outset. Another important functional requirement is the acquisition, processing, and mining of data. This data is used to create digital models and inform decision-making for manufacturing operations. The collected data originates from various sources, such as machines, operators, sensors, suppliers, markets, and users. With the rapid increase in networked components within the system, decision-making activities for manufacturing operations often involve vast datasets, commonly referred to as big data [30].

- **Structural Level:** It is an organized representation of the combination, linkage, and arrangement relationships among the mechanical structures that bring about manufacturing functions, principles, and behaviors. The interconnectedness of the mechanical structure serves as the basis for the transfer and transformation of material, energy, information, and motion within the manufacturing system. This model typically encompasses the definition of topology, planning of layout, design of fixtures/carriers, and the creation of buffers/caches within the smart manufacturing system (SMS) [31].
- **Behavioral Level:** It serves as a systematic portrayal of the organized depiction of mechanical motion transmission and the conversion of different motion patterns, along with their interconnectedness. In the context of the SMS, the interplay of material, energy, and information is predominantly evident through the interactions of mechanical motion behavior. This model conventionally encompasses the movement of equipment, work-in-progress (WIP), material conveyance/handling, and process engagement within the SMS. The foundational components of motion behavior primarily encompass attributes like force, displacement, velocity, acceleration, jerk, angular displacement, angular velocity, angular acceleration, angular jerk, snap, and their higher derivatives.
- **Control Level:** This level addresses the arrangement, functioning, or computation of a process using statistical or engineering techniques in order to regulate the process output. It typically encompasses the design of electrical and pneumatic systems, networking of controls and sensors, designing of equipment controls, and coordination of multiple processes [32].
- **Performance Level:** Variations in design choices lead to variations in the efficiency of manufacturing execution. This model typically involves assessing and improving system performance in terms of metrics such as efficiency, flexibility, reconfigurability, robustness, adaptivity, extensibility, and resilience.

7.4.5 Digital twin-enabled predictive maintenance in smart manufacturing

DTs are transforming the manufacturing business by providing unprecedented visibility into the manufacturing process. Manufacturers may test multiple situations, foresee failures, and make adjustments in real time by constructing virtual replicas of physical assets. Manufacturers can make sensible decisions about improvements to the physical object or process by analyzing data from the DT. DTs twins provide an accurate virtual equivalent in real-time, giving maintenance personnel a comprehensive assessment of equipment performance and health. DTs can detect potential flaws before they arise, allowing manufacturers to take preventive remedies.

Predictive maintenance aims to anticipate the point at which a component or system will cease to function correctly. This prediction is measured by factors such as the remaining usable life (RUL), a Health Indicator, or equipment efficiency. Using RUL and other machine data, companies can predict when to plan maintenance for a component or system. With the emergence of IoT and DTs, businesses are emphasizing the creation of a Predictive Maintenance model that fine-tunes the maintenance cycle, finding a middle ground between corrective and preventative maintenance strategies. Predictive maintenance embraces a just-in-time approach to component replacement [33].

Conventional predictive maintenance lacks a comprehensive understanding of equipment's real-time condition and attributes due to limited insights from condition monitoring. Visual fault information in predictive maintenance is restricted, leaving employees without complete fault details. The DT, presented through a visualization interface, showcases a virtual model of the equipment. This model interacts with real-time data from the actual equipment, effectively conveying real-time status and malfunction details to operators in a clear and intuitive manner. The numerous requirements, such as problem detection and maintenance decision-making, will become much more obvious and more reliable [34,35].

Predictive maintenance with a digital twin cannot handle all use cases or business problems. Here are some significant qualifying factors to consider during use case qualification. The problem must have a target or outcome to anticipate. It should include a record of the equipment's operational history, which should include both positive and negative consequences. The history should be relevant and of high enough quality to serve the use case. Finally, the company should have domain specialists who understand the problem thoroughly.

7.4.5.1 *Predictive maintenance of solar power systems - a case study*

Predictive maintenance assists engineers in precisely determining the best timing for equipment upkeep. This approach reduces downtime and prevents equipment breakdowns by enabling maintenance scheduling based on real-time needs rather than a fixed timetable. Often, it's either too costly or unfeasible to generate the required malfunction scenarios for directly training a predictive maintenance algorithm on the actual machine. To tackle this challenge, a solution involves using operational data gathered from a fully functioning machine to refine a physical 3D model and generate a digital twin. This digital twin can then be employed to devise a predictive maintenance detection algorithm, which can be deployed on the controller of the real equipment. The entire process can be automated, facilitating rapid adjustment to changing conditions, different materials, and equipment setups [36,37].

For example, in solar power systems, predictive maintenance can analyze data from individual solar panels to identify any anomalies or efficiency losses. By detecting issues early on, maintenance teams can strategically plan repair or replacements in advance, effectively reducing downtime and optimizing efficiency in energy production.

In a solar PV (photovoltaic) installation, various factors can disrupt the actual performance, causing it to deviate from the expected performance. If failures or unforeseen degradation issues remain undetected, they can result in reduced energy generation and potential loss of component warranty due to manufacturer changes. When a single cell within a panel is compromised, it adversely affects the overall system's power output and efficiency. Unfortunately, without measuring performance, it becomes impossible to identify and evaluate these issues.

Furthermore, considering the turnover within the solar energy industry's component manufacturers, it is important to promptly address potential warranty concerns so as to optimize the return on investment. Failure to detect failures or unexpected degradation issues can lead to diminished energy generation, loss of electricity credits, and the risk of voiding component warranties due to manufacturer changes.

Given the scale of this problem and the existing gaps in current solutions, PV system owners require an impartial, readily available predictive maintenance tool at the system level. This tool would enable them to optimize their return on investment and minimize the time required to make warranty claims in PV installations [38,39]. The predictive maintenance of a solar power system is shown in Figure 7.3.

Prognostic Insights The practice of creating digital duplicates of power systems and their specific components is utilized for running simulations that can predict potential failures and identify required maintenance measures.

Real-Time Monitoring By utilizing live monitoring, individuals can remotely detect and analyze irregular operations, glitches, as well as the origins of issues.

Remote Diagnostics Examining historical performance data of the system or its individual components enables the enhancement of power system efficiency and the optimization of processes.

Conceptual Development The potential to explore hypothetical possibilities, which may not be easily analyzed through their physical counterparts due to practical limitations, is made possible.

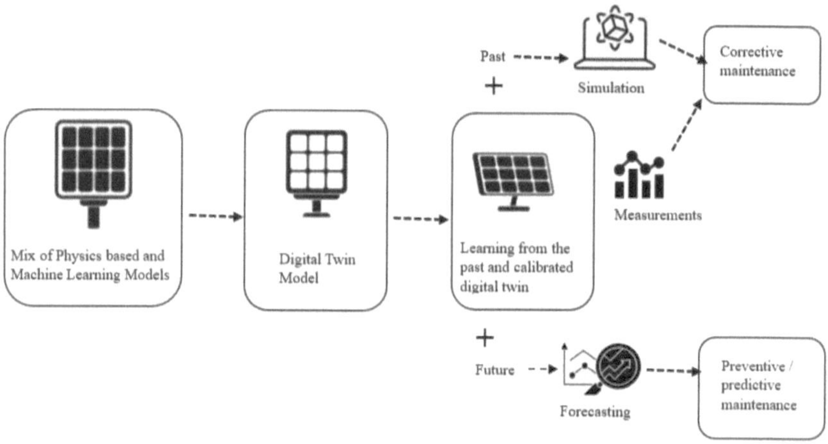

Figure 7.3 Predictive maintenance of a solar power system [39].

7.4.6 Challenges of AI in digital twins' technology in intelligent manufacturing

a. **Cost:** The implementation of AI technologies can come with a high cost, and numerous smaller businesses might not have the means to invest in such technologies due to resource limitations.
b. **Skill:** A shortage of proficient workers skilled in AI technologies can pose a challenge. Numerous manufacturers might face difficulty in locating employees possessing the required expertise to successfully integrate and sustain AI technologies.
c. **Data Security and Privacy:** With the increasing accumulation and analysis of data through AI technologies, manufacturers need to proactively safeguard this data and adhere to privacy regulations.

During the industrial product design process, without digital assistance, creating a product would entail numerous cycles, leading to resource depletion and delivery time impact. In the development of a tightly integrated industrial production line, it is vital to enhance overall efficiency by streamlining and harmonizing different elements like equipment, materials, quality checks, and manual assembly, all based on precise timing. Conventional planning methods rely solely on manual simulations or testing within an actual production environment [40].

7.5 CONCLUSION

Smart manufacturing using DT technology driven by AI holds immense potential for revolutionizing the manufacturing industry. The key technologies employed in smart manufacturing design and implementation, such as

DTs, enable significant benefits and address various challenges. Digital twin technology allows for the virtual representation and synchronization of physical manufacturing systems, providing real-time insights and analysis. This enables operators to optimize processes, enhance productivity, and ensure the safety of operations through advanced testing and operator training. The application of digital twins in smart manufacturing brings numerous benefits. It facilitates efficient and flexible manufacturing processes by enabling real-time monitoring, predictive maintenance, and optimization of production systems. Simulating and visualizing complex manufacturing operations helps detect and resolve issues before they occur, thereby reducing downtime and costs. Additionally, digital twins improve decision-making by providing valuable insights based on data collected from multiple sources, leading to informed and proactive actions. However, the adoption of smart manufacturing using DT technology also presents certain challenges. Integration with existing systems, data management, and cybersecurity are areas that require careful attention. Additionally, the complexity of implementing and maintaining DT systems and the need for specialized expertise pose challenges in terms of cost and resource allocation. Despite these challenges, the potential benefits and advancements offered by smart manufacturing using DT technology outweigh the difficulties. As industries embrace this transformative approach, smart manufacturing will continue to evolve, enabling enhanced productivity, efficiency, and competitiveness in the manufacturing sector. With ongoing research and technological advancements, digital twin-based smart manufacturing is poised to reshape the future of the industry, driving innovation and sustainable growth.

REFERENCES

1. L. Li, B. Lei, C. Mao, Digital twin in smart manufacturing, *Journal of Industrial Information Integration*, 26, 2022, 100289, https://doi.org/10.1016/j.jii.2021.100289. (https://www.sciencedirect.com/science/article/pii/S2452414X21000868)
2. Y. Lu, C. Liu, K. I-Kai Wang, H. Huang, X. Xu, Digital Twin-driven smart manufacturing: Connotation, reference model, applications and research issues, *Robotics and Computer-Integrated Manufacturing*, 61, 2020, 101837, https://doi.org/10.1016/j.rcim.2019.101837. (https://www.sciencedirect.com/science/article/pii/S0736584519302480)
3. M. Attaran, Bilge gokhan celik, digital twin: Benefits, use cases, challenges, and opportunities, *Decision Analytics Journal*, 6, 2023, 100165, https://doi.org/10.1016/j.dajour.2023.100165
4. S. Ramakrishna, T. C. Khong, T. K. S. Leong, Manufacturing, *Procedia Manufacturing*, 12, 2017, 128–131, https://doi.org/10.1016/j.promfg.2017.08.017
5. K. Haricha, A. Khiat, A. Bahnasse, Y. Issaoui, O. Hassan, Towards smart manufacturing: Implementation and benefits, *Procedia Computer Science*, 177, 2020, 639–644. https://doi.org/10.1016/j.procs.2020.10.091

6. P. Zheng, H. Wang, Z. Sang, et al., Smart manufacturing systems for Industry 4.0: Conceptual framework, scenarios, and future perspectives, *Frontiers of Mechanical Engineering*, 13, 2018, 137–150. https://doi.org/10.1007/s11465-018-0499-5
7. L. Ante, Digital twin technology for smart manufacturing and industry 4.0: A bibliometric analysis of the intellectual structure of the research discourse, *Manufacturing Letters*, 27, 2021, 96–102, https://doi.org/10.1016/j.mfglet.2021.01.003. (https://www.sciencedirect.com/science/article/pii/S2213846321000031).
8. Digital Twins: Components, Use Cases, and Implementation Tips, https://www.altexsoft.com/blog/digital-twins/ [accessed on May 10, 2023].
9. S. Vaidya, P. Ambad, S. Bhosle, Industry 4.0 - a glimpse, *Procedia Manufacturing*, 20, 2018, 233–238. https://doi.org/10.1016/j.promfg.2018.02.034LicenseCCBY-NC-ND 4.0
10. M. Javaid, A. Haleem, R. P. Singh, R. Suman, Artificial intelligence applications for industry 4.0: A literature-based study, *Journal of Industrial Integration and Management*, 07, 2022, 83–111. https://doi.org/10.1142/S2424862221300040
11. G. Lampropoulos, K. Siakas, T. Anastasiadis, Internet of things in the context of industry 4.0: An overview, *International Journal of Entrepreneurial Knowledge*, 7, 2019, 1. https://dspace.vsp.cz/handle/ijek/103
12. Q. Qi, F. Tao, A smart manufacturing service system based on edge computing, fog computing, and cloud computing, *IEEE Access*, 7, 2019, 86769–86777. https://doi.org/10.1109/ACCESS.2019.2923610
13. Z. Huang, Y. Shen, J. Li, M. Fey, C. Brecher, A survey on AI-driven digital twins in industry 4.0: Smart manufacturing and advanced robotics, *Sensors*, 21, 2021, 6340. https://doi.org/10.3390/s2119634
14. Industry 4.0 Technologies, https://www.businessprocessincubator.com/content/industry-4-0-technologies/ [accessed on May 12, 2023].
15. K. Nagorny, P. Lima-Monteiro, J. Barata, A. Colombo, Big data analysis in smart manufacturing: A review, *International Journal of Communications, Network and System Sciences*, 10, 2017, 31–58. https://doi.org/10.4236/ijcns.2017.103003
16. T. -H. Chien, C. -Y. Siao, R. -G. Chang, A smart simulation system for manufacturing, *2020 IEEE Eurasia Conference on IOT, Communication and Engineering (ECICE)*, Yunlin, Taiwan, 2020, pp. 171–173, https://doi.org/10.1109/ECICE50847.2020.9301939
17. Q. Qi, F. Tao, T. Hu, N. Anwar, A. Liu, Y. Wei, L. Wang, A. Y. C. Nee, Enabling technologies and tools for digital twin, *Journal of Manufacturing Systems*, 58, part B, 2021, 3–21, https://doi.org/10.1016/j.jmsy.2019.10.001
18. Digital Twins for Smart Manufacturing, https://www.mathworks.com/videos/digital-twins-for-smart-manufacturing-1576049506471.html [accessed on May 10 2023].
19. E. Barkmeyer Jr., E. Wallace, Reference architecture for smart manufacturing part 1: Functional models, *Advanced Manufacturing Series (NIST AMS), National Institute of Standards and Technology*, Gaithersburg, MD, [online], 2016, https://doi.org/10.6028/NIST.AMS.300-1, https://tsapps.nist.gov/publication/get_pdf.cfm?pub_id=917362 [accessed on June 30 2023]

20. G. Moiceanu, G. Paraschiv, Digital twin and smart manufacturing in industries: A bibliometric analysis with a focus on industry 4.0, *Sensors (Basel)*, 22, 4, 2022, 1388. https://doi.org/10.3390/s22041388. PMID: 35214289; PMCID: PMC8963121.
21. J. Wilhelm, C. Petzoldt, T. Beinke, M. Freitag, Review of digital twin-based interaction in smart manufacturing: enabling cyber-physical systems for human-machine interaction, *International Journal of Computer Integrated Manufacturing*, 34, 10, 2021, 1031–1048, https://doi.org/10.1080/0951192X.2021.1963482
22. J. Leng, D. Wang, W. Shen, X. Li, Q. Liu, X. Chen, Digital twins-based smart manufacturing system design in industry 4.0: A review, *Journal of Manufacturing Systems*, 60, 2021, 119–137. https://doi.org/10.1016/j.jmsy.2021.05.011
23. Y. H. Son, others, Past, present, and future research of digital twin for smart manufacturing, *Journal of Computational Design and Engineering*, 9, 1, 2022, 1–23. https://doi.org/10.1093/jcde/qwab067
24. Z. Lv, S. Xie, Artificial intelligence in the digital twins: State of the art, challenges, and future research topics [version 2; peer review: 2 approved], *Digital Twin*, 1, 2022, 12. https://doi.org/10.12688/digitaltwin.17524.2
25. B. He, K. J. Bai, Digital twin-based sustainable intelligent manufacturing: A review, *Advances in Manufacturing*, 9, 2021, 1–21. https://doi.org/10.1007/s40436-020-00302-5
26. Z. Ren, J. Wan, P. Deng, Machine-learning-driven digital twin for lifecycle management of complex equipment, *IEEE Transactions on Emerging Topics in Computing*, 10, 1, 2022, 9–22.
27. Artificial Intelligence (AI) Improving the Manufacturing Industry and Its Challenges, https://www.planettogether.com/blog/artificial-intelligence-ai-improving-the-manufacturing-industry-and-its-challenges#:~:text=One%20of%20the%20biggest%20challenges,with%20expertise%20in%20AI%20technologies [accessed on June 20 2023].
28. Pros and Cons of AI In Manufacturing, https://www.datasciencecentral.com/pros-and-cons-of-ai-in-manufacturing/ [accessed on June 09, 2023].
29. R. Nujoom, A. Mohammed, Q. Wang, A sustainable manufacturing system design: A fuzzy multi-objective optimization model, *Environmental Science and Pollution Research*, 25, 2018, 24535–24547. Available from: https://www.researchgate.net/publication/351862149_Digital_twins-based_smart_manufacturing_system_design_in_Industry_40_A_review [accessed on June 29 2023].
30. J. Kombaya Touckia, N. Hamani, L. Kermad, Digital twin framework for reconfigurable manufacturing systems (RMSs): Design and simulation, *International Journal of Advanced Manufacturing Technology*, 120, 2022, 5431–5450. https://doi.org/10.1007/s00170-022-09118-y
31. Y. Wang, X. Kang, Z. Chen, A survey of digital twin techniques in smart manufacturing and management of energy applications, *Green Energy and Intelligent Transportation*, 1, 2, 2022, 100014. https://doi.org/10.1016/j.geits.2022.100014.
32. S. M. Zayed, G. Attiya, A. El-Sayed, et al., An efficient fault diagnosis framework for digital twins using optimized machine learning models in smart industrial control systems, *International Journal of Computational Intelligence Systems*, 16, 2023, 69. https://doi.org/10.1007/s44196-023-00241-6

33. Y. You, C. Chen, F. Hu, Y. Liu, Z. Ji, Advances of digital twins for predictive maintenance, *Procedia Computer Science*, 200, 2022, 1471–1480. https://doi.org/10.1016/j.procs.2022.01.348. (https://www.sciencedirect.com/science/article/pii/S187705092200357X)
34. D. Zhong, Z. Xia, Y. Zhu, J. Duan, Overview of predictive maintenance based on digital twin technology, *Heliyon*, 9, 4, 2023, e14534. https://doi.org/10.1016/j.heliyon.2023.e14534. PMID: 37025897; PMCID: PMC10070392. https://www.ncbi.nlm.nih.gov/pmc/articles/PMC10070392/ [accessed on Jun 18, 2023].
35. S. Mihai, W. Davis, D. Hung, R. Trestian, M. Karamanoglu, B. Barn, R. Prasad, H. Venkataraman, H. Nguyen, A digital twin framework for predictive maintenance in industry 4.0, *2020 International Conference on High Performance Computing & Simulation*, Barcelona, Spain, 2021
36. H. Boyes, T. Watson, Digital twins: An analysis framework and open issues, *Computers in Industry*, 143, 2022, 103763. https://doi.org/10.1016/j.compind.2022.103763. (https://www.sciencedirect.com/science/article/pii/S0166361522001609)
37. P. Aivaliotis, K. Georgoulias, G. Chryssolouris, The use of digital twin for predictive maintenance in manufacturing, *International Journal of Computer Integrated Manufacturing*, 32, 11, 2019, 1067–1080. https://doi.org/10.1080/0951192X.2019.1686173
38. K. Alexopoulos, N. Nikolakis, G. Chryssolouris, Digital twin-driven supervised machine learning for the development of artificial intelligence applications in manufacturing, *International Journal of Computer Integrated Manufacturing*, 33, 5, 2020, 429–439. https://doi.org/10.1080/0951192X.2020.1747642
39. Decoding Digital Twin for Solar Power Plants, https://smarthelio.com/decoding-digital-twin-for-solar-plants/ [accessed on June 29 2023]
40. Y. Xu, Y. Sun, X. Liu, et al., A digital-twin-assisted fault diagnosis using deep transfer learning, *IEEE Access*, 7, 2019, 19990–19999.

Chapter 8

Machine learning and deep learning in cyber security

B. Ragavi, S. Divya, M.R. Mano Jemila, and Ahmed A. Elngar

8.1 INTRODUCTION

Cybersecurity is the process of protecting vital sensitive data against digital threats and it is also known as security in information systems. This system is projected to prevent intimidation of network systems and real-time applications, whether they are initiated inside or outside of an organization. Since our society has moved into digital technologies, there are different approaches used in cybersecurity for protecting information from hackers and cybercriminals. As a result of the COVID-19 outbreak across the world, many industries and IT sectors are compelled to reevaluate their standard working circumstances and stay at home due to transitioning into the entire digital world [1,2]. It leads to increasing data and sensitive document uploading in the cloud. So, many industries and IT sectors are searching for innovative and novel ways to prevent hackers and data breaches and Figure 8.1 represents different technologies used in cybersecurity for controlling the security measures. Moreover, in recent days cybersecurity experts and professionals are exploring new innovative techniques to recognize the various attacks in which cyber criminals will have unavoidable concerns that need to be fixed right away. In real-time cybersecurity, an emerging concern is the integration of preparedness to effectively address attacks on computer network security.

Additionally, as a result of cybersecurity security attacks and breaches frequently occur, particularly in industrial sectors which results in a data breach including the compromise of confidential information, loss of customer trust, negative impact on public relations, business disruptions, and significant financial losses. Furthermore, loss of confidential data including corruption, blackmail, technology theft, and even hazards to national security so it is more important to enhance the security level in real-time applications. Additionally, types of security are classified under different types, and it is discussed below:

DOI: 10.1201/9781003534761-8

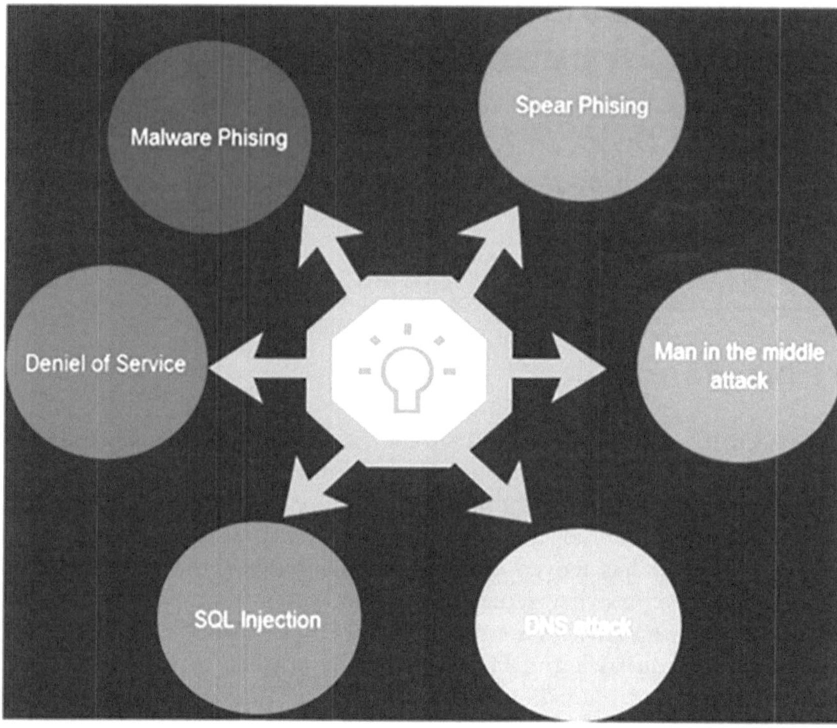

Figure 8.1 Overview of cybersecurity.

8.2 TYPES OF CYBERSECURITY

The following are the types of cybersecurity which is also represented in Figure 8.2.

8.2.1 Network security

A set of technologies in network security ensures the efficacy and integrity of technology that are used in computer networks from unauthorized access, hackers, and disruption. To control or prevent the security hazards, security architecture has been used. It is a specific tool used to protect the network and its application [3]. This architecture has been instrumental in devising highly effective network security strategies that utilize scalable and automated defense mechanisms, presenting themselves as a comprehensive collection of security guidelines. The administrator should choose and implement the set of security guidelines for each protective layer. Furthermore, to enhance the security application, hardware and software tools are combined to enforce the network security and prevent unauthorized access between two networks.

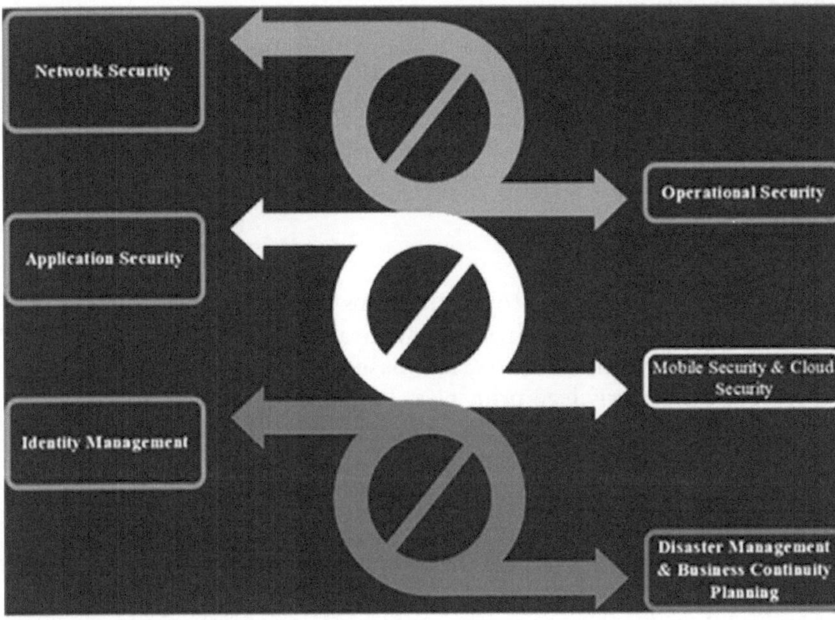

Figure 8.2 Types of cybersecurity.

8.2.2 Application security

The process involves developing, testing, and securing application features to prevent and protect against security vulnerabilities, such as unauthorized access and modification. It has a corrective control application to reduce the effect of attacks and other vulnerabilities. It includes protecting data in the system through the device's intrusion from unwanted threats. This protection can be given continuous security by updating the apps to ensure and secure from hackers. Successful security is instigated by the designer in the initial design stage, programming stage, validation, risk modeling, etc. before a programming device is deployed.

8.2.2.1 Data security

Data security is a vital component of cybersecurity, which centers on safeguarding sensitive information from unauthorized access, disclosure, alteration, and destruction. Effective data security measures are essential to maintain the confidentiality, integrity, and availability of data becomes increasingly crucial, especially with the continuous evolution of cyber threats. Here are some key considerations and practices for ensuring data security in cybersecurity.

8.2.2.2 Encryption

The process of encrypting data to render it incomprehensible to unauthorized individuals. It involves using cryptographic algorithms to convert plain text into ciphertext. Employing strong encryption algorithms and implementing encryption at rest, in transit, and in use can help safeguard sensitive data.

8.2.2.3 Access control

Deploying strong access control mechanisms to guarantee that data is only accessible by authorized individuals with appropriate permissions. This encompasses user authentication, role-based access controls (RBAC), and adherence to least privilege principles. Regularly reviewing and updating access privileges helps mitigate the risk of unauthorized access.

8.2.2.4 Data classification

Data is classified according to its sensitivity and importance, and security controls and protections are assigned based on the designated classification levels. By following this approach, organizations can effectively prioritize their security efforts and allocate resources accordingly.

8.2.2.5 Data loss prevention

Data loss prevention (DLP) solutions help prevent unauthorized data disclosure or leakage. These systems can monitor, detect, and prevent sensitive data from leaving the organization's network. DLP solutions employ techniques such as content filtering, data masking, and data encryption to protect sensitive information.

8.2.2.6 Data backup and disaster recovery

Regularly backup critical data and develop comprehensive disaster recovery plans. This ensures that in the event of a data breach, natural disaster, or system failure, organizations can recover their data and continue operations with minimal disruption.

8.2.2.7 Secure data transmission

Protect data while it is being transmitted over networks or between systems. Implement secure protocols such as SSL/TLS for secure communication and use virtual private networks (VPNs) to establish secure connections.

8.2.2.8 *Regular patching and updates*

By consistently updating systems and software with the latest security patches and updates, organizations can proactively reduce vulnerabilities that attackers may exploit to gain unauthorized access to sensitive data.

8.2.2.9 *Employee training and awareness*

Empower employees with data security best practices, including robust password management, the ability to identify phishing attempts and adherence to data handling policies. This education instills a deep understanding of the significance of data security and the pivotal role each employee plays in upholding it.

8.2.2.10 *Data privacy compliance*

Adhere to pertinent data privacy regulations and standards, such as the General Data Protection Regulation (GDPR) or the California Consumer Privacy Act (CCPA), to guarantee that personal data is collected, processed, and stored in strict accordance with legal requirements.

8.2.2.11 *Incident response and monitoring*

Implement robust incident response plans and establish continuous monitoring of systems and networks. Swiftly detecting and responding to security incidents plays a critical role in minimizing the impact of data breaches and halting further damage, safeguarding the organization's sensitive information [4,5]. Through the implementation of these data security practices, organizations can bolster their overall cybersecurity stance and fortify sensitive information against unauthorized access and misuse. Regular assessments, audits, and updates of data security measures are vital to tackle emerging threats and sustain the continuous protection of valuable data. This includes establishing a robust storage system that ensures data integrity and confidentiality throughout the database and data transfer processes.

8.2.3 Identity management

Identity Management establishes access controls that restrict private information access while allowing employees to read, copy, and modify content pertinent to their responsibilities. This process defines the access levels of individuals within an organization, encompassing a wide range of data from sensitive to company-specific information. It serves as a crucial IT security discipline and framework for managing digital identities, ensuring that only authorized individuals have access to resources and specific activities [6,7,8].

By eliminating unauthorized identities, it effectively prevents unauthorized parties from infiltrating the system while ensuring legitimate parties have appropriate access to resources at the right times. Identity Management extends access to a specific set of individuals, including vendors, customers, contractors, and employees, ensuring a secure and controlled environment for accessing essential resources.

8.2.4 Operational security

Operational security, or OPSEC, plays a pivotal role in cybersecurity, aiming to protect sensitive information from compromise by implementing measures that counter potential threats during various operational activities. It involves implementing practices and measures to protect critical assets, maintain confidentiality, ensure system integrity, and minimize vulnerabilities. Here are key considerations and practices for operational security in cybersecurity:

8.2.4.1 Risk assessment and management

By conducting a thorough risk assessment, organizations can proactively identify potential threats and vulnerabilities in their operational activities. This assessment lays the foundation for developing risk management strategies that prioritize and effectively mitigate the identified risks.

8.2.4.2 Secure configuration management

Implement secure configurations for hardware, software, and network devices. To enhance security measures, organizations should adhere to industry best practices and security guidelines, effectively reducing the attack surface and minimizing potential vulnerabilities.

8.2.4.3 Security awareness and training

Promote a security-conscious culture within the organization by educating employees and stakeholders about operational security best practices. This includes comprehensive training on essential topics like password management, social engineering awareness, and safe browsing habits, empowering everyone to actively contribute to a more secure environment.

8.2.4.4 Incident response planning

Create a robust incident response plan to efficiently address security incidents. This involves defining clear roles and responsibilities, establishing effective communication channels, and conducting regular drills and exercises to evaluate the plan's effectiveness in real-world scenarios.

8.2.4.5 Access control

Strengthen access control measures to maintain authorized personnel's appropriate access privileges. This entails employing multi-factor authentication (MFA), implementing strong passwords, and leveraging RBAC to restrict access to sensitive systems and data, ensuring a higher level of security and confidentiality.

8.2.4.6 Physical security

Safeguard essential physical assets, such as servers, data centers, and critical infrastructure, by implementing robust physical security measures [9]. This includes deploying access controls, video surveillance, and diligent monitoring of all physical entry points, ensuring comprehensive protection against potential threats and unauthorized access.

8.2.4.7 Change management

Establish robust change management processes to control and monitor changes made to systems and configurations. Implement a formalized process for evaluating, testing, and approving changes to minimize the risk of introducing vulnerabilities or disrupting operations.

8.2.4.8 Incident monitoring and logging

Establish monitoring and logging mechanisms to effectively track and identify suspicious activities. Employ security information and event management (SIEM) systems to consolidate and analyze logs from diverse sources, enabling early detection of potential security incidents, and facilitating timely response to emerging threats.

8.2.4.9 Vendor management

Maintain strong security practices among third-party vendors and suppliers by exercising due diligence during their selection process [10]. Set forth explicit security requirements in contracts or service-level agreements to ensure they adhere to appropriate security standards, contributing to a secure and reliable partnership.

8.2.4.10 Continuous improvement and evaluation

Continuously enhance operational security measures by conducting regular assessments and updates that consider emerging threats, industry best practices, and insights gained from security incidents. Conduct periodic audits and evaluations to identify areas for improvement, ensuring compliance with security policies and standards and fostering a proactive approach to security management [11]. Operational security serves as the backbone of an organization's overall security posture. By implementing these practices, organizations can effectively mitigate risks, safeguard critical assets, and uphold the confidentiality, integrity, and availability of systems and data. A proactive and continuous effort is essential to adapt to ever-evolving threats, ensuring operational activities align seamlessly with robust security measures.

8.2.5 Mobile security & cloud security

Mobile security is a fundamental pillar of cybersecurity, centering on safeguarding mobile devices, applications, and data from an array of threats and vulnerabilities. With the rapid expansion in the use of smartphones and tablets, the imperative to ensure the security of these devices has escalated significantly. Here are key considerations and practices for mobile security in cybersecurity:

8.2.5.1 Device protection

Strengthen the security of mobile devices by implementing robust passcodes or biometric authentication methods (e.g., fingerprint or facial recognition) to secure access. Enable automatic device locking and configure remote wiping capabilities in case of loss or theft, providing an added layer of protection for sensitive data.

8.2.5.2 App security

Ensure the security of mobile devices by exclusively downloading apps from reputable sources, such as official app stores. Regularly update apps to maintain the latest security patches. Exercise caution when granting app permissions, only providing necessary access, and thoroughly review app permissions prior to installation.

8.2.5.3 Mobile device management

Adopt mobile device management (MDM) solutions to establish centralized management and security for mobile devices within the organization. These solutions facilitate remote configuration, data encryption, and policy enforcement, ensuring a consistent security posture across all mobile devices.

8.2.5.4 Secure network connections

Steer clear of public Wi-Fi networks due to their inherent insecurity and vulnerability to man-in-the-middle attacks. When connecting to Wi-Fi networks, opt for encrypted connections (e.g., VPN) to safeguard data transmitted between the device and the network, bolstering security and privacy.

8.2.5.5 *Mobile app testing*

Conduct thorough security testing of mobile applications before deployment. This includes analyzing the code for vulnerabilities, testing for potential data leakage, and verifying the implementation of secure communication protocols.

8.2.5.6 *Data encryption*

Enable device-level encryption and secure sensitive data stored on mobile devices through encryption. This precaution guarantees that in the event of device compromise, the data remains safeguarded and indecipherable to unauthorized individuals.

8.2.5.7 *Regular software updates*

Maintain mobile operating systems and applications up to date with the latest security patches and updates. Regular software updates address known vulnerabilities and deliver crucial security enhancements, bolstering the overall security of mobile devices.

8.2.5.8 *Mobile threat detection*

Implement mobile threat detection solutions that can identify and mitigate mobile-specific threats, such as malicious apps, phishing attacks, and mobile malware. These solutions monitor device behavior and network traffic to detect suspicious activities.

8.2.5.9 *User education*

Educate mobile users about mobile security best practices. Train them to recognize and avoid social engineering attacks, phishing attempts, and suspicious app requests. Encourage strong password practices and the use of reputable security apps.

8.2.5.10 *Data backup*

Implement regular data backups for mobile devices to safeguard against loss or damage. Utilize secure backup solutions, such as cloud-based services with robust encryption, to ensure data restoration in the event of device compromise or loss, providing added peace of mind for data protection. Mobile security requires a multi-layered approach to protect against the evolving threats targeting mobile devices. Through the adoption of these practices, organizations and individuals can fortify the security of their mobile devices, applications, and data, effectively protecting against unauthorized access, data breaches, and other vulnerabilities unique to mobile environments.

8.2.6 Disaster recovery and business continuity planning

The concept involves leveraging machine intelligence to enhance various processes, monitoring capabilities, alert systems, and preparedness plans within society, enabling swift and effective responses to any hacking incidents resulting in operational disruptions or data loss. Its policy dictates restarting the lost operations after any calamity happens to the identical operational capacity [12,13]. For example, business continuity and disaster recovery should follow the same principle for any recovery plan. But it should specify the awareness of risk of cyberattack or breach and the following steps are discussed below:

i. The first step includes various professionals across business, IT sectors, and other employees and each member should have clear roles and responsibilities to remove ambiguity [14].
ii. After that risk assessment for cybersecurity will give an outline about all possible risks related to attacks and breaches.
iii. Once we identify all the cyber risks, we should perform a business impact analysis because it will determine the recovery strategy.
iv. Once determining all the strategies, it is important to test the system and it will adapt or review the current plans.

8.3 MACHINE LEARNING IN CYBERSECURITY

In recent times, machine learning has found widespread application across various domains, notably in areas like zero malware detection, threat analysis, anomaly-based intrusion detection for critical infrastructures, and more. Researchers have increasingly emphasized signature-based methods to effectively detect zero attacks, further advancing the field of cybersecurity [15, 16]. In this view, many cybersecurity applications were used for train and test the data to render the effective in the cybersecurity core sector which are discussed below:

8.3.1 Alert management in cybersecurity

Alert Management is essential for handling the large number of alerts generated by modern environments. In this view, it can be employed to address challenges such as filtering, prioritizing, and fusing alerts into more comprehensive events as represented in Figure 8.3. Alert filtering aims to eliminate redundant alerts that may not be relevant or indicate malicious activity and it is achieved by notable success by mitigating the occurrence of deception generated by ML-NIDS (Machine Learning Network Intrusion

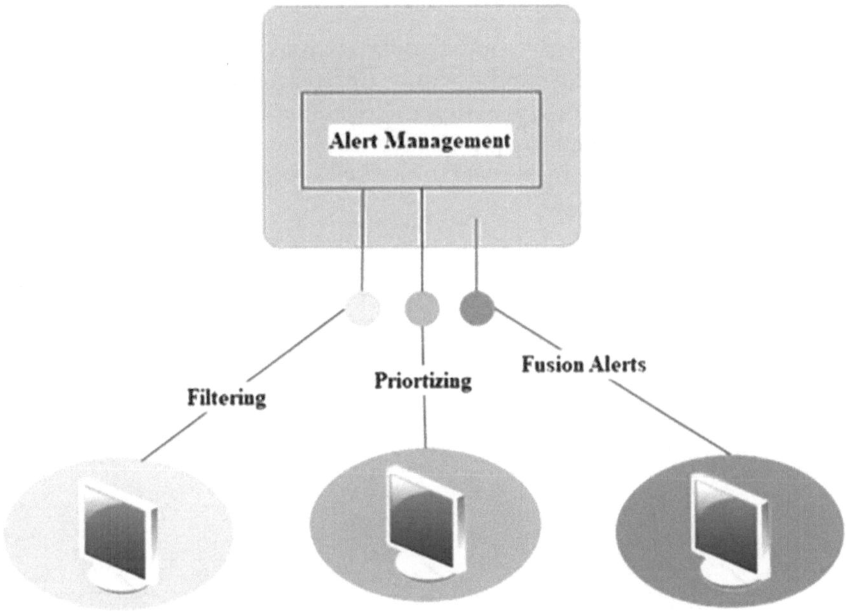

Figure 8.3 Diagram for alert management.

Detection Systems) [17,18]. This method achieved a remarkable 75% reduction in triaging time for deception, exceeding non-ML mechanisms by 45%. Incident triage becomes necessary when security specialists are overwhelmed by the sheer volume of alerts. ML possesses the remarkable ability to autonomously learn crucial ranking criteria even with limited instruction. Recent research has unequivocally showcased its effectiveness, as it successfully generates sensitive alerts at the top position in an impressive 95% of cases. This highlights the potential of ML to significantly enhance alert systems and decision-making processes, bolstering various applications across diverse industries. To manage large quantities of alerts, alert fusion techniques can be employed whereas it involves an aggregation of similar alerts and determining analogies and causal mechanisms relevant to security tasks. For example, clustering techniques have been used to detect protocol and network ports which are often attacked by malicious nodes. It is worth noting that attackers are increasingly utilizing the Remote Desktop Protocol for conducting lateral movement activities. These alert management techniques can be combined to create comprehensive solutions. Moreover, ML-based alert management techniques offer effective solutions to handle the high volume of alerts in modern environments. By filtering, prioritizing, and fusing alerts, these techniques help security professionals focus on the most relevant and critical security incidents.

8.3.2 Source data analysis

As represented in Figure 8.4, source data analysis can able to analyze the data from multiple sources and it will be differentiated into different applications such as operational decision and labeling optimization whereas it analyses the log data and to optimize the labeling (for the Unlabeled Data) efforts and promote the implementation of supervised ML and few applications are mentioned below:

a. Analyze Traffic for identifying the patterns for indicating the attacks occurring inside the network.
b. It is used to monitor the user's behavior.

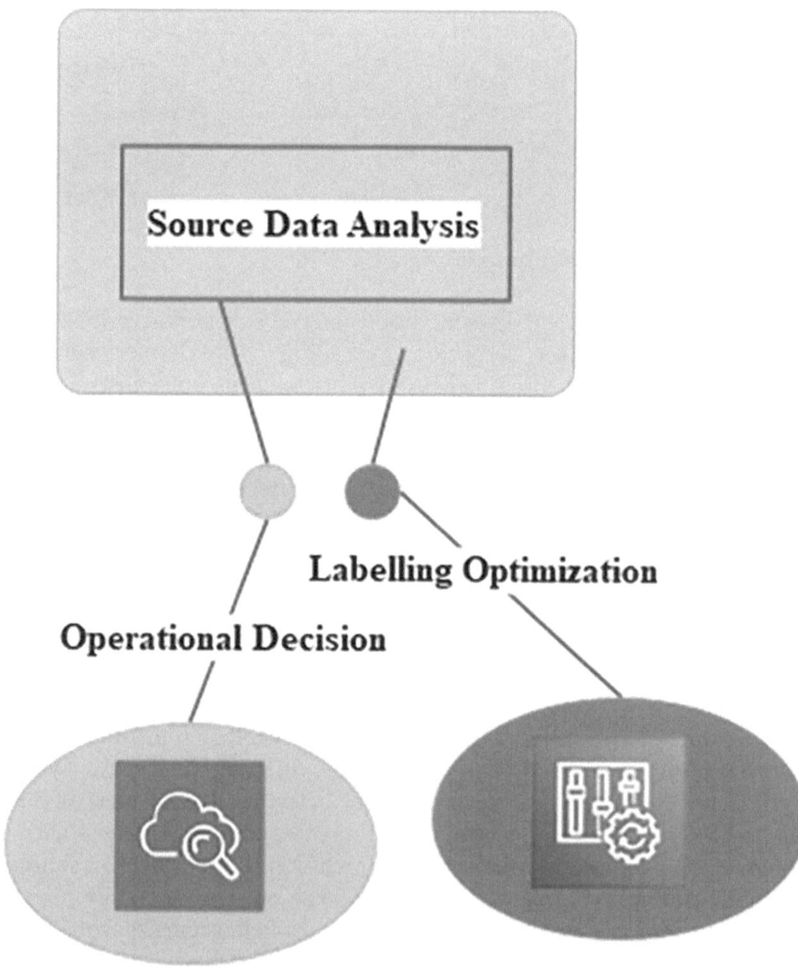

Figure 8.4 Diagram for source data analysis.

c. Easy to detect the threat that occurs inside the system and is also used to identify data exfiltration.
d. Used to Monitor the activities of internal users to easily identify the inside threats and attacks.

8.3.3 Operational decision

In operational decisions, ML shows promise in operational security due to the vast volume of log data in contemporary information systems. High-profile security incidents involving data exfiltration highlighted the significance of log data evaluation. For example, Beehive was an early unsupervised ML system that pitched in extracting knowledge from diverse log data sources such as proxy, DHCP, and VPN servers [19]. It will combine these logs to detect anomalies, detecting incidents that necessitated manual intervention Beehive detected around 800 incidents, with 65% being true security incidents, outperforming non-ML methods significantly. Deep learning, as exemplified by Deep Log, revolutionized the field by eliminating the requirement for manual feature engineering and achieving remarkable outcomes. Notably, it achieved an impressive detection rate of nearly 100% even when trained on just 1% of the available data. This breakthrough underscores the immense potential of deep learning techniques in enhancing the efficiency and accuracy of various data-driven tasks and applications.

8.3.4 Labeling optimization

As an integral part of the data structure, it serves the purpose of linking a specific information object with a set of security attributes, ensuring that the relevant security measures are properly associated and implemented. In this optimization, supervised ML method is used to detect the threats that occur in large amounts of labeled data but it is not possible for practical scenarios such as intrusion system and malware detection. On the other hand, semi-supervised learning and active learning methods are used to optimize the labeling efforts. For example, a botnet detection attained by obtaining an $F1$-score of 0.83 with a limited set of only 2,400 labels.

8.3.5 Risk exposure assessment

By using this exposure assessment, machine learning can enhance cybersecurity, identifying the weak spot and also predicting the likely threats. Also, it will aid in aiding the tasks like penetration testing and estimating compromise indicators. Penetrating testing in ML enhances vulnerability assessment by autonomously attacking security systems and also reducing inspection time by achieving high-speed results. Its potential in penetration testing remains largely unexplored, prompting the development of dedicated

ML-assisted platforms. Estimation of compromised Indicators in ML is used to estimate compromised hosts in a corporate environment by analyzing heterogeneous sources, including host behavior and network data. The most common indicators of compromised hosts are visits to business and travel websites during their working hours [20]. ML combined with honey pots can identify hosts susceptible to botnet malware. For example, nowadays Facebook utilizes machine learning techniques and methodologies to reduce fraudulent accounts by nearly 30% through the correlation of different sources.

8.3.6 Threat intelligence

Machine learning can be utilized with internal corporate data to foresee future attack strategies. For instance, SAGE compresses individual alerts into graphs and demonstrates the workflow of an offensive strategy [21]. It can also disassemble the developing code executables to detect potentially malicious patterns for achieving high accuracy. By combining primary data and secondary data sources, ML solutions can provide more accurate predictions than non-ML approaches. For example, external sources, such as Twitter, Facebook, and social media can be used for open source intelligence, by detecting the security incidents like Mirai botnet and data breaches. ML methods can analyze tweets to study ransomware attacks and predict Common Vulnerability Scores (CVS) earlier than traditional feeds. Dark web data can be leveraged to predict exploitability, with ML methods outperforming common feeds. ML can automate analysis of underground forums, uncovering cybercriminal markets and inferring malicious exploit prices. Patents also exist for ML-based prediction of cyberattacks in modern environments. At the same time, it addresses numerous tasks that complement threat detection.

8.3.7 Intrinsic problem of machine learning in cybersecurity

In this comprehensive overview, we will address the intrinsic problem by breaking it down into three distinct sections, enabling us to explore and discuss the issues comprehensively and systematically. General problems affecting any ML solution in cyber security, Problems specific to in-house ML solutions, Problems associated with adopting commercial-off-the-shelf (COTS) ML products. It is crucial to acknowledge that although certain mitigation measures can be implemented, these problems persist and continue to lack definitive solutions.

8.3.8 General problem of machine learning in cybersecurity

In the context of machine learning, the Independent and Identically Distributed (IID) principle posits that the data utilized to train a model is assumed to be representative of future data that the model will encounter. However, when applied to the field of cybersecurity, the IID principle presents challenges to model deployment due to the dynamic nature of concept drift, the adversarial settings inherent in cyberattacks, and the critical requirement for safeguarding data confidentiality [22]. The application of the IID principle in cybersecurity hinders the smooth deployment of machine learning due to its impact on three essential aspects within this domain: concept drift, the adversarial nature of attacks, and the paramount need to ensure data confidentiality.

8.3.8.1 Concept drift

It refers to the problem of machine learning systems becoming obsolete due to evolving environments and changing data. While domains like computer vision can rely on long-standing training data, cyber security faces continuous challenges. New vulnerabilities, network segments, and evolving attacker strategies introduce concept drift [23]. The performance of ML detectors can degrade when faced with concept drift. To mitigate this issue, a practical solution involves consistently updating ML systems with new labeled data that accurately reflects the evolving patterns and dynamics in the domain of interest. By doing so, the ML detectors can maintain their efficacy and adapt to the ever-changing landscape, ensuring a more robust and reliable defense against emerging threats.

8.3.8.2 Adversarial setting

The field of cyber security acknowledges the inherent presence of adversaries who constantly refine their attack strategies. Adversarial samples target ML systems by making small modifications to data, even imperceptible ones, to compromise ML model predictions. These samples have evaded various ML detectors, including those for botnets and malware, and even commercial products like Google Chrome's phishing detector. Despite attempts to develop countermeasures, there is no universal solution yet, and some methods can even decrease performance. The best defense is a proactive approach of anticipating, evaluating, and countering adversaries before deploying ML models. To emphasize the significance of this threat, it is essential to address and dispel two common misconceptions.

8.3.8.3 Confidentiality

The cybersecurity domain poses challenges for lengthy dependency on ML. Encryption, such as HTTPS replacing HTTP, renders certain ML systems unusable, like a payload-inspecting ML-NIDS. Similarly, encrypted content, like PGP-secured emails, hinders ML analysis, impacting phishing email detectors [24]. Changing data regulations, like GDPR, make it difficult to identify reliable data for long-term use. Confidential and encrypted data also hinder manual verification of ground truth in labeling procedures. Limited availability of publicly disclosed datasets hampers ML system evaluation and adoption.

8.3.9 Future of machine learning in cybersecurity

The primary and crucial challenge in revolutionizing and advancing machine learning in cybersecurity lies in effectively addressing the unique demands and complexities of this field. This challenge is of paramount importance across various industries and sectors, as the ability to develop cutting-edge ML solutions that can combat sophisticated cyber threats is essential for ensuring data protection, privacy, and the overall security of digital systems in today's interconnected world. Collaboration among regulatory organizations, business leaders, engineers, and the research community is important to tackle the difficulties in research practice gaps and it enhances the performance for exploring the novelty in machine learning applications.

8.3.10 Certification (Sovereign entities)

The EU and US highlight trustworthiness in AI, particularly for high-risk scenarios like cyber security. To elevate the standards and address future challenges in machine learning for cybersecurity, regulatory bodies play a crucial role in enforcing procedures that certify ML system performance and robustness. This necessitates the collaborative efforts of four key stakeholders: regulatory authorities, corporate leadership, engineers, and researchers.

8.3.10.1 Performance certification

In machine learning (ML) systems validating the performance through testing is the essential one in cybersecurity. However, the absence of standardized evaluation protocols presents a challenge, especially for COTS products. Existing performance assessments often suffer from biased environments and unfair comparisons that artificially inflate results in favor of specific ML solutions. To ensure meaningful assessments, realistic data distributions and temporal shifts must be considered, while traditional cross-validation techniques should be limited to tuning. Performance validation through statistical tests and the establishment of standardized evaluation protocols would promote fair and pragmatic comparisons, thereby facilitating the deployment of ML. The utmost importance is given to transparency, ensuring that customers of COTS ML systems are furnished with comprehensive details of operations. This includes complete disclosure of the data used for training and testing, the evaluation methodology employed, and transparent reporting of final results. By upholding transparency, customers gain trust and clarity, fostering informed decision-making and promoting accountability in the utilization of ML systems for various applications.

8.3.10.2 Robustness certification

Expanding interest in ML has prompted scientific inquiry into its reliability under adversarial scenarios and exposing vulnerabilities. However, no universal solution has been discovered yet, and some defenses have been broken between their initial preprint release and subsequent peer-reviewed publication. It is crucial to acknowledge that no ML solution is without flaws. Remarkably, the security of ML and its adversarial vulnerabilities are frequently overlooked as critical challenges by stakeholders, as emphasized in a recent audit examining the cyber security perspective of European stakeholders. To deal with the problem, mandatory assessments of adversarial robustness should be incorporated into the evaluation process for any ML-based solution in cyber security. Prior knowledge of the potential assurance risks and their consequences is essential before deploying ML systems. Additionally, transparent disclosure of all assessment details is crucial to ensure accountability and understanding. Enhancing transparency and reliability in machine learning requires regulatory bodies to establish standardized procedures that certify the effectiveness and dependability of ML systems. By implementing such procedures, stakeholders can gain confidence in the performance and security of these systems, leading to more informed decisions and bolstering overall trust in the deployment of ML technologies.

8.3.11 Data availability (executives and legislation authorities)

The effectiveness of ML in cyber security hinges on the availability of high-quality data; however, acquiring suitable and relevant data presents a significant challenge in the field. Open datasets often have limitations like erroneous labels, discontinuance, and small or synthetic environments. These issues can only be partly mitigated, hindering the evaluation and deployment of ML. Solving the data shortage requires collaboration between industrial stakeholders for data sharing and regulatory authorities for actionable policies.

8.3.11.1 Data sharing

Promoting data-sharing practices can address the lack of adequate data in cybersecurity. Some data, like malware samples, are easily shared, but acquiring permission for confidential data is challenging. Anonymization and federated learning have helped overcome privacy concerns. Platforms like EU-OF2CEN and STIX CyBox offer opportunities for ML datasets and threat intelligence sharing but require unbiased data and compliance with regulations.

8.3.11.2 Actionable data regulations

Data regulations protect data owners but limit data gathering and ML development. Interpretation challenges and changing policies hinder long-term projects, but GDPR-compliant platforms and actionable policies can promote data sharing and compliance.

8.3.11.3 Recommendation

To tackle the shortage of adequate data, data sharing originating in their environments should be encouraged by companies while the promotion of such disclosure through proper policies and incentives should be defined by regulatory authorities.

8.3.11.4 Usable security research (scientific community)

The ML-cyber security research trend shows promise but also poses challenges due to the absence of practical results and the failure to consider realistic scenarios. Existing research in this domain reveals two notable issues: the lack of practical results and insufficient consideration of realistic scenarios. These problems warrant further attention to ensure that ML in cybersecurity can be effectively applied in real-world settings and deliver tangible outcomes that address practical challenges.

8.3.11.5 Pragmatic results

Research that aims to make progress or move forward in this context existing state-of-the-art in machine learning frequently lacks real innovations and it is characterized by shortcomings made during investigations, creating uncertainty and impeding actual ML deployments. Poor reproducibility and limited disclosure of experimental details further exacerbate the issue, as researchers struggle to replicate prior research and the process of peer-review processes cannot adequately assess experimental protocols. The inclusion of comprehensive technical resources in submissions is challenging due to page limitations, creating a difficult decision for researchers.

The attainment of enhanced transparency and reliability in machine learning requires regulatory bodies to mandate the establishment and adoption of standardized procedures that certify the performance and robustness of ML systems. By enforcing these procedures, stakeholders can instill confidence in the capabilities and dependability of such systems, promoting more accountable and trustworthy practices in the deployment of ML technologies. The peer-review procedure should encourage and require the provision of documentation for carrying out ML experiments. In addition to the regular review process, a dedicated panel of expert reviewers with advanced technical knowledge can be employed to assess the content simultaneously, ensuring its accuracy and reliability. This supplementary evaluation adds an extra layer of assurance, enhancing the overall quality and credibility of the assessed material.

8.3.12 Realistic security scenarios

The focus on better numbers in research papers often overlooks realistic security scenarios and assumptions made in past work, questioning the practical value of such research. Many papers neglect important factors like concept drift and outdated datasets, leading to near-perfect performance without considering real-world impact. While acquiring public data is challenging, several openly released datasets are available. However, the cyber security setting seems to prioritize evaluating new ML methods on security-related data without sufficient consideration of realistic threat models. Adversarial ML is particularly affected, with attacks assuming powerful adversaries and limited accessibility to data-processing pipelines. This confusion and lack of concern from industrial stakeholders is highlighted in recent surveys and a detailed research study. The real world should be more closely connected to future studies on ML applications for cyber security. Realistic assumptions about the threat model, recent trends in the dataset, and concept drift should be addressed. The future of ML in cybersecurity holds immense promise, driven by continuous advancements and evolution of these technologies. Several key areas are anticipated to witness a significant impact from ML in the realm of cybersecurity.

8.3.12.1 Enhanced threat detection

ML algorithms will become more sophisticated in detecting and mitigating advanced and evolving cyber threats. Through the analysis of vast amounts of data and the identification of patterns and anomalies, ML algorithms have the potential to enhance the accuracy and speed of threat detection significantly. This enables the implementation of proactive defense measures, empowering cybersecurity professionals to stay one step ahead of potential threats and protect digital assets more effectively.

8.3.12.2 Behavioral analytics

ML is poised to assume a vital role in the analysis of user behavior and network activity, adeptly identifying anomalous patterns that could signal potential security breaches or insider threats. This proactive approach to anomaly detection allows cybersecurity teams to swiftly respond to emerging risks, bolstering overall threat detection and safeguarding critical systems and data. By building models based on normal user behavior, ML can identify deviations and trigger alerts when suspicious activity occurs.

8.3.12.3 Adversarial ML

Adversarial attacks are techniques where malicious actors attempt to manipulate or evade ML models. In the future, ML advancements will primarily concentrate on building resilient defenses capable of effectively countering these sophisticated attacks. By fortifying ML models and incorporating adaptive strategies, cybersecurity practitioners can better protect against emerging threats, ensuring a more secure digital landscape [25]. To enhance model resilience and robustness, future efforts will incorporate advanced techniques like adversarial training, model ensembling, and anomaly detection. These methods aim to fortify ML models against adversarial attacks and improve their ability to identify and handle anomalies effectively, bolstering overall cyber security defenses.

8.3.12.4 Automated incident response

ML algorithms will enable the automation of incident response processes. By analyzing and correlating security events in real time, these algorithms can automatically trigger response actions, such as blocking malicious traffic, isolating compromised systems, or alerting security personnel.

8.3.12.5 Explainability and interpretability

Future research in ML for cybersecurity will prioritize the development of techniques to enhance the explainability and interpretability of models. By addressing the challenge of "black box" behavior, these advancements aim to provide security analysts and experts with clearer insights into how the models arrive at their decisions. This transparency will foster trust in the effectiveness of ML models, enabling better-informed decision-making and more confident responses to potential threats.

8.3.12.6 Privacy-preserving techniques

With growing concerns about privacy, ML in cybersecurity will incorporate privacy-preserving techniques. To analyze sensitive data while preserving individual privacy, cutting-edge technologies such as federated learning, differential privacy, and secure multiparty computation will be leveraged. These advanced methodologies ensure that data can be collectively analyzed without exposing sensitive information, striking a crucial balance between data utility and individual privacy in various domains, including cyber security.

8.3.12.7 Threat intelligence and sharing

ML will play a pivotal role in analyzing extensive volumes of threat intelligence data, facilitating rapid and precise identification of emerging threats. This capability empowers cyber security professionals to swiftly respond to evolving risks and proactively defend against potential security breaches, fostering a more resilient and robust cyber security ecosystem. Additionally, ML will facilitate the effective sharing of threat intelligence among organizations, enabling collaborative defense against sophisticated cyberattacks.

8.3.13 Continuous learning and adaptation

ML models will continue to evolve by leveraging continuous learning and adaptive capabilities. ML models will continuously learn from new data, dynamically adapting to evolving threats, and steadily enhancing their performance over time. This constant learning and adaptation render these models highly effective in detecting and mitigating emerging cybersecurity risks, ultimately fortifying digital defenses and safeguarding against evolving threats. As the cyber security landscape evolves, the future of ML remains dynamic and ever-changing. ML will assume a vital role in strengthening defenses, automating processes, and empowering cybersecurity professionals with enhanced capabilities to safeguard digital assets and infrastructure. As threats grow in sophistication, ML's adaptive nature ensures it remains a crucial ally in the ongoing battle against cyber threats, promising a more resilient and proactive cybersecurity ecosystem.

8.3.14 Orchestration in ML

ML is not intended to replace existing systems or human professionals; rather, its purpose is to complement and augment their capabilities by providing an alternative perspective to detect unnoticed occurrences. By analyzing vast amounts of data and identifying patterns that might go unnoticed by traditional methods, ML enhances the overall effectiveness of cybersecurity efforts and empowers human experts to make more informed decisions and respond proactively to emerging threats. However, significant variance is exhibited by ML methods, and all threats cannot be protected against by a single solution. Orchestrating diverse ML solutions, in conjunction with

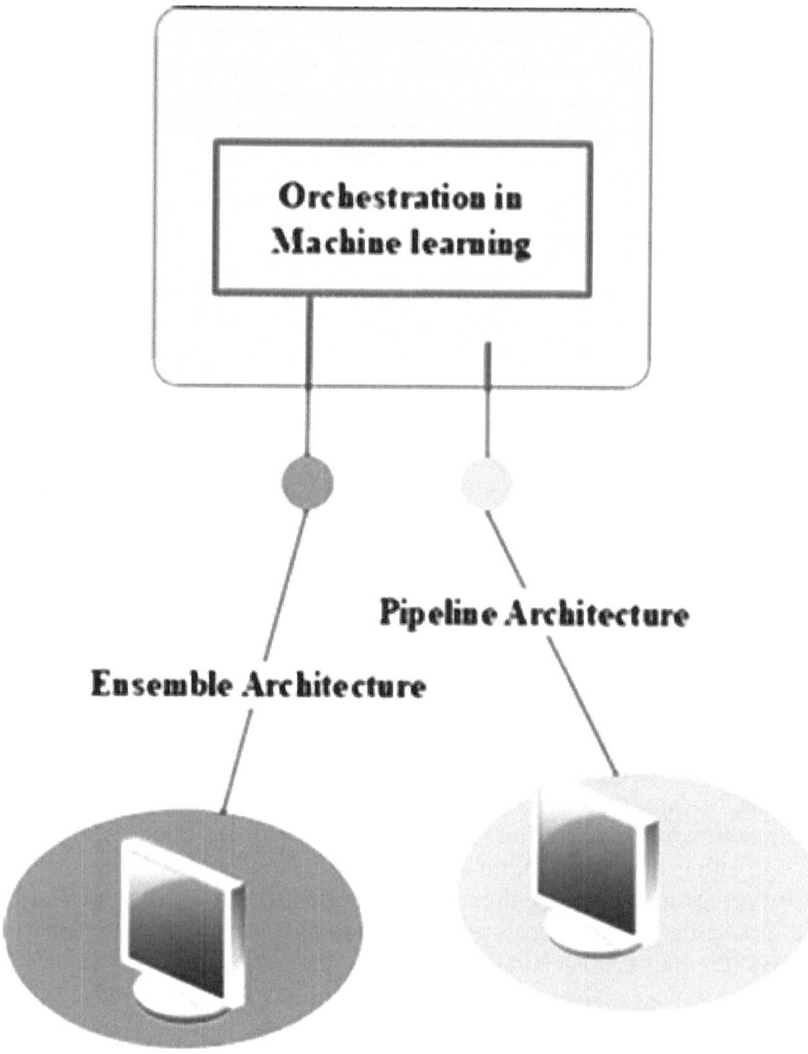

Figure 8.5 Diagram for orchestration in machine learning.

non-ML techniques, can address these issues. ML engineers play a crucial role in coordinating different outputs to extract actionable information, employing ensemble or pipeline architectures based on the system's objectives. Orchestration in ML is represented in Figure 8.5.

8.3.14.1 Ensemble architecture

It is a proficient approach to combine multiple ML models, where each model focuses on a specific problem while analyzing the same data. For example, ensembles of ML models can be utilized in ML-NIDS for intrusion detection, targeting different threats like botnets or DoS attacks. However, engineers face the challenge of lacking standardized feature sets, which can greatly impact the performance of each model in the ensemble.

8.3.14.2 Pipeline architecture

ML models with different inputs and outputs are organized systematically. For example, Model aggregation can be used for threat detection, and its outputs can then be utilized for malware analysis. However, implementing such architectures is challenging due to the issues that occur in networks which have a compounded impact on each individual component.ML engineers and practitioners should emphasize how to make the most of all of these components' practical efficacy, as coordinating complex systems that combine ML and non-ML solutions is advantageous for cyber security.

8.4 DEEP LEARNING FOR CYBERSECURITY

Deep Learning, a subset of artificial intelligence (AI), revolves around computational statistics using neural networks. Its primary focus lies in simulating the human brain's analytical learning capabilities, enabling algorithms to recognize complex patterns and make data-driven decisions akin to human cognition. It has the capability to mimic human brain mechanisms for interrupting the dataset such as text, image, and sound classification. Generally, deep learning is an unsupervised greedy layer-by-layer that enhances and optimizes the deep structure present in deep learning. When compared to ML, deep learning produces more accuracy and it performs well if the data will increase during testing and validation whereas deep learning does not perform the perfect task if the data volume is small. Due to its advanced features and potential for self-learning, deep learning (DL) offers faster and more accurate processing. Furthermore, DL has many matrix operations so that the GPU and its hardware are frequently used for optimizing the matrix operation effectively [26]. DL is already implemented in real-time applications and finding the observation to enhance the trends and technology in the field of cybersecurity. In recent days, cyberattacks and data breaches are more common in digital world and it has led to figure out misconfigured networks. The rise of new smart network technologies has necessitated the development of innovative cybersecurity techniques to address the increasing frequency of network attacks in recent years. To safeguard these advanced networks from sophisticated threats, cybersecurity professionals must continuously adapt and devise novel strategies to ensure the integrity and security of these interconnected systems. Cybersecurity seems to be essential in preventing attack and unauthorized access to key infrastructure. Cybersecurity encompasses a wide array of methods and technologies, categorized into various groups such as software security, information security, network security, disaster recovery, operational security, end-user education, and more. Cyberattacks, akin to modern warfare devoid of physical weapons, wield tremendous destructive power, as they compromise private and business data, disrupt critical processes, create persistent security vulnerabilities, grant unauthorized access to hardware, and severely impact a nation's economy. Hence, the need to counter these threats remains paramount to safeguarding digital infrastructure and maintaining the security and prosperity of nations.

8.4.1 Role of DL in cyber security

DL is a subset of ML, and it possesses the ability to learn and enhance its capabilities autonomously by evaluating computer algorithms, making it a cutting-edge and dynamic approach to safeguarding digital assets. This self-improving nature of the system allows it to adapt rapidly to new threats and challenges, reinforcing its position as a powerful tool in defending against cyber threats and ensuring the security of digital infrastructural technologies can be reverse-engineered, leading to the introduction of bias or vulnerabilities that diminish the effectiveness of their defenses. This highlights the importance of proactive efforts to secure and fortify ML systems against such adversarial attacks, ensuring their reliability and integrity in the face of potential threats. Hackers can exploit ML algorithms to introduce fake data sets and contaminate cyber security systems. However, on the other side, DL eliminates the need for data scientists to manually curate data sets, as the algorithm can automatically learn from vast amounts of data, reducing the burden of manual input and streamlining the learning process. Massive amounts of raw data are processed by DL models so that the cyber security system can be automatically trained. DL neural networks are designed to operate autonomously and independently, functioning without the need for direct supervision or human intervention. This self-sufficiency enables them to learn from data and make decisions without constant human oversight, making DL a powerful and efficient tool in various applications. Large data sets can be used to more quickly and correctly identify very complicated patterns than ML.

8.5 USE CASES WITH REAL-TIME APPLICATION IN CYBERSECURITY

Discussion on use cases with real-time application in cybersecurity as represented in Figures 8.6 is discussed as follows:

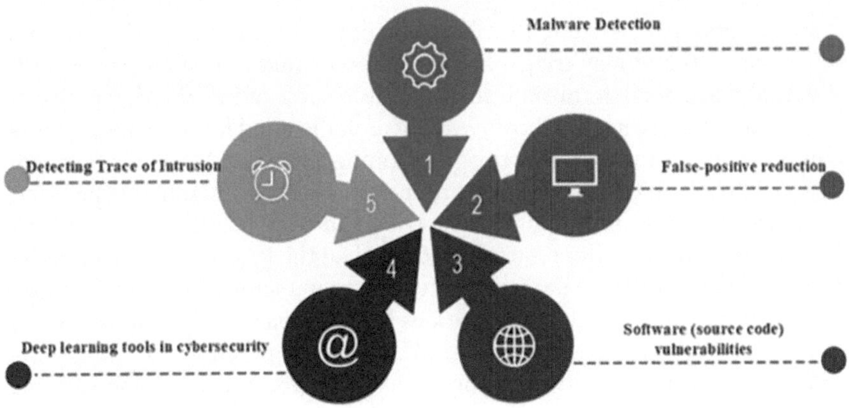

Figure 8.6 Diagram for use cases with real-time application in cybersecurity.

8.5.1 Malware detection

Malware is a program, software, or other code whereas attackers specifically create a design to do something malicious such as stealing confidential data, keeping track of online activities, or even seizing the system. Malware exhibits a diverse range of types, each serving distinct purposes or functionalities. It can be categorized into several classes, including viruses, trojan horses, and adware, each carrying out specific actions or posing unique risks to digital systems and data. Malware can be classified in some cases based on the functionality. For instance, adware shows unwanted ads and forcibly installs software on the computer. At the same time, malware can monitor the network for vulnerable computers using worm techniques, then enter the system and issue commands using backdoor trojan techniques whereas some malware can be identified by its goals. For example, Espionage malware is intended to target a specific organization in order to steal information or gather intelligence. As an illustration, crimeware is made to steal private, confidential, or proprietary information for the purpose of profit. The malware use case is unique from the ransomware use case in the Exabeam Threat Detection Investigation and Response (TDIR) for use case package framework. The malware use case illustrates situations where malicious actors develop harmful software or code with the intention of gaining unauthorized access to a system, causing data corruption, or inflicting harm on the targeted infrastructure. Such malevolent actions underscore the significance of robust cybersecurity measures to defend against these threats and protect sensitive information and digital assets. The ransomware use case specifically outlined that, if the attackers encrypt the business assets.

8.5.2 False-positive reduction

In any cybersecurity system, the likelihood of producing false positives exists, which means that it may trigger alerts for threats that ultimately prove not to be actual risks. Sorting through such a large volume of information to distinguish real threats from false positives can be challenging for human operators. This underscores the importance of developing advanced ML algorithms and intelligent tools to effectively filter and prioritize alerts, enabling cyber security professionals to focus on genuine risks and respond efficiently to potential threats. If the security system identifies then the false positive occurs that doesn't exist or mislabel harmless software as malicious. It can occur for a number of reasons, but it usually comes from an excessively reactive system or an error in the device or solution. ML offers the potential to decrease the number of false positives by training the model to recognize specific patterns. As a result, when a pattern aligns with predefined criteria for a false positive, the model can confidently identify it as such, thereby improving the accuracy of threat detection and reducing the unnecessary burden on cybersecurity professionals to investigate non-threatening incidents.

8.5.3 Software (source code) vulnerabilities

Understanding software vulnerabilities is the key concept for controlling modern security threats. It is the defect in software whereas it allows an attacker to gain the set of processes. An attacker can exploit a software vulnerability to steal or manipulate confidential information and join into the system to a botnet, backdoor, or introduce additional malware. Additionally, after breaking from one computer to another network, the attacker might use the host to break into additional hosts connected to the same network.

8.5.4 DL tools in cybersecurity

DL, often referred to as Deep Neural Network, encompasses a subset of ML techniques. This specialized approach involves using neural networks with multiple layers to process and learn from vast amounts of data, allowing for the extraction of intricate patterns and representations that are essential for solving complex tasks across various domains. It has a capability to learn unsupervised data for solving complex problems. These data are widely used for cybersecurity for protecting the business sectors from various threats such as phishing, spear-phishing, Intrusion, password attack, denial of service, etc.

8.5.5 Detecting trace of intrusion

An intrusion detection system (IDS) serves as vigilant software, continuously monitoring network traffic to swiftly identify any malicious activity or policy violations. Upon detecting anything suspicious, it promptly updates its information and takes proactive measures to protect against unauthorized users, enhancing the overall security and integrity of the network or system. By using the SIEM system, illegal actions are monitored and frequently recorded centrally and communicated to the administration [27]. By using DL methods, this detection system learns and builds a predictive model that is capable of distinguishing good and bad connections.

a. IDS monitors the traffic occurring in the computer network for detecting suspicious activity.
b. It evaluates the information moving through the network in order to look for patterns and indications of abnormal behavior.
c. The IDS scrutinizes network activity against a predefined set of rules and patterns.
d. It diligently seeks out any behaviors that may indicate potential attacks or intrusions.
e. By analyzing network data, the IDS identifies and flags suspicious activities.
f. Its vigilant approach allows for immediate detection and response to potential security threats.
g. Notifications are sent to the system administrator when the IDS identifies activity matching the predefined rules or patterns.
h. The IDS promptly alerts the administrator of potential security incidents.
i. The received notifications allow the administrator to investigate and respond to the identified threats promptly.
j. This proactive approach enables swift actions to mitigate risks and safeguard the network and systems.

8.6 FUTURE WORK

Advanced Threat Detection ML and DL models will continuously evolve to meet the challenges posed by increasingly sophisticated and evolving cyber threats. Their capacity to learn from diverse data sources and adapt to emerging patterns ensures they remain at the forefront of cybersecurity defense, providing robust and proactive solutions to safeguard digital assets and infrastructure from the ever-changing threat landscape. As these technologies continue to advance, the cybersecurity ecosystem will benefit from more effective, intelligent, and resilient defenses against cyber threats. Through the analysis of extensive data sets, encompassing network traffic, system logs, and user behavior, these models excel at detecting anomalies, zero-day exploits, and previously unseen attack patterns with heightened accuracy. Their ability to process and learn from diverse sources empowers them to identify emerging threats swiftly and accurately, reinforcing their crucial role in bolstering cybersecurity defenses and protecting against the most sophisticated cyberattacks.

8.7 CONCLUSION

In conclusion, DL and ML techniques have emerged as powerful tools in the field of cybersecurity. They offer significant potential for enhancing the detection, prevention, and response capabilities against various types of cyber threats. DL algorithms, such as convolutional neural networks (CNNs) and recurrent neural networks (RNNs), have demonstrated remarkable success in tasks such as malware detection, intrusion detection, and spam filtering. By leveraging their ability to automatically extract complex features from raw data, these models can effectively identify patterns and anomalies in large-scale datasets, leading to improved accuracy and reduced false positives. ML approaches, including supervised learning, unsupervised learning, and reinforcement learning, have also proven valuable in cybersecurity applications. Through the analysis of historical data, these techniques can identify hidden patterns, identify potential vulnerabilities, and predict future cyberattacks. They enable security systems to adapt and learn from new threats, thereby enhancing their ability to detect and respond to emerging risks. However, it is important to acknowledge the challenges and limitations associated with DL and ML in cybersecurity. The reliance on large amounts of labeled training data, the potential for adversarial attacks, and the interpretability of complex models are among the key concerns. Additionally, the dynamic nature of cyber threats requires continuous model updates and fine-tuning to maintain effectiveness. To maximize the benefits of DL and ML in cybersecurity, it is crucial to establish robust data collection and management practices, invest in the development of specialized algorithms and models, and foster collaboration between cybersecurity experts and data scientists. Moreover, ethical considerations, such as privacy protection and bias mitigation, must be carefully addressed to ensure responsible and trustworthy deployment of these technologies. Overall, DL and ML hold tremendous potential in fortifying cybersecurity defenses. By leveraging the power of AI, these techniques can enhance threat detection, enable proactive defense mechanisms, and significantly strengthen the resilience of systems and networks against cyberattacks. As the field continues to advance, further research, development, and collaboration are essential to realize the full potential of DL and ML in the realm of cybersecurity.

REFERENCES

1. Abioye, S. O., Oyedele, L. O., Akanbi, L., Ajayi, A., Delgado, J. M. D., & Bilal, M. A. A. (2021). Artificial intelligence in the construction industry: A review of present status, opportunities and future challenges. *Journal of Building Engineering*, 44, 103299.

2. Lee, D., & Yoon, S. N. (2021). Application of artificial intelligence-based technologies in the healthcare industry: Opportunities and challenges. *International Journal of Environmental Research and Public Health*, 18(1), 271.
3. Yildirim, M. (2021). Artificial intelligence-based solutions for cyber security problems. In Luhach, A. K., & Elçi, A. (Ed.) *Artificial Intelligence Paradigms for Smart Cyber-Physical Systems* (pp. 68–86). IGI Global, Hershey, PA.
4. Sharma, A., Podoplelova, E., Shapovalov, G., Tselykh, A., & Tselykh, A. (2021). Sustainable smart cities: Convergence of artificial intelligence and blockchain. *Sustainability*, 13(23), 13076.
5. Zhang, Z., Ning, H., Shi, F., Farha, F., Xu, Y., Xu, J., & Choo, K. K. R. (2022). Artificial intelligence in cyber security: Research advances, challenges, and opportunities. *Artificial Intelligence Review*, 2022, 1–25.
6. Singh, J., Sajid, M., Gupta, S. K., & Haidri, R. A. (2022). Artificial intelligence and blockchain technologies for smart city. In Tripathi, S. L., Ganguli, S., Kumar, A., & Magradze, T. (Eds.) *Intelligent Green Technologies for Sustainable Smart Cities* (pp. 317–330). Wiley, Hoboken, NJ.
7. Ali, A., Septyanto, A. W., Chaudhary, I., Al Hamadi, H., Alzoubi, H. M., & Khan, Z. F. (2022,) Applied artificial intelligence as event horizon of cyber security. In *2022 International Conference on Business Analytics for Technology and Security (ICBATS)* (pp. 1–7). IEEE, New York.
8. Ahmed, I., Zhang, Y., Jeon, G., Lin, W., Khosravi, M. R., & Qi, L. (2022). A blockchain-and artificial intelligence-enabled smart IoT framework for sustainable city. *International Journal of Intelligent Systems*, 37(9), 6493–6507.
9. Jaiswal, A., Arun, C. J., & Varma, A. (2022). Rebooting employees: Upskilling for artificial intelligence in multinational corporations. *The International Journal of Human Resource Management*, 33(6), 1179–1208.
10. Li, W., Huang, R., Li, J., Liao, Y., Chen, Z., He, G., & Gryllias, K. (2022). A perspective survey on deep transfer learning for fault diagnosis in industrial scenarios: Theories, applications and challenges. *Mechanical Systems and Signal Processing*, 167, 108487.
11. Koay, A. M., Ko, R. K. L., Hettema, H., & Radke, K. (2023). Machine learning in industrial control system (ICS) security: Current landscape, opportunities and challenges. *Journal of Intelligent Information Systems*, 60(2), 377–405.
12. Dalle Pezze, D., Deronjic, D., Masiero, C., Tosato, D., Beghi, A., & Susto, G. A. (2023). A multilabel continual learning framework to scale deep learning approaches for packaging equipment monitoring. *Engineering Applications of Artificial Intelligence*, 124, 106610.
13. Atzeni, D., Ramjattan, R., Figliè, R., Baldi, G., & Mazzei, D. (2023). Data-driven insights through industrial retrofitting: An anonymized dataset with machine learning use cases. *Sensors*, 23(13), 6078.
14. Feng, Y., & Zhou, M. (2023). Robot intelligent communication based on deep learning and TRIZ ergonomics for personalized healthcare. *Personal and Ubiquitous Computing*, 27(3), 989–1001.
15. Ulku, I., & Akagündüz, E. (2022). A survey on deep learning-based architectures for semantic segmentation on 2D images. *Applied Artificial Intelligence*, 36(1), 2032924.
16. Cuomo, S., Di Cola, V. S., Giampaolo, F., Rozza, G., Raissi, M., & Piccialli, F. (2022). Scientific machine learning through physics-informed neural networks: Where we are and what's next. *Journal of Scientific Computing*, 92(3), 88.

17. Zuo, C., Qian, J., Feng, S., Yin, W., Li, Y., Fan, P., & Chen, Q. (2022). Deep learning in optical metrology: A review. *Light: Science & Applications*, 11(1), 39.
18. Zhang, Y., Shi, X., Zhang, H., Cao, Y., & Terzija, V. (2022). Review on deep learning applications in frequency analysis and control of modern power system. *International Journal of Electrical Power & Energy Systems*, 136, 107744.
19. Ali, N., Ghazal, T. M., Ahmed, A., Abbas, S., Khan, M. A., Alzoubi, H. M., & Khan, M. A. (2022). Fusion-based supply chain collaboration using machine learning techniques. *Intelligent Automation and Soft Computing*, 31(3), 1671–1687.
20. Paleyes, A., Urma, R. G., & Lawrence, N. D. (2022). Challenges in deploying machine learning: A survey of case studies. *ACM Computing Surveys*, 55(6), 1–29.
21. Abosata, N., Al-Rubaye, S., Inalhan, G., & Emmanouilidis, C. (2021). Internet of things for system integrity: A comprehensive survey on security, attacks and countermeasures for industrial applications. *Sensors*, 21(11), 3654.
22. George, P. G., & Renjith, V. R. (2021). Evolution of safety and security risk assessment methodologies towards the use of bayesian networks in process industries. *Process Safety and Environmental Protection*, 149, 758–775.
23. Gupta, V., Khera, S., & Turk, N. (2021). MQTT protocol employing IOT based home safety system with ABE encryption. *Multimedia Tools and Applications*, 80(2), 2931–2949.
24. Drees, S. M., Nowostawski, M., Jameel, R., & Mourya, A. K. (2021). Security aspects of blockchain technology intended for industrial applications. *Electronics*, 10(8), 951.
25. Hasan, M. K., Habib, A. A., Shukur, Z., Ibrahim, F., Islam, S., & Razzaque, M. A. (2023). Review on cyber-physical and cyber-security system in smart grid: Standards, protocols, constraints, and recommendations. *Journal of Network and Computer Applications*, 209, 103540.
26. Jankorazov, A., Xolmamatova, D., & Murodboyeva, M. (2023). Enzymes and their industrial application methods. *International Bulletin of Engineering and Technology*, 3(3), 102–107.
27. Sasikumar, A., Vairavasundaram, S., Kotecha, K., Indragandhi, V., Ravi, L., Selvachandran, G., & Abraham, A. (2023). Blockchain-based trust mechanism for digital twin empowered industrial internet of things. *Future Generation Computer Systems*, 141, 16–27.

Chapter 9

Artificial intelligence and media industry

Exploring new avenues and future

Sapan Kumar Gupta, Md. Rashid Iqubal Siddique, Shahid Shadab Hassan, and R. Badlishah Ahmad

9.1 INTRODUCTION

Artificial Intelligence and Media Industry can revolutionize the way we consume and create content. With AI, media companies can personalize their offerings to individual users, improving engagement and retention. Additionally, AI can assist in the creation of content by generating ideas, automating production processes, and even writing articles or scripts. As technology continues to advance, the possibilities for AI in the media industry are endless.

This bonding between the two can be really beneficial for both the consumers and the media companies. Consumers can enjoy a more personalized and tailored content experience, while media companies can increase their efficiency and productivity, ultimately leading to greater profitability. However, it is important to consider the ethical implications of AI in the media industry, such as potential biases and loss of human creativity. Media Industry can use the power of AI to analyse data and make informed decisions, but it is crucial to ensure that these decisions align with ethical and moral values.

Additionally, the implementation of AI in the media industry should be done in a way that promotes transparency and accountability to avoid any negative consequences for society. Every technology can be used for good or bad, and it is up to the media industry to take responsibility for the ethical implications of AI and ensure that it is used in a way that benefits society.

Furthermore, it is important to involve diverse perspectives in the development and implementation of AI to mitigate potential biases and ensure equitable outcomes. This relationship appears very promising for the future of AI and its impact on society, but it requires a collective effort from all stakeholders to ensure that it is used ethically and responsibly. As AI continues to evolve and become more integrated into our daily lives, it is crucial that we prioritize transparency, accountability, and inclusivity in its development and deployment. By doing so, we can maximize the benefits of this powerful technology while minimizing any potential harms.

DOI: 10.1201/9781003534761-9

AI already being used in different media formats and platforms such as streaming services, social media platforms, and online news outlets. It has revolutionized the way content is created, personalized, and delivered to users, enhancing their overall media consumption experience. Additionally, AI is also being utilized in advertising and marketing strategies to target specific audiences and optimize campaign performance.

AI tools like virtual reality and augmented reality are revolutionizing the way users interact with digital content. These tools provide immersive experiences, allowing users to engage with content in a more interactive and realistic manner. They have the potential to transform various industries, such as gaming, education, and healthcare, by offering new possibilities for training, simulation, and entertainment.

Big Data and Cloud Computing concepts are on the way to be utilized in media industry as well. By harnessing the power of Big Data, media companies can analyse vast amounts of user data to gain insights into consumer preferences and behaviour. This information can then be used to personalize content recommendations and target advertisements more effectively. Additionally, Cloud Computing enables media organizations to store and access large amounts of data and media files remotely, allowing for seamless collaboration and efficient content distribution across different platforms.

9.2 RELATED WORK

In an article, Media and Artificial Intelligence by Matthew Gentzkow discussed the growth of online web platforms for news gathering and dissemination [1].

Mathias-Felipe de-Lima-Santos, and Wilson Ceron in their published work Artificial Intelligence in News Media: Current Perceptions and Future Outlook published in MDPI' s Journalism and Media discussed in brief about the Natural Language Processing (NLP), Deep Learning, Machine Learning, Computer Vision, Speech Recognition, Expert System and Robotics [2]. All these AI subordinate tools are available and can be optimally recognised and utilised by media organisations, policymakers, and the public.

It is important to educate ourselves about AI and its potential impact on society, so that we can make informed decisions about its use and regulation. Only through a collaborative effort can we ensure that AI is used in a way that benefits everyone. This kind of collaborations should involve experts from various fields, including computer science, ethics, and law, to ensure that AI is developed and used responsibly. Additionally, ongoing research and development in AI should prioritize ethical considerations and transparency to build trust with the public.

Giuseppe Amato, Italian National Research Council, Frédéric Bimbot IRISA [3] - Institut de Recherche en Informatique et Systèmes Aléatoires, Fabrizio Falchi, Italian National Research Council, Fabien Lotte, National Institute for Research in Computer Science and Control in their published work AI in the media and creative industries in The New European Media platform also advocates the evolution of Big Data and Computing Competencies to ensure that individuals are equipped with the necessary skills to understand and navigate the rapidly evolving technological landscape. By promoting digital literacy and education, we can empower individuals to make informed decisions about their use of AI and other emerging technologies. Ultimately, it is only through a comprehensive and inclusive approach that we can harness the full potential of AI while mitigating its potential risks. These powerful technologies can ensure that individuals and organizations are equipped to navigate the rapidly changing landscape of AI [4,5]. By investing in education and training programmes, we can empower people to use AI in a way that maximizes its potential while minimizing potential harm. Furthermore, policymakers must also play a role in regulating AI to ensure that it aligns with societal values and does not perpetuate existing biases or inequalities.

9.3 PROPOSED WORK

Media Industry can use several AI related technologies such as NLP, machine learning, and computer vision to improve their operations. These technologies can help automate content creation, personalize user experiences, and enhance the accuracy of data analysis [6]. Additionally, AI can assist in detecting fake news and identifying potential copyright infringements to enhance their operations and services. For instance, machine learning algorithms can be used to personalize content recommendations for users based on their viewing history and preferences. Additionally, NLP can be utilized to improve the accuracy of automated closed captioning and transcription services for video content.

AI is promising a bright future for media companies. With AI technologies, media organizations can automate tasks such as content creation, editing, and distribution, saving time and resources. AI can also help in analysing audience engagement and sentiment analysis, providing valuable insights for content optimization. Moreover, AI-powered chatbots and virtual assistants [7] can enhance customer service by providing instant responses and personalized recommendations to users. Overall, the integration of AI in media industry holds great potential for improving efficiency and delivering a more tailored user experience.

Furthermore, AI can also assist in analysing user behaviour and preferences to further refine recommendations and improve the overall user experience. In addition, AI can automate the process of content creation by generating personalized articles, videos, and social media posts based on user preferences and trends. This not only saves time and resources for media companies but also ensures that the content produced is highly relevant and engaging for the target audience.

In conclusion, using AI to personalize content has become a crucial strategy for media companies to stay competitive in today's digital landscape. By leveraging AI, media companies can analyse vast amounts of data to gain insights into consumer behaviour and preferences [8]. This allows them to tailor their content to individual users, delivering a more personalized and enjoyable experience. Additionally, AI-powered content personalization can help media companies build stronger relationships with their audience, fostering loyalty and increasing customer satisfaction.

It can also lead to higher engagement and conversion rates. By understanding the preferences and behaviour of individual users, media companies can create targeted advertisements and recommendations that are more likely to resonate with their audience. This can result in higher click-through rates and conversions, ultimately driving revenue growth for the company. Moreover, AI-powered content personalization [9] can help media companies stay ahead of competitors by continuously adapting and improving their offerings based on real-time data and feedback from users.

International media already recognize the value of AI-powered content personalization and have started implementing it in their strategies. For example, streaming platforms like Netflix and Spotify use AI algorithms to analyse user preferences and suggest personalized content recommendations. This not only enhances user experience but also increases customer loyalty and engagement [10]. Additionally, AI can also assist media companies in analysing data from various sources such as social media, surveys, and website analytics to gain insights into audience behaviour and preferences, enabling them to make informed decisions about content creation and distribution. Figure 9.1 shows the key AI applications in the newsrooms.

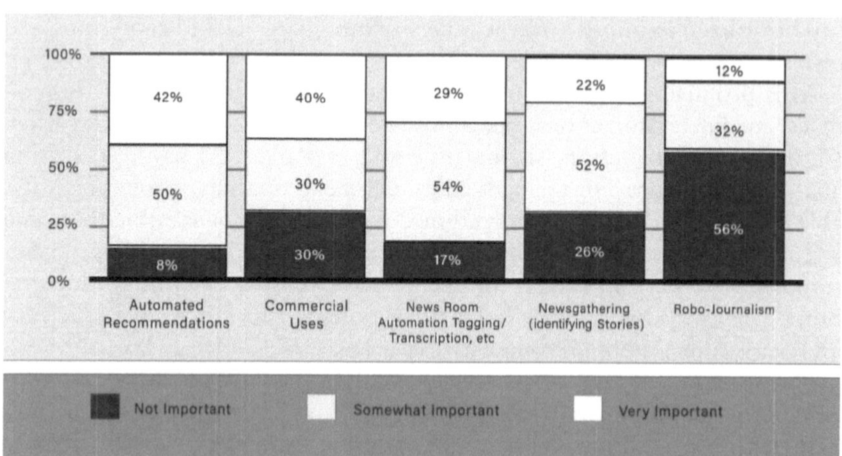

Figure 9.1 Key AI applications in the newsrooms.

In Indian media we recently experienced AI news anchor which is a prime example of how AI can revolutionize the media industry. With AI news anchors, news channels can provide 24/7 coverage without the need for human anchors. This not only saves costs but also ensures that news is delivered promptly and accurately to viewers. These anchores can read news bulletins in multiple languages, without any human intervention [11]. This is just one example of how AI is revolutionizing the media industry and transforming the way news is delivered to audiences around the world.

As AI technology continues to advance, we can expect to see even more innovative applications in media and entertainment. One such application could be the use of AI to personalize news [12] content for individual viewers based on their interests and preferences. Additionally, AI could also be used to fact-check news stories in real-time, helping to combat the spread of misinformation and fake news.

For instance, AI-powered chatbots are already being used to engage with audiences and provide personalized content recommendations. Additionally, AI algorithms [13] can analyse vast amounts of data to identify trends and patterns, helping media companies make more informed decisions about what content to produce and distribute. The potential for AI in the media industry is truly limitless.

Moreover, AI can also be used to automate tasks such as video editing and transcription, saving time and resources for media companies. This technology can also assist in the creation of more engaging and interactive content, such as virtual reality experiences or personalized news articles [14]. AI can also assist in content creation by generating automated news articles and video summaries, freeing up journalists to focus on more in-depth reporting. Furthermore, AI- powered voice assistants and smart speakers are changing the way audiences consume news, providing a hands-free and personalized experience.

AI can also assist in creating and editing content, such as generating captions and subtitles for videos or even writing news articles. With the help of AI, media companies can streamline their workflows and increase efficiency, allowing them to focus on producing high-quality content. Overall, the integration of AI [15] in the media industry has the potential to revolutionize how news is reported and consumed. In addition, AI can also help media companies to analyse data and predict trends, enabling them to tailor their content to the interests of their audience. This can lead to increased engagement and loyalty from viewers or readers. As AI technology continues to advance, it is likely that we will see even more innovative ways in which it can be used in the media industry. However, there are concerns about the impact of AI on the accuracy and objectivity of news reporting. It is important for media companies to ensure that AI is used ethically and transparently, with human oversight to prevent biases and errors.

NLP is a key component of AI that can help media companies analyse and understand large amounts of text data. By using NLP, media companies can gain insights into the preferences and behaviours of their audience [16], allowing them to tailor content to better meet the needs and NLP is one area where AI is already being used to improve the accuracy and efficiency of tasks such as transcription and translation. Additionally, AI can also be used to personalize content for individual users based on their interests and behaviour, further enhancing the user experience. Moreover, AI-powered NLP can also help media companies automate the process of content moderation and flagging inappropriate content [17]. This can save time and resources while ensuring that the platform remains safe and user-friendly for all users.

Another area where AI is making a significant impact is computer graphics. AI-powered algorithms can generate highly realistic and detailed images, videos, and animations, which can be used in various industries such as gaming, film, and advertising [18]. Furthermore, AI can also help to automate the process of creating visual content by suggesting design elements and layouts based on user preferences and historical data. AI uses computer graphics to create virtual environments and characters that can interact with users in real-time, enhancing the overall user experience [19]. As AI continues to evolve, we can expect even more advanced and sophisticated computer graphics applications that will revolutionize the way we interact with digital content.

Speech Recognition [20] is another area where AI has made significant advancements. With the help of machine learning algorithms, speech recognition technology can accurately transcribe spoken words into text. This has many practical applications, such as in virtual assistants like Siri and Alexa, as well as in automated customer service systems [21]. As AI continues to improve in this area, we can expect even more seamless and natural interactions between humans and machines. Speech recognition can help in media industry, where it can be used to automatically.generate captions for videos and TV shows, making them more accessible to people with hearing impairments [22]. Additionally, it can be used in language translation services, allowing for real-time translation of spoken words between different languages. enabling closed captioning and subtitling for the hearing impaired.

Additionally, it can be used in language translation services, making communication easier between individuals who speak different languages. Moreover, speech recognition technology has also found its way into virtual assistants such as Siri, Alexa, and Google Assistant, enabling users to interact with their devices using voice commands [23]. This has revolutionized the way we interact with technology and made it more accessible to people with disabilities or those who prefer hands-free interactions.

Another application of speech recognition technology is in the field of expert systems, where it can be used to analyse and interpret spoken language in order to provide personalized recommendations or advice [24].

This has the potential to greatly improve the efficiency and accuracy of expert systems, as well as make them more user-friendly for individuals who may not be familiar with technical jargon or complex interfaces. Expert systems that incorporate speech recognition technology can also be used in a variety of industries, including healthcare, finance, and customer service, to provide more personalized and efficient assistance to users [25]. As this technology continues to advance, we can expect to see even more innovative applications of speech recognition in the future.

Robotics is another area where speech recognition technology can be utilized to enhance human-robot interaction and improve the functionality of robots in various settings. For instance, robots equipped with speech recognition capabilities can be used in manufacturing plants to receive verbal commands from workers [26], or in homes to assist individuals with disabilities or elderly individuals with daily tasks. Robotics is part of AI, and as AI continues to advance, we can expect even more sophisticated speech recognition technology to be integrated into robots, allowing for more natural and intuitive communication between humans and machines. Additionally, speech recognition technology can also be used in virtual assistants, such as Siri or Alexa, to provide users with a more personalized and efficient experience [27].

Robotics is a machine that can be programmed to perform various tasks, including those that are dangerous or difficult for humans to do [28]. With the help of AI and speech recognition technology, robots can become even more versatile and helpful in assisting individuals with disabilities or elderly individuals with their daily needs. As technology continues to evolve, we can expect robots to become even more advanced and capable of performing complex tasks, ultimately improving the quality of life for many people [29]. Figure 9.2 shows the components of AI solutions.

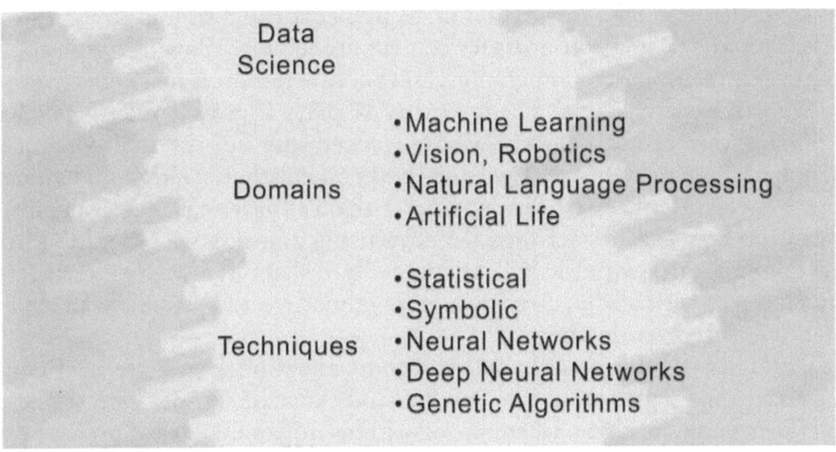

Figure 9.2 Components of AI solutions.

9.4 FUTURE SCOPE

Artificial intelligence and media are two different domains but when it comes together, there is always a scope of innovation and transformation [30]. The integration of artificial intelligence in media has the potential to revolutionize the way content is created, distributed, and consumed. By leveraging AI algorithms and machine learning techniques, media companies can analyse vast amounts of data to gain valuable insights, personalize user experiences, and optimize advertising strategies.

This convergence opens up exciting possibilities for enhanced storytelling, targeted marketing campaigns, and improved audience engagement [31]. They have the scope of revolutionizing the way we consume and interact with information. With the integration of artificial intelligence into media, we can expect more personalized content recommendations, enhanced data analysis for targeted advertising, and even AI-generated content creation [32]. This collaboration has the potential to greatly enhance user experiences and transform the media landscape as we know it.

By leveraging artificial intelligence, media platforms can analyse user preferences and behaviours [33] to deliver tailored content recommendations. This not only improves user satisfaction but also increases the effectiveness of targeted marketing campaigns. Additionally, AI-powered content creation can generate highly relevant and engaging material, further enhancing audience engagement and driving innovation in storytelling techniques. Ultimately, the integration of AI into media has the potential to reshape the industry by providing more personalized experiences and revolutionizing the way information is consumed and shared.

The scope is endless, as AI continues to advance and evolve. From personalized recommendations on streaming platforms to AI-generated news articles [34], the possibilities for enhancing user experiences are limitless. Furthermore, the integration of AI in media can also lead to increased efficiency and cost-effectiveness for content production, allowing media companies to streamline their operations and allocate resources more effectively.

We can expect a great future where AI plays a pivotal role in revolutionizing the media industry. With advancements in NLP and computer vision, AI can create hyper- personalized content that caters to individual preferences and interests. This not only enhances user engagement but also opens up new avenues for targeted advertising and revenue generation. As AI continues to evolve, it will undoubtedly reshape the way we consume and interact with media, creating a more immersive and personalized experience for users worldwide.

Examples of AI-powered content recommendations can be seen in popular streaming platforms like Netflix and Amazon Prime Video. These platforms use machine learning algorithms to analyse user preferences and viewing history, allowing them to suggest personalized content that matches the interests of Indian viewers. Additionally, AI can also help media companies understand user engagement by analysing metrics such as click-through rates, time spent on articles or videos, and social media interactions. This data can then be used to optimize content strategies and improve overall user experience.

Future applications of AI in the media industry include the development of virtual reality and augmented reality experiences, which can enhance storytelling and provide immersive content for viewers [35]. AI can also be used to automate tasks such as video editing and captioning, making the production process more efficient. These advancements in AI technology have the potential to revolutionize the way we consume and interact with media in the future.

Future of Indian media involves a significant integration of AI technologies, ultimately transforming the way news and entertainment are produced and consumed. With AI technologies, news and entertainment in India can be produced and consumed in a more efficient and personalized manner. AI algorithms can analyse vast amounts of data to identify trends, preferences, and user behaviour, enabling media companies to deliver tailored content to their audience. Additionally, AI-powered chatbots and virtual assistants can enhance the user experience by providing instant and accurate information, making media consumption more interactive and engaging.

Prospects for AI in the Indian media industry include the potential for automated content creation, where AI algorithms can generate news articles or produce videos based on real-time data. This could significantly reduce production costs [36] and increase the speed at which news is delivered to the audience. Furthermore, AI can also play a crucial role in combating fake news by fact-checking information and verifying sources, ensuring that users have access to reliable and trustworthy content.

As AI becomes more sophisticated, it is expected to revolutionize the way media is created, consumed, and monetized. This could lead to personalized content recommendations, targeted advertising, and even the creation of entirely new forms of media [37] that were previously unimaginable. Overall, AI has the potential to reshape the media industry and create exciting opportunities for both content creators and consumers alike.

Apart from media, we can see the influence of AI in all aspect of life including healthcare and security [38–40].

9.5 CONCLUSION

To conclude, AI technology has the potential to not only enhance the user experience but also improve the efficiency and effectiveness of media production and distribution processes. With its ability to automate tasks and streamline workflows, AI can help media companies save time and resources, ultimately leading to a more innovative and dynamic media landscape. Also, the future of media is likely to be heavily influenced by AI technology, as it continues to evolve and advance.

REFERENCES

1. Bakshy, E., Messing, S., Adamic, L. A. (2015). Exposure to ideologically diverse news and opinion on Facebook. *Science* 348(6239), 1130–1132. https://doi.org/10.1126/science.aaa1160
2. Bandy, J., Diakopoulos, N. (2021). More accounts, fewer links: How algorithmic curation impacts media exposure in twitter timelines. *Proceedings of the ACM on Human Computer Interaction* 5(CSCW1), 1–28. https://doi.org/10.1145/3449152
3. Beckett, C. (2019). New Powers, New Responsibilities: A Global Survey of Journalism and Artificial Intelligence. https://blogs.lse.ac.uk/polis/2019/11/18/new-powers-new-responsibilities
4. Bender, E. M., Gebru, T., McMillan-Major, A., Shmitchell, S. (2021). On the dangers of stochastic parrots: Can language models be too big? *Proceedings ACM Conference on Fairness, Accountability, and Transparency* 21, 610–623. https://doi.org/10.1145/3442188.3445922
5. Bergstrom, C., Joseph, B.-C. (2019). Information gerrymandering in social networks skews collective decision-making. *Nature* 573, 40–41. https://doi.org/10.1038/d41586-019-02562-z
6. Boididou, C., Middleton, S. E., Jin, Z., Papadopoulos, S., Dang-Nguyen, D. T., Boato, G., Kompatsiaris, Y. (2018). Verifying information with multimedia content on twitter. *Multimedia Tools and Applications* 77(12), 15545–15571. https://doi.org/10.1007/s11042-017-5132-9
7. Broussard, M., Diakopoulos, N., Guzman, A. L., Abebe, R., Dupagne, M., Chuan, C. H. (2019). Artificial intelligence and journalism. *Journalism & Mass Communication Quarterly* 96(3), 673–695. https://doi.org/10.1177/1077699019859901
8. Bruns, A. (2019). *Are Filter Bubbles Real?* John Wiley and Sons, Amsterdam
9. Burel, G., Farrell, T., Mensio, M., Khare, P., Alani, H. (2020). Co-spread of misinformation and fact- checking content during the COVID-19 pandemic. *International Conference on Social Informatics*, Pisa, Italy, pp. 28–42.
10. Chen, J., Dong, H., Wang, X., Feng, F., Wang, M., He, X. (2020). Bias and debias in recommender system: A survey and future directions. *CoRR*. arXiv: 2010. 03240
11. Ciampaglia, G. L., Nematzadeh, A., Menczer, F., Flammini, A. (2018). How algorithmic popularity bias hinders or promotes quality. *Scientific Reports* 8(1), 15951. https://doi.org/10.1038/s41598-018-34203-2
12. Cieri, C., Maxwell, M., Strassel, S., Tracey, J. (2016). Selection criteria for low resource language programs. *Proceedings of the Tenth International Conference on Language Resources and Evaluation*, vol. LREC'16, pp. 4543–4549. European Language Resources Association (ELRA), Paris, France.
13. Commission European. (2018). Communication from The Commission to The European Parlament, The Council, The European Economic and Social Committee and The Committee of the Regions- Tackling online disinformation. A European Approach. https://eur-lex.europa.eu/legal-content/EN/TXT/?uri=CELEX:52018DC0236
14. Conotter, V., Obrien, J. F., Farid, H. (2012). Exposing digital forgeries in ballistic motion. *IEEE Transactions on Information Forensics and Security* 7, 02. https://doi.org/10.1109/TIFS.2011.2165843

15. Costera, M. I. (2020). Journalism, audiences and news experiences. In: Wahl-Jorgensen, K., Hanitzsch, T. (eds.) *The Handbook of Journalism Studies*. Routledge, New York. https://doi.org/10.4324/9781315167497-25
16. Das, R., Ytre-Arne, B. (eds.) (2018). *The Future of Audiences*. Palgrave Macmillan, London. https://doi.org/10.1007/978-3-319-75638-7
17. Dawson, A., Hirt, M., Scanlan, J. (2016). The Economic Essentials of Digital Strategy. McKinsey Quarterly. https://www.mckinsey.com/capabilities/strategy-and-corporate-finance/our-insights/the-economic-essentials-of-digital-strategy
18. Del Vicario, M., Bessi, A., Zollo, F., Petroni, F., Scala, A., Caldarelli, G., Stanley, H. E., Quattrociocchi, W. (2016). The spreading of misinformation online. *Proceedings of the National Academy of Sciences* 113(3), 554–559.
19. Devlin, J., Chang, M. W., Lee, K., Toutanova, K. (2019). Bert: Pre-training of deep bidirectional transformers for language understanding. *Proceedings Conference of the North American Chapter of the Association for Com putational Linguistics* 5, 5.
20. Diakopoulos, N. (2019). Towards a design orientation on algorithms and-automation in news production. *Digit Journal* 7(8), 1180–1184. https://doi.org/10.1080/21670811.2019.1682938
21. Diakopoulos, N. (2019). *Automating the News: How algorithms are Rewriting the Media*. Harvard University Press, Cambridge. https://doi.org/10.4159/9780674239302
22. Diakopoulos, N. (2020). Computational news discovery: Towards design considerations for editorial orientation algorithms in journalism. *Digit Journal* 8(7), 1–23. https://doi.org/10.1080/21670811.2020.1736946
23. Diakopoulos, N., Trielli, D., Lee, G. (2021). Towards understanding and supporting journalistic practices using semi-automated news discovery tools. *Proceedings of the ACM (PACM): Human-Computer Interaction (CSCW)* 5, 1–30.
24. Draper, N. A., Joseph, T. (2019). The corporate cultivation of digital resignation. *New Media & Society* 21(8), 1824–1839. https://doi.org/10.1177/1461444819833331
25. Ekstrand, M. D., Burke, R., Diaz, F. (2019). Fairness and discrimination in recommendation and retrieval. *Proceedings of ACM Conference on Recommender Systems* 2019, 1403–1404. https://doi.org/10.1145/3331184.3331380
26. Elahi, M., Jannach, D., Skjærven, L., Knudsen, E., Sjøvaag, H., Tolonen, K., Holmstad, Ø., Pipkin, I., Throndsen, E., Stenbom, A., Fiskerud, E., Oesch, A., Vredenberg, L., Trattner, C. (2021). Towards responsible media recommendation. *AI Ethics* 2, 103–114. https://doi.org/10.1007/s43681-021-00107-7
27. Elahi, M., Kholgh, D. K., Kiarostami, M. S., Saghari, S., Rad, S. P., Tkalcic, M. (2021). Investigating the impact of recommender systems on user-based and item-based popularity bias. *Information Processing & Management* 58, 102655. https://doi.org/10.1016/j.ipm.2021.102655
28. Ferrer, X., van Nuenen, T., Such, J. M., Coté, M., Criado, N. (2021). Bias and discrimination in AI: A cross-disciplinary perspective. *IEEE Technology and Society Magazine* 40, 72–80.

29. Hoffmann, H., Kohl, C. (2018). Marktüberblick zum Einsatz von künstlicher Intelligenz in Medien und Publishing. KI-Report Eine Übersicht zum Angebot im deutschsprachigen Raum. https://hspartner.de/jsp292/site/hauptmenue/publikationen/studien/ki
30. Lovrinovic, C., & Volland, H. (2019). The Future Impact of Artificial Intelligence on the Publishing Industry. Gold Finch, Frankfurter Buchmesse, 2019
31. Blasi, D., Anastasopoulos, A., & Neubig, G. (2022). Systematic Inequalities in Language Technology Performance across the World's Languages. In *Proceedings of the 60th Annual Meeting of the Association for Computational Linguistics*, pp. 5486–5505.
32. Castro, D., & Steinberg, M. (2017). Blocked: Why some companies restrict data access to reduce competition and how open APIs can help. Center for Data Innovation. SSRN, 2017. http://dx.doi.org/10.2139/ssrn.3108763
33. Freeman, R. B. (2014). "Who Owns the Robots Rules the World", IZA World of Labor, May 2014. https://wol.iza.org/uploads/articles/5/pdfs/who-owns-the-robots-rules-the-world.pdf
34. González Otero, B., Quintais, J. P. (2018). "Before the Singularity: Copyright and the Challenges of Artificial Intelligence", Kluwer Copyright Blog, September 25, 2018. https://copyrightblog.kluweriplaw.com/2018/09/25/singularity-copyright-challenges-artificial-intelligence
35. Guadamuz, A. (2017). Artificial intelligence and copyright. *WIPO Magazine* 2017, 14–19. https://www.wipo.int/wipo_magazine/en/2017/05/article_0003.html
36. Röder, M. (2018). Artificial intelligence and copyright: The rise of the machine. *Presentation at the IFRRO International Conference*, Athens, 21 October 2018.
37. Rognstad, O.-A. (2018). Artificial intelligence and copyright ownership. *Presentation at the Conference EU Copyright, Quo Vadis? From the EU Copyright Package to the Challenges of Artificial Intelligence*, Brussels, 25 May 2018.
38. Khadidos, A. O., Alyoubi, K. H., Mahato, S., Khadidos, A. O., Mohanty, S. N. (2023). Computer aided detection of major depressive disorder (MDD) using electroencephalogram signals. *IEEE Access* 11, 41133–41141. https://doi.org/10.1109/ACCESS.2023.3262930
39. Tyagi, N., Bhushan, B. (2023). Natural language processing (NLP) based innovations for smart healthcare applications in healthcare 4.0. In: Ahad, M. A., Casalino, G., Bhushan, B. (eds) *Enabling Technologies for Effective Planning and Management in Sustainable Smart Cities*. Springer, Cham. https://doi.org/10.1007/978-3-031-22922-0_5
40. Bhushan, B., Kumar, A., Agarwal, A. K., Kumar, A., Bhattacharya, P., Kumar, A. (2023). Towards a secure and sustainable internet of medical things (IoMT): Requirements, design challenges, security techniques, and future trends. *Sustainability* 15, 6177. https://doi.org/10.3390/su15076177

Chapter 10

Machine learning feature selection method for anatomic brain MRI in dementia

A comparative study

Anindya Nag, Ayontika Das, Biva Das, Moupriya Das, and Anupam Kumar Bairagi

10.1 INTRODUCTION

The brain is an incredibly intricate and essential organ within the human body. It plays a crucial role in various cognitive functions such as imagination, problem-solving, critical thinking, deliberation, and memory. Memories and information acquired over time are stored within the brain's memory systems and can be recalled later. These stored experiences and knowledge significantly shape individuals' personalities and identities [1]. However, the devastating effects of dementia, particularly the loss of memory and the inability to recognize familiar objects and places can be distressing. Alzheimer's disease (AD) is responsible for the majority of dementia cases. Therefore, its significance lies in its potential to predict and detect dementia at its earliest stages [2].

Dementia is a global affliction that impacts individuals worldwide. It is a condition characterized by the disruption of normal functioning in specific populations of brain cells, leading to a combination of symptoms [3]. Typically, it affects the brain regions responsible for cognitive functions, memory, and communication. Figure 10.1 illustrates the various types of dementia, categorized based on their underlying causes, distinguishing between neurodegenerative and non-neurodegenerative forms [4].

AD is widely recognized as the primary ethology of dementia [5]. Dementia can be classified as early-onset when diagnosed in individuals under 65 and late-onset in older adults aged 65 and above [6]. Projections from the World Alzheimer Report 2015 indicate a significant increase in the global dementia population, with an estimated 131.5 million people living with dementia by 2050, compared to 74.7 million in 2030 [7]. Visual evidence is critical in numerous scientific disciplines, particularly medical imaging, which has become a powerful tool for studying brain processes. Medical professionals often rely on neuroimaging techniques such as MRI to visualize the brain's anatomy and functionality when diagnosing

DOI: 10.1201/9781003534761-10

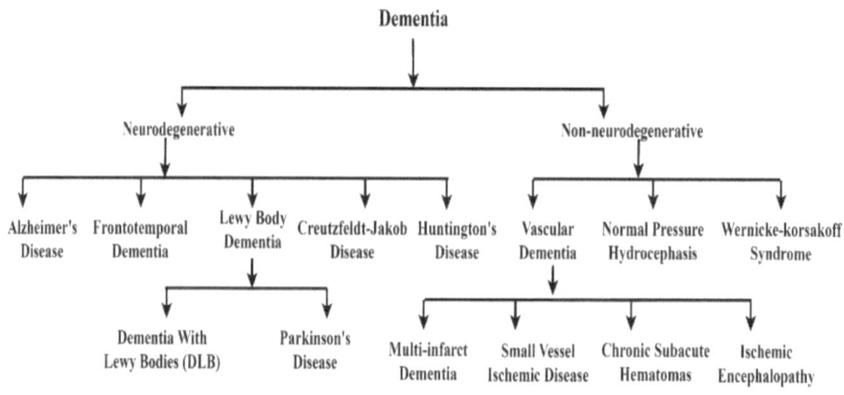

Figure 10.1 Significant types of dementia.

brain-related conditions [8]. In cases of dementia caused by AD, diagnosis involves a comprehensive evaluation of symptoms by a physician, who may also recommend additional laboratory tests, brain imaging, or memory assessments. Magnetic resonance images (MRI) scans can reveal brain abnormalities in individuals with moderate cognitive impairment, namely mild cognitive impairment (MCI), has the potential to serve as a predictive indicator for the probability of acquiring AD. Physicians typically look for specific indicators in MRI scans, such as a reduction in the size of specific brain regions or other anomalies, particularly affecting the temporal and parietal lobes [9]. Figure 10.2 demonstrates changes in body structure over 2 years, featuring the top row depicting a healthy male control subject (baseline age: 84.8), the middle row depicting a female MCI subject (baseline age: 71.8) who later developed AD, and the bottom row depicting a male AD patient (baseline age: 77.5) [10].

The combination of machine learning (ML) classification algorithms with feature selection approaches has facilitated the development of automated techniques for classifying brain illnesses and determining the most informative set of predictions [11]. Several ML methods, including support vector machines (SVMs), adaptive neuro-fuzzy inference systems, and neural networks, have been utilized in study investigations that centre on the evaluation of dementia through imaging assessments. The objective of these investigations is to achieve precise diagnosis of Alzheimer's and vascular dementia through the utilization of a meticulously chosen collection of MRI features [12].

10.1.1 AD diagnostic

Alzheimer's is a degenerative neurological disorder that progressively impairs memory, critical thinking, and the ability to perform even simple

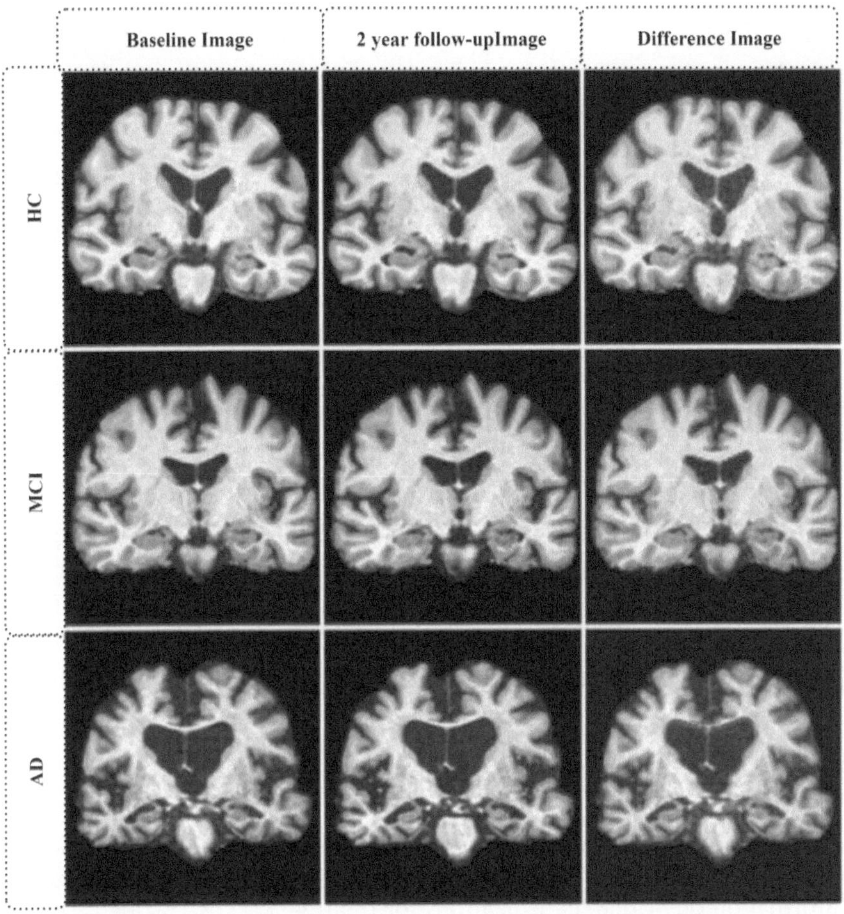

Figure 10.2 Three coronal section MRI images (brain-extracted) of subjects from the AD neuroimaging initiative (NI) cohort.

tasks [13]. It is the leading cause of dementia among individuals aged 65 and above. Dementia is distinguished by substantial reductions in cognitive abilities, such as memory, thinking, judgement, and language, to the point that they impede daily functioning [14]. It is not a standalone illness but a constellation of symptoms often accompanying various other medical conditions. A recent study published in the American Journal of Psychiatry explored the potential use of an olfactory test to monitor the onset of AD [15]. Figure 10.3 illustrates brain and neuronal physiology.

AD is subject to the effect of multiple risk factors, encompassing genetics, age, vascular diseases, head injuries, infections, and environmental variables, such as trace metals and heavy metals as depicted in Figure 10.4 [16].

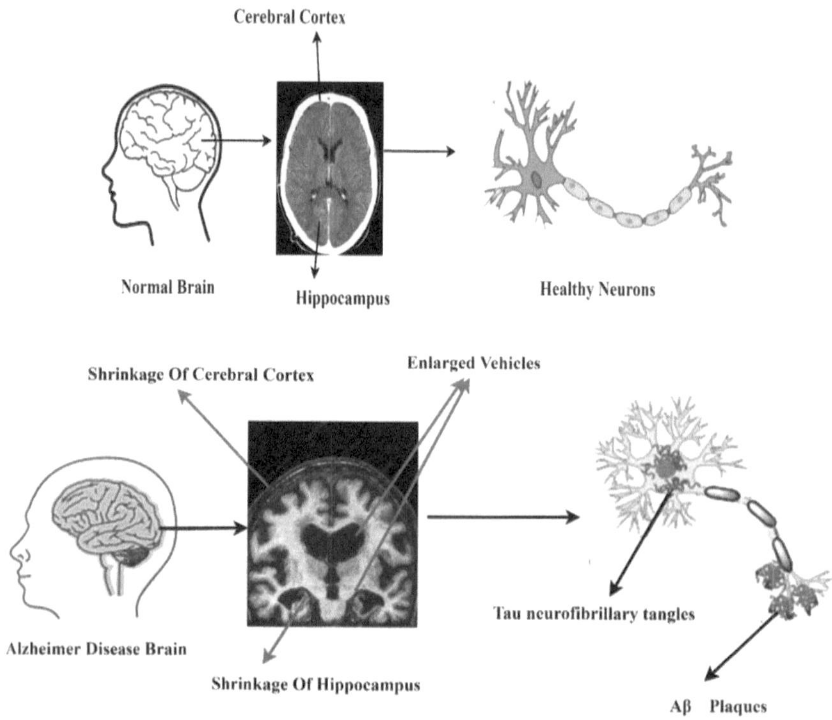

Figure 10.3 (a) Healthy brain and (b) AD brain neurophysiology.

The pathological changes associated with AD, including Aβ accumulation and synaptic loss, are still not fully comprehended [17]. Table 10.1 provides information on available medications for treating AD [18,19].

10.1.2 Fundamentals of ML

ML is a scientific discipline that empowers computers to acquire knowledge from data. Through the application of algorithms, ML uncovers inherent patterns in data, aiming to enhance decision-making and predictive capabilities [20]. ML techniques can be categorized into various types based on their objectives and data characteristics, including supervised, unsupervised, semi-supervised, reinforcement, transduction, and meta-learning. Among these techniques, supervised learning (SL) stands out as a prevalent ML approach, leveraging known input and output data to train models [21]. This training equips the models to address classification, regression, and forecasting tasks. In the context of SL, the correct solutions to problems are pre-established. The ML algorithm identifies data patterns linked to these solutions, enabling it to make predictions. The accuracy of these forecasts is enhanced by incorporating corrections made by an operator [22].

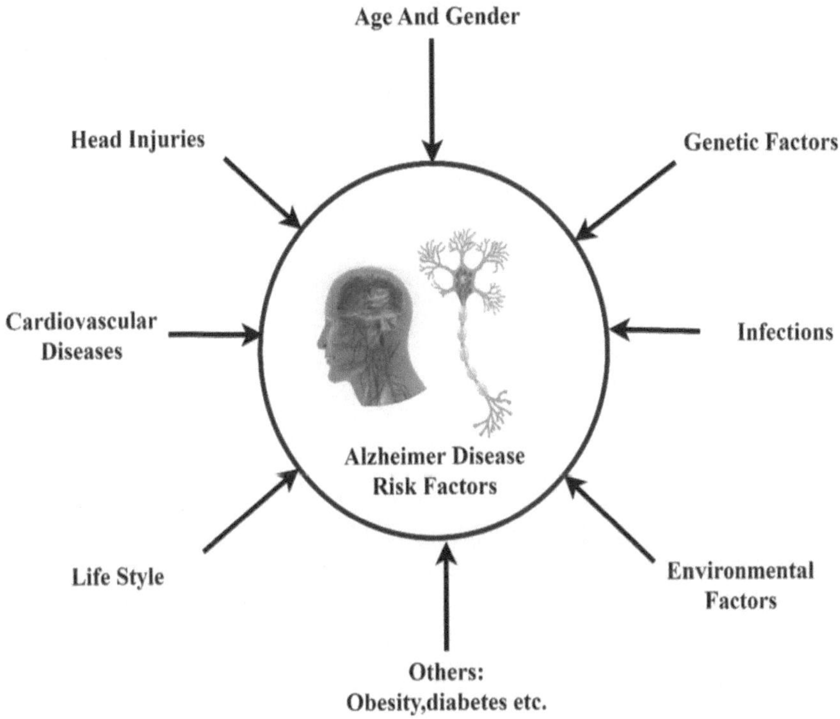

Figure 10.4 The risk factor of AD.

Table 10.1 Popular medications for the treatment of AD

Popular meditation	*Phase of treatment*	*Target pathway*
Memantine	Moderate to severe	N-Methyl-D aspartate Receptor
Tacrine	Mild to moderate	AChE Inhibitor
Galantamine	Mild to moderate	AChE Inhibitor
Rivastigmine	Mild to moderate	AChE Inhibitor
Donepezil	Effective in all phases	AChE Inhibitor

This iterative process continues until the algorithm achieves optimal prediction accuracy.ML algorithms are categorized into four types: unsupervised, supervised, semi-supervised, and reinforcement learning (RL) [23].

10.1.2.1 Types of machines leaning

Below is a brief overview of various learning methods and their respective applications, while Figure 10.5 illustrates the different types of ML approaches.

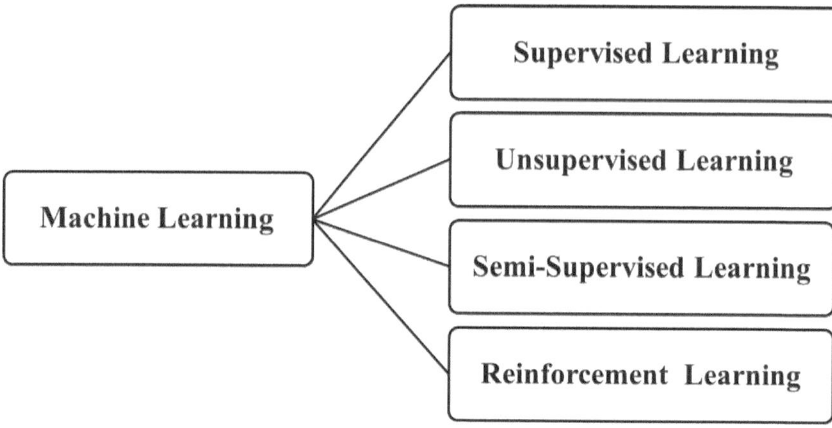

Figure 10.5 Types of machine learning.

- **Supervised Learning:** The primary objective of SL in ML is to train a mathematical function that establishes a mapping between input and output, utilizing a set of sample pairs consisting of input and corresponding output [24]. This process is referred to as SL. It involves learning a function using a set of training examples and labelled training data. SL is typically employed when the objective is to obtain specific outcomes from a given set of inputs, often in a task-oriented context [25].
- **Unsupervised Learning (UL):** UL, as a data-driven approach, involves the analysis of unlabelled datasets without human intervention [26]. It encompasses applications such as extracting generative features, uncovering meaningful patterns and structures, clustering observations, and purposefully investigating them. The routine activities conducted at UL encompass various computational tasks including as clustering, feature learning, dimensionality reduction, density estimation, anomaly detection, and the identification of association rules.
- **Semi-Supervised Learning (SSL):** SSL combines both the SL and UL approaches mentioned earlier, as it operates with labelled and unlabelled data [27]. It falls in between the realms of "unsupervised learning" and "supervised learning." SSL is valuable because labelled data can be scarce in real-world scenarios, while unlabelled data is abundant. The objective of an SSL model is to provide a more accurate estimation compared to using only the labelled data available to the model.
- **Reinforcement Learning:** RL is a technique within the field of ML that adopts an environment-driven approach, enabling software agents and machines to autonomously determine the optimal course of action in various scenarios [28]. The objective of RL is to leverage knowledge gained from interactions with the environment to maximize value and minimize risk [29].

Table 10.2 Types of ML

Types of ML	*Data set type*	*Example*
SL	The data is labelled with different classifications and outcomes, allowing the algorithm to make more accurate predictions.	This category encompasses a variety of ML techniques, including RF, Bayesian networks, and ANN.
UL	Since this data needs labels, it can be called unlabelled or unannotated. The program identifies significant correlations within the datasets.	This category encompasses a range of clustering approaches, such as principal component analysis (PCA), k-means, and hierarchical clustering.
SSL	This learning method can be utilized for datasets containing labelled and unlabelled columns.	This encompasses labelled and unlabelled outcomes and classifications, primarily employed in image and voice recognition applications.
RL	A technique rooted in human psychology utilizes reward-based criteria.	Its usage is more prevalent in diagnostic imaging, analytics-driven disease screening, and medication selection.

ML encompasses a diverse array of approaches, summarized in Table 10.2, showcasing the breadth of techniques within the field [30,31].

10.1.3 Different ML techniques

In SL, the computer has a labelled dataset containing input and output parameters [32]. When presented with a new dataset, the SL algorithm analyses it based on the labelled data and generates the appropriate output. On the other hand, UL does not have a labelled dataset, and the algorithm is designed to discover patterns and categorize information independently. RL algorithms are designed to actively seek optimal solutions, using a reward and punishment system to guide their actions and ultimately achieve the desired outcome [33]. Logistic regression (LR), SVM, Random Forest, cluster analysis, artificial neural networks (ANNs), and CNN are among the popular ML approaches extensively utilized in cardiac imaging and diagnostics [34] (Figure 10.6).

- **Logistic Regression:** The SL method, LR, utilizes the sigmoid function (logarithmic) as its fundamental building block, which maps real numbers to values between 0 and 1. During the training phase, the algorithm receives an N-sample training dataset, where each instance contains X properties and a Y label that determines its category. Through the training process, the system generates a model that can be utilized to label new data that was not included in the training set [35].

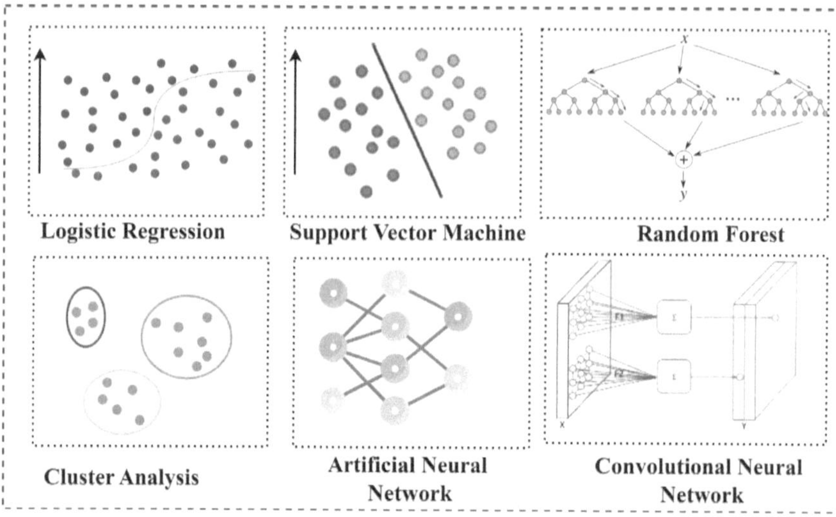

Figure 10.6 The most widely used ML methods in cardiac imaging are being employed.

- **Support Vector Machine:** SVM is a widely used and advanced ML method specialising in data organization. The algorithm operates based on the principle of margin calculation, where artificial boundaries are established between different classes. The goal is to maximize the separation between classes, ensuring the minor possible classification error in the process [36].
- **Random Forest:** Random Forest is a regression technique that leverages ensemble learning by building multiple Decision Trees during training. RF is a ML algorithm that is particularly suitable for classification tasks due to its ability to choose the final output based on the majority class selected by the ensemble of trees. The RF provides the average prediction obtained from all the trees for regression tasks [37].
- **Clustering Analysis:** Unsupervised cluster analysis is a method used to identify subgroups within a given dataset. Clustering algorithms group objects based on their proximity, determining the degree to which an item belongs to a particular cluster by measuring its distance from the cluster centre [38].
- **Artificial Neural Network:** ANN, inspired by the functioning of neurons in the human brain, mimics their behaviour. Neurons are cellular structures found in the brain, and understanding neural networks requires knowledge of neuron theory, including components like dendrites, nuclei, soma, and axon. Similarly, ANN operates at three levels: input layer, hidden layer (analogous to soma and axon), and output layer. The input layer receives data, which is then processed by

the hidden layer, and the output layer transmits the calculated output. ANN falls into three main categories: RL, SL, and UL [39].

- **Convolutional Neural Network (CNN)**: CNNs, an extension of ANNs, incorporate convolutional operations to modify the values of nodes in a layer based on their proximity to nodes in the preceding layer. This approach is mainly designed for image processing tasks, where the spatial relationships among pixels are crucial for accurate predictions. CNNs leverage both the advantages and limitations of ANNs. Notably, these models differ significantly from their predecessors as they directly take images as input without requiring explicit feature extraction steps [40].

This section provides an overview of the commonly used ML techniques in cardiac imaging and diagnostics, explaining their benefits and limitations in a clear and accessible manner. For a more comprehensive understanding, it is recommended to see Table 10.3 [41].

Table 10.3 Overview of ML methods

Techniques	*Description*	*Advantages*	*Disadvantages*
LR	Expanding linear regression to generate a binary classification as the ultimate result.	• Simple to execute • Efficient in terms of processing speed	• This statement does not align with the non-linearity model framework.
SVM	Identifies the optimal boundary between classes	• Exceptional accuracy • Swift classification of newly identified objects.	• Choosing the appropriate kernel may not be immediately evident. • Requires significant computational resources.
RF	Generates a sequence of hierarchical decision queries performed on the input and output data.	• Exhibits robust performance • Calculates derived variable measures	• Requires significant computational resources • Potential issue of overfitting
ANN	Models achieve intricate classification tasks by sequentially subjecting input data to a series of non-linear transformations within a network.	• Effectively deal with noisy data • Detect non-linear relationships	• Training time is slow • Interpretation is challenging

(Continued)

Table 10.3 (Continued) Overview of ML methods

Techniques	*Description*	*Advantages*	*Disadvantages*
CNN	And adapted for the processing and categorization of image data.	• Flexible design that can be tailored to the application's needs • Able to learn optical characteristics directly from the images	• Shares the same limitations as ANNs
Clustering	Unsupervised exploration of subgroups within the input feature space	• Effective for identifying subgroups when the labels for the larger groups are known in advance. • Straightforward and efficient	• The results may vary depending on the scale and initialization used. • Quantifying the exact number of subgroups can be challenging.

10.1.4 Utilizing ML for dementia diagnosis

The determination of the participants' dementia status at the time of their demise was made through the evaluation of interviews and evaluations that were conducted throughout the last years of their lives. The procedure entailed employing the comprehensive Geriatric Mental State-Automated Geriatric Examination for Computer-Assisted Taxonomy diagnostic algorithm and making use of references from the Diagnostic and Statistical Manual of Mental Disorders (third edition - revised) [42]. In addition to considering informant interviews done after the participant's death and examining the cause of death, these factors were also taken into consideration. The categorization of participants as non-dementia individuals at the time of their death was determined based on two criteria. Firstly, participants who had not received a dementia diagnosis during their final interview within 6 months prior to their passing were classified as non-dementia individuals. Secondly, participants whose last and subsequent retrospective interviews revealed the absence of dementia were also defined as non-dementia individuals. The researchers utilized Bayesian analysis methods to estimate the probability of dementia in instances when the final interview took place over 6 months prior to the individual's death, and there were no indications of dementia in the interview records or retrospective informant interviews (RINI) [43].

Figure 10.7 illustrates the essential sequential process of computer-aided dementia diagnosis utilizing ML techniques [44].

Neuroimaging-based dementia diagnosis is a complex process influenced by various factors contingent on the imaging techniques employed. ML offers

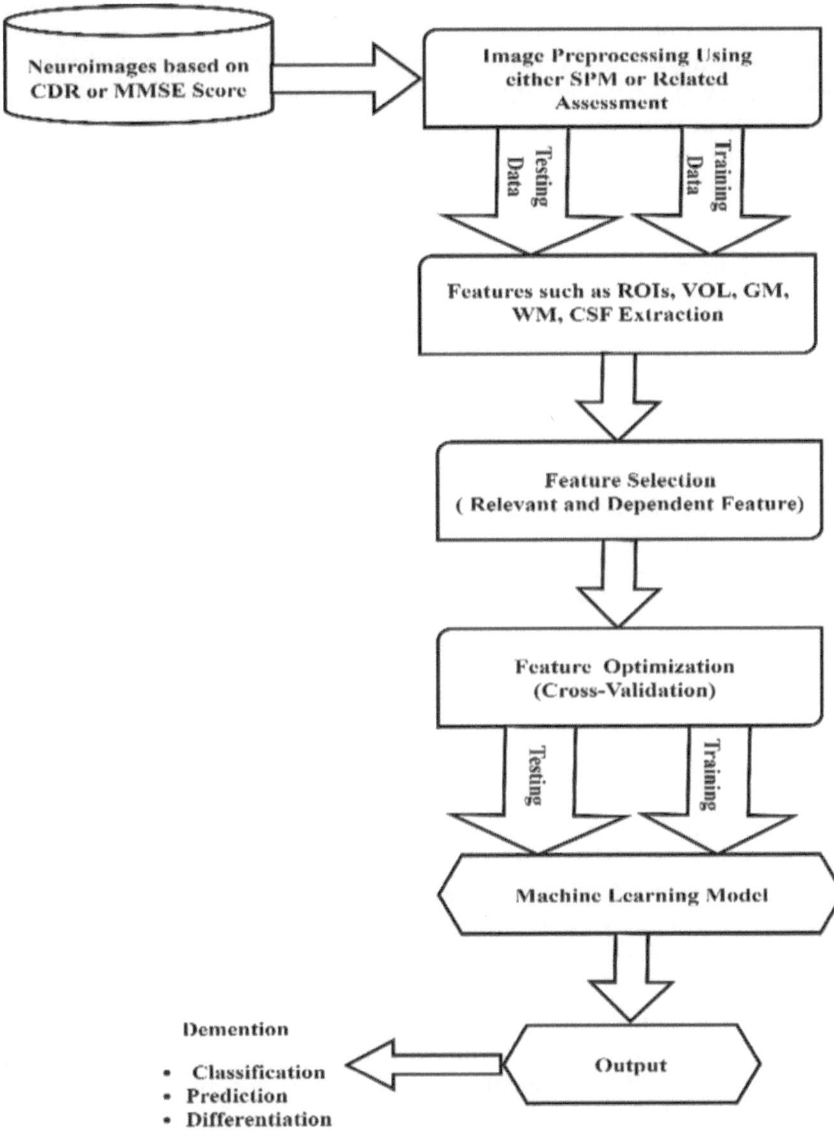

Figure 10.7 The basic computer-aided dementia diagnosis utilizing Machine Learning.

the potential to aid clinical diagnostic decisions by automating dementia classification and prediction through computer-aided diagnosis (CAD) techniques and neuroimaging data. The foundational steps of computer-aided dementia diagnosis using ML are illustrated in Figure 10.6. Initial image preprocessing, carried out using tools like SPM or SeeCAT [44], hinges on parameter selection. For extracting pertinent features and selecting relevant ones, diverse statistical methods are utilized, with cross-validation optimization ensuring method generation reliability. Subsequently, ML methods, including SVM [45], are harnessed for tasks such as dementia classification, prediction, or differentiation. Within this section, we systematically delve into the cutting-edge ML approaches applied in the realm of dementia research [46].

10.1.5 Deep learning techniques for dementia diagnosis

Deep learning, a subset of ML, has emerged as a powerful tool in the realm of dementia diagnosis. This approach involves training neural networks with multiple layers to learn intricate patterns and features from complex medical data automatically [47]. In the context of dementia, deep learning methods can analyse a diverse range of data sources, including neuroimaging scans, clinical assessments, and genetic information. CNNs excel at processing images, allowing for the detection of subtle brain abnormalities indicative of dementia. Recurrent neural networks (RNNs) can analyse sequential data, such as temporal changes in cognitive scores [48]. Moreover, hybrid architectures and attention mechanisms enhance the interpretability of model predictions. The fusion of deep learning and dementia diagnosis holds great potential for early detection, accurate classification, and personalized treatment strategies, ushering in a new era of precision medicine in neurodegenerative disorders [49].

Figure 10.8 depicts the architecture of a deep learning CNN employed for hierarchical feature extraction and subsequent classification, contributing to the prognosis of dementia. This approach leverages the innate ability of CNNs to automatically learn intricate features from input data, enabling a multi-level comprehension of the information [50]. The final stage of this process involves the application of a SoftMax classifier which assigns probabilities to various classes, thereby enhancing the accuracy and reliability of dementia prognosis. This integration of deep learning and classification techniques holds the potential to improve dementia prediction significantly, advancing the field of medical decision-making and tailored treatment strategies [51]. A novel approach involves utilizing a deep learning CNN for hierarchical feature extraction and subsequent classification aimed at prognosticating dementia. This methodology capitalizes on the inherent capability of CNNs to automatically learn intricate features from raw data, facilitating a multi-level understanding of the input information. The final step in this framework employs a SoftMax classifier, which assigns probabilities to different classes, enabling accurate dementia prognosis. This amalgamation of deep learning and classification holds the potential to enhance the precision and reliability of dementia prediction, paving the way for more informed medical decisions and personalized interventions [52].

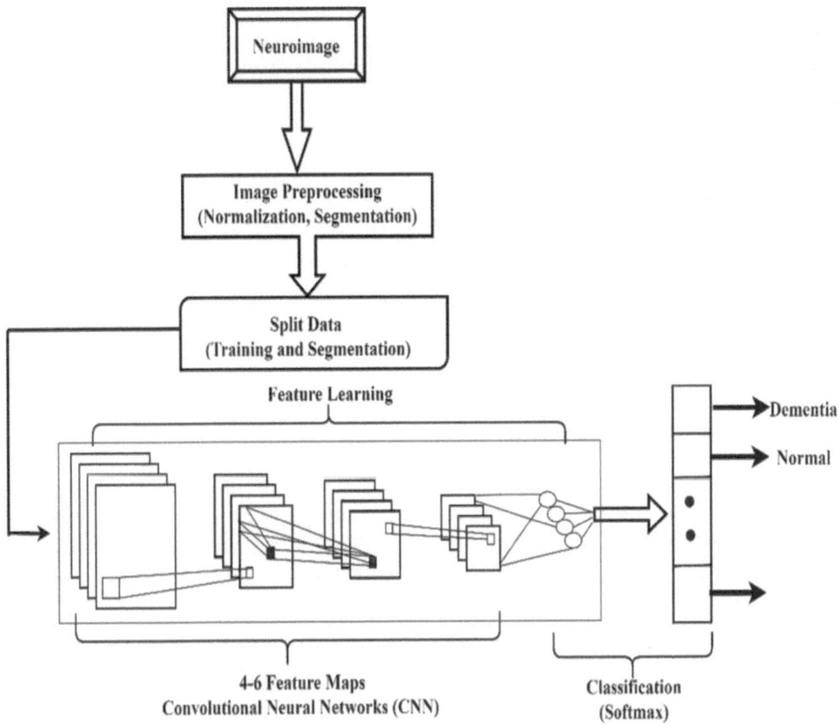

Figure 10.8 Employing a deep learning convolutional neural network for gradual feature extraction and prognosis of dementia through softmax classification.

The study initiates by providing a comprehensive overview of the diagnostic criteria for AD, its associated risk factors, and commonly prescribed medications for treatment. In Section 10.2 provides major challenges utilizing ML for feature selection in anatomic brain MRI. In Section 10.3, the authors assess the existing methods described in the literature, examining their strengths and limitations. Section 10.4 offers a comparative analysis of various ML approaches employed in the field. Based on the findings, the report presents a conclusive assessment and puts forth recommendations for future research in Section 10.5.

10.2 MAJOR CHALLENGES

In this Section authors outline the primary obstacles encountered in the domain of employing ML techniques for the purpose of feature selection in anatomic brain MRI scans for the diagnosis of dementia. The field of utilizing ML for feature selection in anatomic brain MRI for dementia diagnosis presents both challenges and promising future prospects [53].

One significant challenge is handling MRI data's high-dimensional and complex nature while selecting informative features to contribute to an accurate diagnosis. Moreover, ensuring robustness and generalizability across diverse patient populations is crucial. As the field advances, addressing these challenges will likely involve developing innovative algorithms that can effectively capture subtle yet critical patterns in brain images. Furthermore, the integration of multimodal data and the incorporation of advanced deep learning techniques hold the potential to refine feature selection and enhance diagnostic accuracy. These efforts could ultimately lead to more reliable and early detection of dementia, revolutionizing the way we diagnose and understand neurodegenerative diseases [54].

10.2.1 Classification of dementia: deciphering cognitive decline

Dementia classification is a pivotal pursuit in the realm of medical diagnosis, aiming to untangle the complexities of cognitive decline. This endeavour involves employing various ML algorithms to accurately distinguish between individuals with different stages of dementia and those without the condition [55]. The challenges in this domain are manifold, ranging from handling the heterogeneity of data sources (including clinical, imaging, and genetic data) to mitigating the impact of class imbalances. Moreover, the interpretability of models and the ability to generalize across diverse patient populations are crucial considerations. Looking ahead, the fusion of cutting-edge techniques such as deep learning and multimodal data integration holds significant promise in improving the accuracy and early detection capabilities of dementia classification [56]. The ultimate goal is to enhance patient care, enable personalized interventions, and advance our understanding of neurodegenerative diseases.

10.2.2 Neuroimaging clinical and statistical analysis for dementia diagnosis

This section focuses on the clinical and statistical analysis of neuroimaging techniques for the purpose of diagnosing dementia. The primary imaging modalities and analytical techniques that have been developed for the purpose of evaluating the initial progression, ageing, and diseased phases of the brain. To enhance the process of diagnosing dementia by employing a comprehensive approach that integrates clinical and statistical approaches [57]. Early detection, subsequent intervention, and ongoing assistance hold paramount importance for individuals in PR symptomatic or predementia phases, aiming to decelerate degeneration and sustain a high quality of life. Biomedical imaging is a tool for visualizing deteriorative histological alterations, encompassing features like hypo-metabolism, amyloid plaques, and

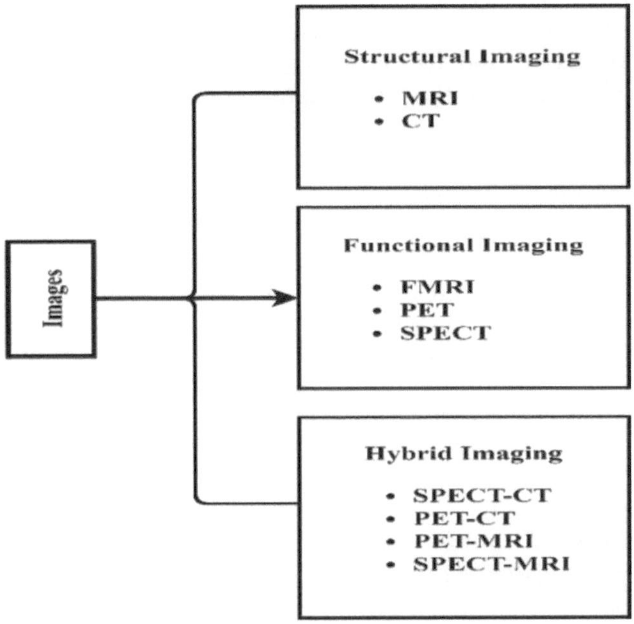

Figure 10.9 Illustrates various modalities used in the field of biomedical imaging.

atrophy. Specifically, these imaging techniques can furnish quantitative biomarkers indicative of specific pathologies and prognoses for dementia [58]. Figure 10.9 illustrates diverse imaging modalities, such as MRI, PET, and SPECT images (discussed in subsequent sections), which offer advantages in capturing brain irregularities and neural susceptibilities related to dementia. Various publicly accessible neuroimaging databases are available, including the Alzheimer's Disease Neuroimaging Initiative (ADNI) established in 2004, the Open Access Series of Imaging Studies (OASIS) first introduced in 2007, the Australian Imaging, Biomarker & Lifestyle Flagship Study of Ageing (AIBL) initiated in 2006, and CAD Dementia, which involves structural brain MRI scans. These repositories have been extensively employed in numerous studies, significantly bolstering dementia-related research [59].

10.3 LITERATURE OF REVIEW

The subsequent section of this chapter consists of a literature review. This literature review examines the current corpus of literature on the ML Feature Selection Method for Anatomic Brain MRI in Dementia.

Saeed et al. [9] introduced a pre-trained deep learning model called ResNet50, specifically designed to automatically extract features from

MRI scans for AD diagnosis. The authors further evaluated the performance of the CNN model by comparing it with traditional methods such as SoftMax, SVM, and RF, using various metrics, including accuracy. The results revealed that the proposed framework outperformed the existing state-of-the-art approach, achieving 85.7% and 99% accuracy for models trained on the MRI ADNI dataset (ResNet50 + SoftMax: 99%, ResNet50 + SVM: 92% and ResNet50 + RF: 85.7%).

Badshah et al. [39] presents a comprehensive approach to enhance the classification accuracy and reduce the computation time of brain MRI tumour images, involving preprocessing, feature extraction, selection, and classification phases, using various ML algorithms, achieving significant improvements over existing methods in terms of accuracy and efficiency.

Vijithananda et al. [40] focuses on utilizing diffusion-weighted MRI images of human brain tumours to extract demographic and texture features, implementing ML techniques for the differentiation of malignant and benign tumours. Using feature selection based on ANOVA f-test, a final Random Forest ML model achieved 90.41% accuracy in distinguishing tumour types, suggesting the potential of these features to aid non-invasive brain tumour diagnosis decisions.

Rajab et al. [42] focused on developing ML techniques for identifying AD pathologies associated with dementia by selecting relevant attributes. They utilized a cohort of 186 participants from the Cognitive Function and Ageing Study (CFAS) to employ ML methods for feature ranking and classification. The study aimed to compare neuropathological characteristics and assess their diagnostic performance objectively. The highest-performing classifier for dementia achieved 79% sensitivity, 69% specificity, and 75% accuracy by utilizing the top seven neuropathology variables.

Rallabandi et al. [43] developed an automated ML classifier utilizing MRI to differentiate between healthy controls (HCs), Parkinsonism dementia, AD, and MCI. The accuracy of various classifiers, including discrete tree, Random Forest, and multilayer perceptron, was evaluated using different approaches. The study revealed that naive Bayesian (NB) and SVM demonstrated the highest performance in 5-fold cross-validation, indicating their effectiveness in accurately classifying the mentioned conditions.

Marioni et al. [45] introduced a method for diagnosing AD using structural MRI data by combining subcortical and cortical variables from a cohort of 308 participants in the ADNI dataset. The study conducted three binary categorisation experiments: AD vs MCI, AD vs lMCI, and AD versus HC. The proposed method employed a Feature Selection approach incorporating PCA and the Recursive Feature Elimination technique to reduce dimensionality and simultaneously select the most relevant features.

Multiple experiments were conducted on the dataset, and the results demonstrated that SVM achieved the highest accuracy levels when classifying AD cases against lMCI, HC, and MCI.

Li et al. [47] developed three classifiers utilizing ML techniques to predict and infer the brain regions affected in the onset and progression of AD using MRI data. The authors constructed three distinct classifiers, namely SVM, 3D Deep Residual Network (ResNet), and 3D Very Deep Convolutional Network (VGGNet), and systematically compared them. The SVM, VGGNet, and ResNet classifiers achieved accuracies of 90%, 95%, and 95%, respectively, in effectively differentiating individuals with AD from elderly control subjects.

Herzog et al. [53] presented a technique that enables data processing using readily available hardware. The method involved analysing structural changes in the brain by utilizing asymmetry characteristics derived from MRIs in the ADNI database and disease classification using ML. Supervised ML algorithms, specifically SVM and CNNs, were employed to differentiate between normal cognition (NC), Early MCI (EMCI), and NC and AD, achieving accuracy in the classification process.

Thapa et al. [54] proposed a model that incorporates feature selection to enhance the model's performance by enabling the learning of valuable patterns from the available data. The study demonstrates the potential of integrating MRI characteristics with neuropsychological test results, such as the Mini-Mental State Examination (MMSE), to create a more favourable decision-making environment for the early identification of AD. Experimental results show that including MMSE and other characteristics significantly improves AD classification.

Bansal et al. [55] proposed a method for classifying dementia using MRI scans from the ADNI database. The method utilizes the Bag of Feature approach to extract robust features efficiently. Experimental results showed that the suggested technique achieved superior classification accuracy. The combination of the BOF method and SVM yielded a remarkable success rate of 93%.

Lin et al. [56] demonstrates the feasibility of an economical and sensitive approach for predicting brain amyloid-β burden, a key factor in AD progression. By harnessing the power of ML and combining multimodal MRI techniques, the research successfully predicted amyloid levels based on voxel-based morphometry and fractional anisotropy measurements. These findings highlight a promising avenue for early Alzheimer's diagnosis through the correlation between structural and diffusion MRI parameters and amyloid accumulation in older individuals.

Ahmed et al. [57] comprehensively surveys recent advancements in using neuroimaging and ML, particularly deep learning, for accurate and timely automated diagnosis of dementia, highlighting challenges in identifying various types of dementia beyond AD. The authors provided a comprehensive overview of the latest neuroimaging techniques, including MRI, PET, and SPECT, as well as ML and deep learning methodologies for extracting relevant features and classifying different types of dementia.

Shafi et al. [58] proposes a ML approach for the detection of brain tumours using MRI images, encompassing preprocessing, feature extraction, selection, and a SVM classifier, resulting in a high-performance classification method with sensitivity, specificity, precision, and accuracy all reaching 98.8%, thus offering a significant advancement for accurate tumour detection in neuro-medicine diagnosis.

Zheng et al. [59] aimed to differentiate between vascular dementia (VaD) and AD using multi-parameter features derived from structural MRI through ML techniques. Ninety-three patients (58 AD and 35 VaD) were included, with brain tissue segmentation providing volumetric measurements for feature extraction. Significantly different features between VaD and AD were selected, reduced dimensionality using LASSO, and employed in a SVM classifier, achieving an encouraging performance with sensitivity, specificity, and accuracy of 82.65%, 87.17%, and 84.35%, respectively, highlighting the potential of this CAD method for clinical decision-making in distinguishing between the two conditions.

The comparison table of the Literature of Review is presented in Table 10.4, providing a comprehensive overview of the reviewed studies.

10.4 COMPARISON ANALYSIS OF APPROACH

This review section conducts a comparative analysis of the literature reviewed thus far, considering various datasets and the reported accuracy in previous studies. A comparative analysis was conducted to evaluate several ML methodologies. The findings indicate that the SVM strategy, implemented without utilising any pre-trained CNN deep learning model, demonstrated superior performance in terms of accuracy (97%), specificity (98%), and sensitivity (95%). The ResNet50 + SoftMax model demonstrated a notably elevated prediction accuracy of 99% [9]. The specificity of the test is 98% and the sensitivity is 99%. After careful examination, it is determined that the SVM algorithm without any pre-trained CNN deep learning model demonstrates superior performance in the classification of AD vs lMCI, AD vs HC, and AD vs eMCI, achieving accuracies of 95.83%, 97.83%, and 97.87%, respectively [45]. The CFAS Data Set was utilised to achieve a minimum accuracy of 75% by employing the SVM-RBF + LR method. Additionally, the specificity was found to be 69% and the sensitivity was determined to be 79% [42]. Figure 10.10 depicts a graphical representation of this analysis.

Table 10.4 Comparison of literature of review

Authors	*Techniques used*	*Outcomes*	*Methodology*	*Accuracy*	*Specificity*	*Sensitivity*
Saeed et al. [9]	Pre-trained CNN deep learning model ResNet50 + (SoftMax/SVM/RF) (Data Set: Alzheimer's Disease Neuroimaging Initiative (ADNI))	Outperforming most state-of-the-art models, the ResNet50 + SoftMax model achieved a significantly higher prediction accuracy of 99%.	Utilizes a pre-trained CNN deep learning model, ResNet50, as an automatic feature extraction method for diagnosing Alzheimer's disease through MRI images, comparing its performance with conventional machine learning techniques such as SoftMax, SVM, and RF, demonstrating the superiority of the proposed model with accuracy ranging from 85.7% to 99% on the MRI ADNI dataset.	99%	98%	99%
Badshah [39]	SVM, KNN, ELM, LDA	Involving feature extraction, selection, and classification using S-K-E-L Classifier, P-S-D-G Classifier, and S-P-S-O Classifier achieved highly favourable results in terms of accuracy and computational efficiency, outperforming various state-of-the-art algorithms in the classification of brain MRI images of tumours.	This research presents a comprehensive methodology for brain tumour classification in MRI images, involving preprocessing, feature extraction, feature selection using statistical measures, and the utilization of combined machine learning classifiers, including a novel S-P-S-O Classifier, which collectively achieves high accuracy levels of up to 99% and outperforms other state-of-the-art algorithms in comparative evaluations.	99%	–	–

(*Continued*)

Table 10.4 (Continued) Comparison of literature of review

Authors	*Techniques used*	*Outcomes*	*Methodology*	*Accuracy*	*Specificity*	*Sensitivity*
Vijithananda [40]	Machine Learning (ML) techniques	This study demonstrates that the extraction of demographic and texture features from MRI apparent diffusion coefficient (ADC) images, coupled with ML techniques, can lead to a high-performance model with 90.41% accuracy for distinguishing malignant from benign brain tumours, offering a non-invasive diagnostic approach to assist in decision-making and potentially reduce the need for invasive procedures like brain biopsies.	In this prospective study, demographic and texture features were extracted from MRI apparent diffusion coefficient (ADC) images of brain tumours, followed by feature selection using ANOVA *f*-test, and subsequently, ML techniques were applied, particularly the Random Forest classifier, to differentiate between malignant and benign brain tumours, achieving a accuracy in the final model.	90.41%	–	–
Rajab et al. [42]	SVM-RBF + LR (Data Set: Cognitive Function and Ageing Study (CFAS))	The classifier for dementia achieved an accuracy of 75% by utilizing the top seven neuropathology characteristics in terms of ranking.	This study utilizes machine learning techniques for feature ranking and classification to objectively compare neuropathological features associated with dementia, assessing their significance using a cohort from the CFAS, and highlights the potential of machine learning in identifying key indices of Alzheimer-related pathologies for dementia classification.	75%	69%	79%

(Continued)

Table 10.4 (Continued) Comparison of literature of review

Authors	*Techniques used*	*Outcomes*	*Methodology*	*Accuracy*	*Specificity*	*Sensitivity*
Rallabandi et al. [43]	NB and SVM	Upon comparing multiple approaches using 5-fold cross-validation, the authors observed that NB and SVM classifiers achieved the highest accuracy of 78%.	A ML classifier is developed to differentiate between healthy controls, individuals with Parkinsonism dementia, Alzheimer's disease, and mild cognitive impairment using T1-weighted MRI images through a process involving 3D tissue segmentation, fractal dimension calculation, grey matter density estimation, local gyrification index computation, anatomical lobar parcellation, and training/testing with various classification algorithms.	78%	77%	77%
Marioni et al. [45]	SVM, KNN, NB and RF	The dataset testing revealed that SVM is highly effective in classifying AD vs IMCI, AD vs HC, and AD vs eMCI, achieving accuracies of 95.83%, 97.83%, and 97.87%, respectively.	The results of the dataset testing indicate that SVM exhibit a high level of effectiveness in accurately categorizing AD vs. late mild cognitive impairment (IMCI), AD vs. HC, and AD vs. eMCI.	97%	98%	95%
Li et al. [47]	VGGNet, SVM and ResNet	For classifying AD participants against old control subjects, SVM classifiers achieved an accuracy of 90%, while VGGNet and ResNet classifiers attained accuracies of 95% each.	In the task of categorizing individuals with AD in comparison to a control group of elderly adults, SVM classifiers demonstrated a classification accuracy of 90%. On the other hand, VGGNet and ResNet classifiers achieved classification accuracies of 95% apiece.	95%	96%	95%

(Continued)

Table 10.4 (Continued) Comparison of literature of review

Authors	*Techniques used*	*Outcomes*	*Methodology*	*Accuracy*	*Specificity*	*Sensitivity*
Herzog et al. [53]	SVM and CNN (Data Set: ADNI)	The evaluated supervised machine learning methods and CNN demonstrate an accuracy of 92.5% and 75.0% in distinguishing between NC and EMCI and an accuracy of 93.0% and 90.5% in distinguishing between NC and AD, respectively.	The study proposes a data processing pipeline utilizing MRI data from the ADNI database to extract brain asymmetry features and employs supervised machine learning algorithms and convolutional neural networks to achieve high accuracy in classifying different stages of dementia (normal cognition vs. early mild cognitive impairment, and normal cognition vs. Alzheimer's Disease), offering a cost-effective approach for dementia classification applicable to other brain degenerative disorders with asymmetry changes.	90%	92%	89%
Thapa et al. [54]	SVM with Radial basis function (RBF) (Data Set: ADNI)	The suggested method, when employed alongside other types of MRI and multimodal approaches, exhibits improvements in accuracy (91.3%), sensitivity (92%), and specificity (90%).	Involves employing feature selection techniques to identify essential diagnostic features by combining neuropsychological scores, such as the mini-mental state examination (MMSE), with MRI-derived features to enhance the decisional space for early Alzheimer's disease diagnosis and significantly improve classification accuracy.	91%	90%	92%

(*Continued*)

Table 10.4 (Continued) Comparison of literature of review

Authors	*Techniques used*	*Outcomes*	*Methodology*	*Accuracy*	*Specificity*	*Sensitivity*
BansaL et al. [55]	BoF + SVM (Dataset :ADNI)	Combining Bag-of-Features (BoF) with SVM presents a potent approach to enhance algorithm efficiency by simultaneously reducing computational costs and increasing throughput.	Employed a SVM classification approach utilizing speeded-up robust features extracted through the Bag-of-Features (BOF) technique to distinguish between demented, mildly cognitively impaired, and normal control categories using magnetic resonance images (MRI) from the ADNI database, yielding a competitive classification accuracy of 93%.	93%	81%	84%
Lin et al. [56]	Integration of ML techniques and multimodal MRI data	Developed a cost-effective method using machine learning and multimodal MRI to predict brain amyloid-β burden, demonstrating a strong correlation between structural and diffusion MRI indices and amyloid burden, thus offering the potential for early Alzheimer's disease diagnosis.	Developed a predictive model using regression machine implementations to estimate brain amyloid-β burden in older individuals by integrating voxel-based morphometry and fractional anisotropy measures from multimodal MRI data, demonstrating significant correlations and identifying key brain regions for prediction.	–	–	–

(*Continued*)

Table 10.4 (Continued) Comparison of literature of review

Authors	*Techniques used*	*Outcomes*	*Methodology*	*Accuracy*	*Specificity*	*Sensitivity*
Ahmed et al. [57]	ML and deep learning technologies for feature extraction and classification methods (Dataset :ADNI, OASIS, AIBL, CADDementia)	This study provided a systematic elucidation of contemporary ML techniques, with a specific focus on deep learning approaches, for the timely identification of dementia.	Comprehensively survey recent advancements in automated diagnostic systems for dementia using medical image analysis and ML algorithms, focusing on neuroimaging procedures, particularly emphasizing the application of deep learning techniques for early detection of various types of dementia beyond Alzheimer's disease.	97.5%	–	–
Zheng et al. [59]	Demonstrate the SVM with RBF	Demonstrates that utilizing multi-parameter features extracted from structural MRI and applying machine learning techniques, specifically a SVM classifier, can achieve promising accuracy, sensitivity, and specificity (84.35%, 82.65%, and 87.17% respectively) in distinguishing between vascular dementia (VaD) and Alzheimer's Disease (AD), showcasing the potential for computer-aided diagnosis using these methods.	This study employs automated brain tissue segmentation and machine learning techniques, including feature selection using LASSO and a SVM classifier, to analyse multi-parameter volumetric measurements extracted from structural MRI for differential diagnosis between vascular dementia (VaD) and Alzheimer's disease (AD), achieving encouraging classification performance with an SVM using radial basis function (RBF) kernel.	84.35%,	87.17%,	82.65%,

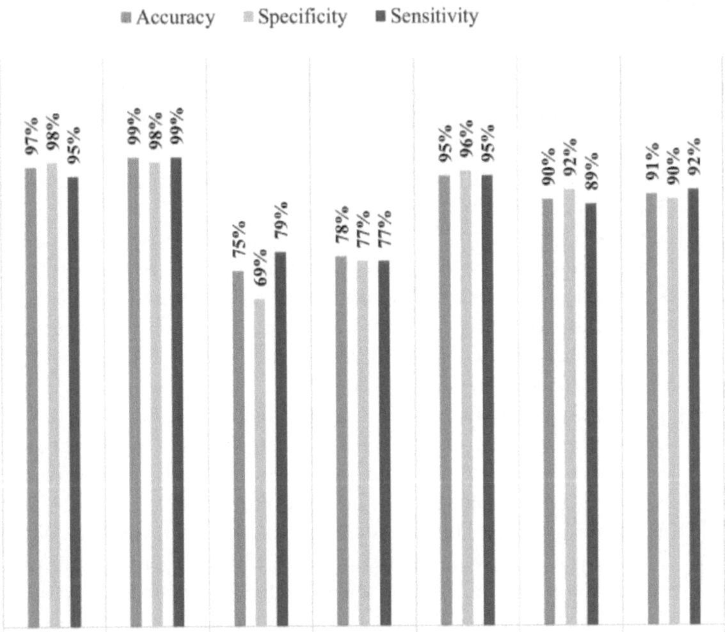

	R. E. Marioni et al., (2023)	A. Saeed et al., (2022)	Rajab et al., (2022)	Rallabandi et al., (2021)	Li et al., (2021)	Herzog et al., (2021)	Thapa et al., (2020)
Accuracy	97%	99%	75%	78%	95%	90%	91%
Specificity	98%	98%	69%	77%	96%	92%	90%
Sensitivity	95%	99%	79%	77%	95%	89%	92%

Figure 10.10 Comparison analysis of approaches.

10.5 CONCLUSION AND FUTURE SCOPE

In this study, the authors examined the influence of the subject sample on classification accuracy and selected features during ML analysis of MRI data. This investigation is of significant importance as ML analysis is increasingly being utilized in brain imaging, and understanding the accuracy and reproducibility of these analyses is crucial. The findings of this study support the use of advanced ML techniques in anatomical neuroimaging while emphasizing the need for caution in interpreting ML results. Comparative analysis of various ML approaches revealed that the SVM (without any pre-trained CNN deep learning model) strategy exhibited the highest levels of accuracy (97%), specificity (98%), and sensitivity (95%). In the work that would be done in the future, authors would be extracting a new collection of traits that are intended to be more sensitive to the early identification of dementia. Second, to improve the effectiveness of dementia detection methods, it is necessary to weed out unnecessary characteristics from previously collected feature sets.

REFERENCES

[1] Gaikar, R., Azad, A., Schieda, N., & Ukwatta, E. (2023). Fully automated cascaded approach for renal mass detection on T2 weighted MRI images. *Medical Imaging 2023: Computer-Aided Diagnosis*, 12465, 2653286. https://doi.org/10.1117/12.2653286

[2] Mendez, M. F. (2017). Early-onset Alzheimer disease. *Neurologic Clinics*, 35(2), 263–281. https://doi.org/10.1016/j.ncl.2017.01.005

[3] Khamseh, F., Hadjati, G., Nahvinejad, H., Nouparast, Z., Salehi, M., Noroozian, M., Foroughan, M., Alaedini, F., Saffarifard, A., Farin, F., Khansari, M., Pourseid Mohammad, M., & Ahmadi, Z. (2023). Comparing the diagnostic value of four dementia tests in the amnestic and healthy elderly. *Basic and Clinical Neuroscience Journal*, 14(1), 137–142. https://doi.org/10.32598/bcn.2022.1745.1

[4] Young, C. B., Johns, E., Kennedy, G., Belloy, M. E., Insel, P. S., Greicius, M. D., Sperling, R. A., Johnson, K. A., Poston, K. L., & Mormino, E. C. (2023). Apoe effects on regional tau in preclinical Alzheimer's disease. *Molecular Neurodegeneration*, 18(1), 4. https://doi.org/10.1186/s13024-022-00590-4

[5] Apostolova, L. G., Dutton, R. A., Dinov, I. D., Hayashi, K. M., Toga, A. W., Cummings, J. L., & Thompson, P. M. (2006). Conversion of mild cognitive impairment to Alzheimer disease predicted by hippocampal atrophy maps. *Archives of Neurology*, 63(5), 693. https://doi.org/10.1001/archneur.63.5.693

[6] Dening, T., & Sandilyan, M. B. (2015). Dementia: Definitions and types. *Nursing Standard*, 29(37), 37–42. https://doi.org/10.7748/ns.29.37.37.e9405

[7] Szablewski, L. (2022). Human gut microbiota in health and Alzheimer's disease. *Advances in Alzheimer's Disease*, 62(2), 549–560. https://doi.org/10.3233/aiad220015

[8] Wu, Y.-T., Beiser, A. S., Breteler, M. M., Fratiglioni, L., Helmer, C., Hendrie, H. C., Honda, H., Ikram, M. A., Langa, K. M., Lobo, A., Matthews, F. E., Ohara, T., Pérès, K., Qiu, C., Seshadri, S., Sjölund, B.-M., Skoog, I., & Brayne, C. (2017). The changing prevalence and incidence of dementia over time - current evidence. *Nature Reviews Neurology*, 13(6), 327–339. https://doi.org/10.1038/nrneurol.2017.63

[9] AlSaeed, D., & Omar, S. F. (2022). Brain MRI analysis for Alzheimer's disease diagnosis using CNN-based feature extraction and machine learning. *Sensors*, 22(8), 2911. https://doi.org/10.3390/s22082911

[10] Ledig, C., Schuh, A., Guerrero, R., Heckemann, R. A., & Rueckert, D. (2018). Structural brain imaging in Alzheimer's disease and mild cognitive impairment: Biomarker analysis and shared morphometry database. *Scientific Reports*, 8(1), 9. https://doi.org/10.1038/s41598-018-29295-9

[11] Shilaskar, S., & Ghatol, A. (2013). Feature selection for medical diagnosis: Evaluation for cardiovascular diseases. *Expert Systems with Applications*, 40(10), 4146–4153. https://doi.org/10.1016/j.eswa.2013.01.032

[12] Verma, A. K., Pal, S., & Kumar, S. (2019). Prediction of skin disease using ensemble data mining techniques and feature selection method-a comparative study. *Applied Biochemistry and Biotechnology*, 190(2), 341–359. https://doi.org/10.1007/s12010-019-03093-z

[13] Castellazzi, G., Cuzzoni, M. G., Cotta Ramusino, M., Martinelli, D., Denaro, F., Ricciardi, A., Vitali, P., Anzalone, N., Bernini, S., Palesi, F., Sinforiani, E., Costa, A., Micieli, G., D'Angelo, E., Magenes, G., & Gandini Wheeler-Kingshott, C. A. (2020). A machine learning approach for the differential diagnosis of Alzheimer and vascular dementia fed by MRI selected features. *Frontiers in Neuroinformatics*, 14, 25. https://doi.org/10.3389/fninf.2020.00025

[14] Khundrakpam, B. S., Tohka, J., & Evans, A. C. (2015). Prediction of brain maturity based on cortical thickness at different spatial resolutions. *NeuroImage*, 111, 350–359. https://doi.org/10.1016/j.neuroimage.2015.02.046

[15] Mohr, H., Wolfensteller, U., Frimmel, S., & Ruge, H. (2015). Sparse regularization techniques provide novel insights into outcome integration processes. *NeuroImage*, 104, 163–176. https://doi.org/10.1016/j.neuroimage.2014.10.025

[16] Dexte, M. (2017). Brain atrophy rates in normal aging and Alzheimer disease. *Neuroscience and Neurological Surgery*, 1(1), 01–02. https://doi.org/10.31579/2578-8868/009

[17] Sandhya, A., & Kannayiram, G. (2018). Alzheimer's disease therapeutic approaches. *Asian Journal of Pharmaceutical and Clinical Research*, 11(7), 17. https://doi.org/10.22159/ajpcr.2018.v11i7.25104

[18] Devanand, D. P., Michaels-Marston, K. S., Liu, X., Pelton, G. H., Padilla, M., Marder, K., Bell, K., Stern, Y., & Mayeux, R. (2000). Olfactory deficits in patients with mild cognitive impairment predict Alzheimer's disease at follow-up. *American Journal of Psychiatry*, 157(9), 1399–1405. https://doi.org/10.1176/appi.ajp.157.9.1399

[19] Armstrong, R. (2019). Risk factors for Alzheimer's disease. *Folia Neuropathologica*, 57(2), 87–105. https://doi.org/10.5114/fn.2019.85929

[20] Anand, P., & Singh, B. (2013). A review on cholinesterase inhibitors for Alzheimer's disease. *Archives of Pharmacal Research*, 36(4), 375–399. https://doi.org/10.1007/s12272-013-0036-3

[21] Breijyeh, Z., & Karaman, R. (2020). Comprehensive review on Alzheimer's disease: Causes and treatment. *Molecules*, 25(24), 5789. https://doi.org/10.3390/molecules25245789

[22] Tripathi, S., & Mazumder, P. M. (2020). Apple cider vinegar (ACV) and their pharmacological approach towards Alzheimer's disease (AD): A review. *Indian Journal of Pharmaceutical Education and Research*, 54(2s), 62. https://doi.org/10.5530/ijper.54.2s.62

[23] Sidey-Gibbons, J. A., & Sidey-Gibbons, C. J. (2019). Machine learning in medicine: A practical introduction. *BMC Medical Research Methodology*, 19(1), 4. https://doi.org/10.1186/s12874-019-0681-4

[24] Zhao, Q., Liu, Z., Adeli, E., & Pohl, K. M. (2021). Longitudinal self-supervised learning. *Medical Image Analysis*, 71, 102051. https://doi.org/10.1016/j.media.2021.102051

[25] Argentiero, A., Muscogiuri, G., Rabbat, M. G., Martini, C., Soldato, N., Basile, P., Baggiano, A., Mushtaq, S., Fusini, L., Mancini, M. E., Gaibazzi, N., Santobuono, V. E., Sironi, S., Pontone, G., & Guaricci, A. I. (2022). The

applications of artificial intelligence in cardiovascular magnetic resonance-A comprehensive review. *Journal of Clinical Medicine*, 11(10), 2866. https://doi.org/10.3390/jcm11102866

[26] Ray, S. (2019). A quick review of machine learning algorithms. *2019 International Conference on Machine Learning, Big Data, Cloud and Parallel Computing (COMITCon)*, Faridabad, India. https://doi.org/10.1109/comitcon.2019.8862451

[27] Chetna, D., & Shailja, S. (2023). A study on aspect of artificial neural networks for machine learning. *International Journal of Advanced Research in Science, Communication and Technology*, 2023, 270–274. https://doi.org/10.48175/ijarsct-8594

[28] Shruti, B. (2014). A survey: Research summary on neural networks. *International Journal of Research in Engineering and Technology*, 03(15), 385–389. https://doi.org/10.15623/ijret.2014.0315076

[29] Shameer, K., Johnson, K. W., Glicksberg, B. S., Dudley, J. T., & Sengupta, P. P. (2018). Machine learning in cardiovascular medicine: Are we there yet? *Heart*, 104(14), 1156–1164. https://doi.org/10.1136/heartjnl-2017-311198

[30] Johnson, K. W., Torres Soto, J., Glicksberg, B. S., Shameer, K., Miotto, R., Ali, M., Ashley, E., & Dudley, J. T. (2018). Artificial intelligence in cardiology. *Journal of the American College of Cardiology*, 71(23), 2668–2679. https://doi.org/10.1016/j.jacc.2018.03.521

[31] Mothkur, R., & Poornima, K. M. (2018). Machine learning will transfigure medical sector: A survey. *2018 International Conference on Current Trends towards Converging Technologies (ICCTCT)*, Coimbatore, India. https://doi.org/10.1109/icctct.2018.8551134

[32] Singh, V., & Kaushik, V. D. (2022). *Concepts of data mining and process mining. In Process Mining Techniques for Pattern Recognition.* Yadav, V., Dubey, A.K., Singh, H.P., Dubey, G., & Suryani, E. (pp. 1–17). CRC Press. https://doi.org/10.1201/9781003169550-1

[33] Sarker, I. H., Kayes, A. S., Badsha, S., Alqahtani, H., Watters, P., & Ng, A. (2020). cybersecurity data science: An overview from machine learning perspective. *Journal of Big Data*, 7(1), 5. https://doi.org/10.1186/s40537-020-00318-5

[34] Khoshaba, F., Kareem, S., Awla, H., & Mohammed, C. (2022). Machine learning algorithms in bigdata analysis and its applications: A review. *2022 International Congress on Human-Computer Interaction, Optimization and Robotic Applications (HORA)*, Ankara, Turkey. https://doi.org/10.1109/hora55278.2022.9799848

[35] Song, X., Liu, X., Liu, F., & Wang, C. (2021). Comparison of machine learning and logistic regression models in predicting acute kidney injury: A systematic review and meta-analysis. *International Journal of Medical Informatics*, 151, 104484. https://doi.org/10.1016/j.ijmedinf.2021.104484

[36] Habehh, H., & Gohel, S. (2021). Machine learning in healthcare. *Current Genomics*, 22(4), 291–300. https://doi.org/10.2174/1389202922666210705124359

[37] Waring, J., Lindvall, C., & Umeton, R. (2020). Automated machine learning: Review of the state-of-the-art and opportunities for healthcare. *Artificial Intelligence in Medicine*, 104, 101822. https://doi.org/10.1016/j.artmed.2020.101822

[38] Battleday, R. M., Peterson, J. C., & Griffiths, T. L. (2021). From convolutional neural networks to models of higher-level cognition (and back again). *Annals of the New York Academy of Sciences*, 1505(1), 55–78. https://doi.org/10.1111/nyas.14593

[39] Badshah, S. H., Farhatullah, G. Z., Hassan, M. A., Junaid, H., Sohail, M., Mahbob, M. A., Ahamad, I., & Ullah, N. (2023). An efficient classification techniques for brain tumor using features extraction and statistic methods, with machine learning algorithms. *Lecture Notes on Data Engineering and Communications Technologies*, 757–775. Springer, Berlin. https://doi.org/10.1007/978-3-031-24475-9_62

[40] Vijithananda, S. M., Jayatilake, M. L., Hewavithana, B., Gonçalves, T., Rato, L. M., Weerakoon, B. S., Kalupahana, T. D., Silva, A. D., & Dissanayake, K. D. (2022). Feature extraction from MRI ADC images for brain tumor classification using machine learning techniques. *BioMedical Engineering OnLine*, 21(1), 6. https://doi.org/10.1186/s12938-022-01022-6

[41] Bueff, A., & Belle, V. (2023). Logic+reinforcement learning+deep learning: A survey. *Proceedings of the 15th International Conference on Agents and Artificial Intelligence*, Lisbon, Portugal. https://doi.org/10.5220/0011746300003393

[42] Rajab, M. D., Jammeh, E., Taketa, T., Brayne, C., Matthews, F. E., Su, L., Ince, P. G., Wharton, S. B., & Wang, D. (2022). Assessment of Alzheimer-related pathologies of dementia using machine learning feature selection. *Alzheimer's Research & Therapy*, 15(1), 47. https://doi.org/10.1101/2022.04.28.22274107

[43] Subramanyam Rallabandi, V. P., & Seetharaman, K. (2021). Machine learning-based classification of dementia types: MRI Study. *2021 International Conference on Artificial Intelligence and Smart Systems (ICAIS)*, Coimbatore, India. https://doi.org/10.1109/icais50930.2021.9395957

[44] The Medical Research Council Cognitive Function and Ageing Study (MRC CFAS). (1998). Cognitive function and dementia in six areas of England and Wales: The distribution of MMSE and prevalence of GMS organicity level in the MRC CFA study. *Psychological Medicine*, 28(2), 319–335. https://doi.org/10.1017/s0033291797006272

[45] Marioni, R. E., Matthews, F. E., & Brayne, C. (2010). The association between late-life cognitive test scores and retrospective informant interview data. *International Psychogeriatrics*, 23(2), 274–279. https://doi.org/10.1017/s1041610210001201

[46] Bron, E. E., et al. (2021). Cross-cohort generalizability of deep and conventional machine learning for MRI-based diagnosis and prediction of Alzheimer's disease. *NeuroImage: Clinical*, 31, 102712. https://doi.org/10.1016/j.nicl.2021.102712

[47] Li, Q., & Yang, M. Q. (2021). Comparison of machine learning approaches for enhancing Alzheimer's disease classification. *PeerJ*, 9, 10549. https://doi.org/10.7717/peerj.10549

[48] Prakash, S. S., Harshitha, K., Charitha, A. O., Janvitha, C., & Indu, K. (2023). Computer-aided diagnosis using machine learning techniques. *2023 3rd International Conference on Smart Data Intelligence (ICSMDI)*, Trichy, India. https://doi.org/10.1109/icsmdi57622.2023.00075

[49] Ramineni, V., & Kwon, G.-R. (2023a). Diagnosis of alzheimer's disease using wrapper feature selection method. *Korean Institute of Smart Media*, 12(3), 30–37. https://doi.org/10.30693/smj.2023.12.3.30

[50] Alkabawi, E. M., Hilal, A. R., & Basir, O. A. (2017). Feature abstraction for early detection of multi-type of dementia with sparse auto-encoder. *2017 IEEE International Conference on Systems, Man, and Cybernetics (SMC)*, Banff, AB, Canada. https://doi.org/10.1109/smc.2017.8123168

[51] Bándi, P., Balkenhol, M., van Dijk, M., Kok, M., van Ginneken, B., van der Laak, J., & Litjens, G. (2023). Continual learning strategies for cancer-independent detection of lymph node metastases. *Medical Image Analysis*, 85, 102755. https://doi.org/10.1016/j.media.2023.102755

[52] Rodrigues, J. F., Brandoli, B., & Amer-Yahia, S. (2020). DermaDL: Advanced convolutional neural networks for automated melanoma detection. *2020 IEEE 33rd International Symposium on Computer-Based Medical Systems (CBMS)*. Virtual Event. https://doi.org/10.1109/cbms49503.2020.00101

[53] Herzog, N. J., & Magoulas, G. D. (2021). Brain asymmetry detection and machine learning classification for diagnosis of early dementia. *Sensors*, 21(3), 778. https://doi.org/10.3390/s21030778

[54] Thapa, S., Singh, P., Jain, D. K., Bharill, N., Gupta, A., & Prasad, M. (2020). Data-driven approach based on feature selection technique for early diagnosis of Alzheimer's disease. *2020 International Joint Conference on Neural Networks (IJCNN)*. Virtual Event. .https://doi.org/10.1109/ijcnn48605.2020.9207359

[55] Bansal, D., Khanna, K., Chhikara, R., Dua, R. K., & Malhotra, R. (2020). Classification of magnetic resonance Images using bag of features for detecting dementia. *Procedia Computer Science*, 167, 131–137. https://doi.org/10.1016/j.procs.2020.03.190

[56] Lin, L., Wang, X., Wang, X., Li, X., Sun, Y., Du, W., Chen, G., & Han, Y. (2020). Machine learning regression analysis predicts brain amyloid-β burden: A multimodal neuroimaging research study. *Alzheimer's & Dementia*, 16(S5), 041799. https://doi.org/10.1002/alz.041799

[57] Ahmed, M. R., Zhang, Y., Feng, Z., Lo, B., Inan, O. T., & Liao, H. (2019). Neuroimaging and machine learning for dementia diagnosis: Recent advancements and future prospects. *IEEE Reviews in Biomedical Engineering*, 12, 19–33. https://doi.org/10.1109/rbme.2018.2886237

[58] Shafi, A. S., Hasan, Md. M., Molla, M. M., Alam, M. K., & Islam, Md. T. (2022). Automatic brain tumor detection using feature selection and machine learning from MRI images. *Lecture Notes in Electrical Engineering*, Pahang, Malaysia, 751–762. https://doi.org/10.1007/978-981-16-8690-0_66

[59] Zheng, Y., Guo, H., Zhang, L., Wu, J., Li, Q., & Lv, F. (2019). Machine learning-based framework for differential diagnosis between vascular dementia and Alzheimer's disease using structural MRI features. *Frontiers in Neurology*, 10, 1097. https://doi.org/10.3389/fneur.2019.01097

Chapter 11

Study of bio-nanofluid flow via vertical surface

A machine learning approach

Priyanka Chandra, Raja Das, Balakrishnama Manohar, and K. Madhusudhan Reddy

11.1 INTRODUCTION

The rate of heat transfer characteristics has attracted a growing interest from many scientists due to their huge industrial features, such as in mechanical, electrical, optical, and cooling instruments. The rate of heat transmission plays an essential role in energy depositing. As a result, researchers are currently concentrating on the study of nanofluids, which are a blend of nanoparticles with a size of 100 nm and higher thermophysical properties than usual fluids. The term "nanofluid" was first coined by Choi & Eastman [1], Buongiorno [2] later used these nanofluids to create a mathematical structure by incorporating Brownian motion and thermophoretic variables. Recently, Prabakaran et al. [3] investigated the mixed convective flow of CNTs/Al_2O_3 nanofluid in water via a heated elastic plate. Aziz et al. [4] investigated the Lorentz and Coriolis forces in nanoparticles. More nanofluid advances can be found in [5–7].

The investigation of bioconvection problems receives a lot of attention since it has applications in biosensors, the pharmaceutical sector, microbially improved oil recovery, biotechnology, and medical sciences. Bioconvection occurs due to the upward motion of microorganisms that are denser than water. Self-propulsion distinguishes microorganisms from passive particles such as solutes or sediments. This self-propulsion frequently leads to complex dynamic behaviour and distribution patterns. Kuznetsov and Avramenko [8] demonstrated the bioconvection of nanoparticles in the presence of gyrotactic bacteria over a convectively heated horizontal plate. Uddin et al. [9] used a well-known numerical methodology known as the Chebyshev collocation method to investigate the rheological properties of Newtonian fluid in the presence of microorganism species. Some of the important contributions to bioconvection are provided in references [10–13]. Recently, Elbashbeshy & Asker [14] used the Runge-Kutta method to explain the gyrotactic microbes effect on nanofluid across a stretched plate having variable fluid parameters.

Although an accurate analytical solution to a problem may not always be attainable, the scientist employs a variety of numerical and semi-numerical

DOI: 10.1201/9781003534761-11

approaches to solve the problem, such as the Keller-Box method [15] (the spectral relaxation method [16]), the homotopic perturbation method [17], and many others [17–19]. All the above-mentioned numerical methods for solving diverse problems have advantages and drawbacks. As a result, a stochastic computer framework is developed to solve complex nonlinear problems. The improvement of stochastic solvers helps to obtain solutions to very complex real-life problems in terms of ODEs or PDEs. Stochastic computing solvers possess a wide range of applications, including finance problems [20], plasma physics [21], Emden-Fowler equations [22], fluid dynamics, nuclear physics [23], biological investigations [23,24,25]. Many academics have recently implemented the stochastic numerical approach for fluid flow systems [13, 26–30]. Several techniques based on artificial intelligence are employed by researchers [31–34]. These computational tasks are carried out using the MATLAB® and Mathematica infrastructures.

In this chapter, a unique application of the novel hybrid machine learning algorithm designed by using the finite difference method (FDM) and the backpropagated Levenberg-Marquardt algorithm (LMNN) is presented. Gyrotactic microorganisms have become increasingly common in biomedical and industrial applications, according to the literature review on bioconventional fluid flow that was mentioned above. In response to this development, a study was written on the magnetohydrodynamic (MHD) nanofluid flow across a vertically inclined stretching surface in a porous media with gyrotactic microorganisms. The principal objective is to provide results as a function of input using ANN, in contrast to numerical methods, which only provide results at discrete points. PDEs are used to express the flow problem's mathematical foundation. The Lie symmetric transformations are then used to transform the system of PDEs into ODEs. The Levenberg–Marquardt algorithm with backpropagation in a neural network (LMNN) is used to solve the proposed flow problem. To create a reference dataset for the proposed LMNN, the FDM is applied by varying four significant parameters of the proposed geometry: Lewis number, Brownian motion, Thermophoresis parameter, and bioconventional Lewis number. Each of these fluid flow situations uses the testing, training, and validation techniques created with LMNN with the goal to determine an approximate solution and evaluate it against benchmark outcomes. Convergence graphs of the mean squared error-based merit function, state transition, regression statistics, and histogram error are used to determine the performance of the proposed LMNN.

11.2 MATHEMATICAL FORMULATION

The study investigates the mixed convection boundary layer flow model of a water-based nanofluid flowing across an inclined vertical stretching surface embedded in porous media with gyrotactic microorganisms.

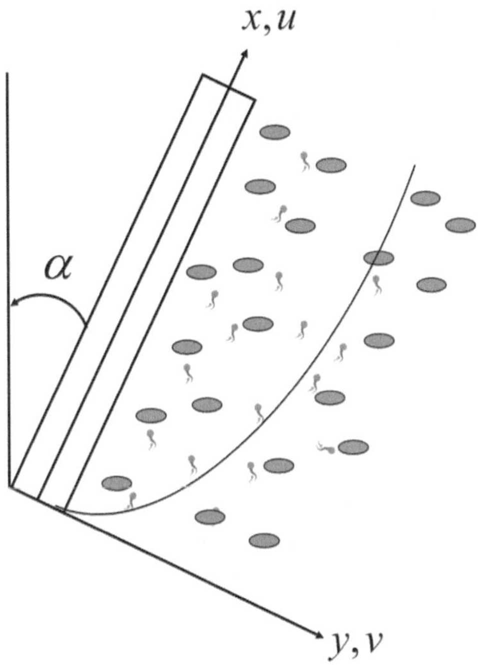

Figure 11.1 Flow geometry.

The problem is modelled using Cartesian coordinates, as shown in Figure 11.1. The system is chosen so that the *Y*-axis measures the distance perpendicular to the plate and the *X*-axis measures the entire length of the plate. With the velocity of $U_w = qx^n$ the inclined flat surface, it is vertically stretched forward in the *X* direction, where q and n are arbitrary constants. The flow model is represented by the following set of equations [14] (Elbashbeshy & Asker, 2022):

$$\frac{\partial u}{\partial x}+\frac{\partial v}{\partial y}=0 \tag{11.1}$$

$$\rho_{f\infty}\left(u\frac{\partial u}{\partial x}+v\frac{\partial v}{\partial y}\right)-\frac{\mu\cdot\partial^2 u}{\partial y^2}+\frac{\mu\cdot u}{K}$$

$$=\left[(1-C_\infty)\rho_{f\infty}\beta_f(T-T_\infty)-(\rho_p-\rho_{f\infty})g^*(C-C_\infty)\right. \tag{11.2}$$

$$\left.-(\rho_{n\infty}-\rho_{f\infty})g^*\gamma(D-D_\infty)\right]\cos(\alpha)$$

$$u\frac{\partial T}{\partial x}+v\frac{\partial T}{\partial y}=D_B\frac{\partial T}{\partial y}\frac{\partial C}{\partial y}\tau+\frac{v}{\mu}\frac{\partial^2 T}{\partial y^2}+\frac{D_T}{T_\infty}\left(\frac{\partial T}{\partial y}\right)^2\tau \tag{11.3}$$

$$u\frac{\partial C}{\partial x}+v\frac{\partial C}{\partial y}=\frac{D_T}{T_\infty}\frac{\partial^2 T}{\partial y^2}+D_B\frac{\partial^2 C}{\partial y^2} \tag{11.4}$$

$$u\frac{\partial D}{\partial x}+v\frac{\partial D}{\partial y}+\frac{b_c W_C}{(C_w-C_\infty)}\frac{\partial}{\partial y}\left(D\frac{\partial C}{\partial y}\right)=D_n\frac{\partial^2 D}{\partial y^2} \tag{11.5}$$

along with boundary conditions:

$$\begin{aligned}&\text{at } y=0: u=U_w=qx^n,\quad v=0,\quad T=T_w,\quad C=C_w,\quad D=D_w\\&\text{as } y\to\infty: u\to 0,\quad v\to 0,\quad T\to T_\infty,\quad C\to C_\infty,\quad D\to D_\infty\end{aligned} \tag{11.6}$$

Each of the above five field equations, Eqs. (11.1)–(11.5), corresponds to the conservation of total mass, momentum, thermal energy, nanoparticle concentration, and microbial density, respectively, where u and v stand for the velocity's x and y components, respectively. The letters T, C, and D stand for the fluid's temperature, concentration of nanoparticles, and density of motile microorganisms, respectively. g^* denotes gravitational acceleration. K denotes the permeability parameter. n denotes the index power, β_f denotes fluid's volume expansion coefficient, μ denotes the kinematic viscosity, γ shows the microbe's average volume, τ denotes the effective heat capacitance, b_c represents the chemotaxis constant, W_c is the constant maximum swimming velocity of the cell, D_n denotes the diffusivity of microbes, D_B and D_T represents the Brownian diffusion coefficient and the thermophoresis coefficient, respectively.

The stream function $\psi(x;y)$ is assumed in such a way that it satisfies the continuity Eq. (11.1).

$$u=\frac{\partial\psi}{\partial y},\quad v=-\frac{\partial\psi}{\partial x} \tag{11.7}$$

To render the equations nondimensional and allow for a numerical solution, the following transformation is employed:

$$\begin{aligned}&\eta=\sqrt{\frac{U_w}{vx}\frac{n+1}{2}}y,\psi=q\sqrt{\frac{2v}{n+1}}U_w x f(\eta),\theta(\eta)=\frac{T-T_\infty}{T_w-T_\infty},\\&\phi(\eta)=\frac{C-C_\infty}{C_w-C_\infty},\chi(\eta)=\frac{D-D_\infty}{D_w-D_\infty}\end{aligned} \tag{11.8}$$

η is employed to simplify the model and $f(\eta)$ is dimensionless function. $\theta(\eta)$, $\phi(\eta)$ *and* $\chi(\eta)$ represents the dimensionless temperature, concentration, and microorganism density, respectively. The Lie symmetric transformations are applied to governing PDEs in Eqs. (11.1)–(11.6), to transform them into nonlinear ODEs using Eq. (11.8) in Eqs. (11.2)–(11.5) we get:

$$f'''+ff''-\left(\frac{2n}{n+1}\right)f'^2+\frac{2}{n+1}[\theta-R_b\chi-Nr\phi]Ri\cos(\alpha)-\lambda f'=0 \quad (11.9)$$

$$\theta''+\Pr\left[f\theta'+Nb\phi'\theta'+Nt(\theta')^2\right]=0 \quad (11.10)$$

$$\phi''+\frac{Nt}{Nb}\theta''+Lef\varphi'=0 \quad (11.11)$$

$$\chi''+\Pr L_bf\chi'-\mathrm{Pe}\phi'[\chi+\chi']-\mathrm{Pe}\sigma\phi'=0 \quad (11.12)$$

with boundary conditions:

$$\begin{gathered}\eta=0:f=0;\quad f'=1;\quad \theta=1;\quad \phi=1;\quad \chi=1;\\ \eta\to\infty:f'\to 0;\quad \theta\to 0;\quad \phi\to 0;\quad \chi\to 0\end{gathered} \quad (11.13)$$

where R_b and Ri are the bioconvection Rayleigh number and Richardson number, respectively. λ is the porosity parameter, Pr *is* the Prandtl number, Pe denotes the Peclet number, σ represents the motile parameter, Nb, Nt, and Nr represent the Brownian motion parameter, the thermophoresis parameter, and the buoyancy ratio parameter, respectively. L_b and Le represent the bioconvection Lewis number and the Lewis number, respectively. The above parameters are defined as:

$$R_b=\frac{(\rho_{m\infty}-\rho_\infty)\gamma\Delta D}{(1-C_\infty)\rho_\infty\beta_f\Delta T},\quad Ri=\frac{Gr_x}{\mathrm{Re}_x^{\ 2}}=\frac{g^*\beta_f l^2\Delta T}{U_0^2x},$$

$$\mathrm{Nb}=\frac{\tau D_B\Delta C}{vT_\infty},\quad \sigma=\frac{D_\infty}{D_w-D_\infty},$$

$$\mathrm{Nr}=\frac{(\rho_p-\rho_\infty)\Delta C}{(1-C_\infty)\rho_\infty\beta_f\Delta T},\quad \lambda=\frac{v}{Kqx^{n-1}},\quad \mathrm{Nt}=\frac{\tau D_T\Delta T}{vT_\infty},$$

$$\Pr=\frac{v}{\alpha},\quad Le=\frac{v}{D_B\Delta C},\quad L_b=\frac{D_B}{D_n},$$

The coefficient of skin friction C_f, the local Nusselt number Nu_x, the local Sherwood number Sh_x, and the motile microbe density number Dn_x are defined as

$$C_f={}^{2\tau_w}\!\Big/\!\left(\rho U_w^2\right),\quad Nu_x={}^{xq_w}\!\Big/\!(k\Delta T),$$

$$Sh_x={}^{xq_m}\!\Big/\!(D_B\Delta C),\quad Dn_x={}^{xq_n}\!\Big/\!(D_n\Delta M) \quad (11.14)$$

where q_w represents the heat flux, q_m denotes the mass flux, q_n is the motile microorganisms surface flux respectively and τ_w represents the shear stress and are given by

$$q_w = -k\left(\frac{\partial T}{\partial y}\right)_{y=q},\ q_m = -D_B\left(\frac{\partial C}{\partial y}\right)_{y=q},\ q_n = -D_n\left(\frac{\partial D}{\partial y}\right)_{y=q},\ \tau_w = -\mu\left(\frac{\partial u}{\partial y}\right)_{y=q}$$

Eq. (11.14) is changed by the similarity transformation into the following form:

$$\frac{C_f\sqrt{\mathrm{Re}_x}}{2} = -f''(0),\quad \frac{Nu_x}{\sqrt{\mathrm{Re}_x}} = -\theta'(0),\quad \frac{Sh_x}{\sqrt{\mathrm{Re}_x}} = -\varphi'(0),\quad \frac{Dn_x}{\sqrt{\mathrm{Re}_x}} = -\chi'(0) \tag{11.15}$$

where $Re_x = {U_0 x^2}/{vl}$ denotes the local Reynolds number.

11.3 SOLUTION METHODOLOGY

The solution process comprises two steps. The modified ODE system specified in Eqs. (11.9)–(11.14) is first solved using the FDM by altering the Lewis number (Le), Brownian motion (Nb), thermophoresis parameter (Nt), and Bioconvection Lewis number (L_b). This produces the reference data set for LMNN. Later, using the MATLAB® code, the proposed backpropagated Levenberg-Marquardt scheme (LMNN) based on machine learning is implemented. The entire domain is divided into 20 elements to provide 21 discrete solutions, which are then used by Levenberg-Marquardt backpropagation with neural networks to produce the complete functions of inputs. The suggested LMNN structure for the fluid flow model is shown in Figure 11.2. It shows the neural network's architectural arrangement, with one input η and four outputs: velocity profile $f'(\eta)$, temperature profile $\theta(\eta)$, concentration of nanoparticle profile $\phi(\eta)$, and microbe density profile $\chi(\eta)$. To implement the hidden layer concept, LMNN uses the activation function and appropriate weight modifications. The stated LMNN efficacy is validated using regression analysis, histogram investigations, and examination of the outcome using absolute errors analysis (AEA).

Regression analysis, histogram investigations, and examination of the outcome with AEA are all used to validate the indicated LMNN effectiveness. The reference data set is divided into three distinct subsets (80%), testing (10%), and validation data (10%) in order to execute the intended LMNN. The complete work flow of the study is shown in Figure 11.3.

11.3.1 Finite difference method

The FDM is one of the numerical techniques used to deal with boundary value and initial value issues in ODEs and PDEs. In the finite difference

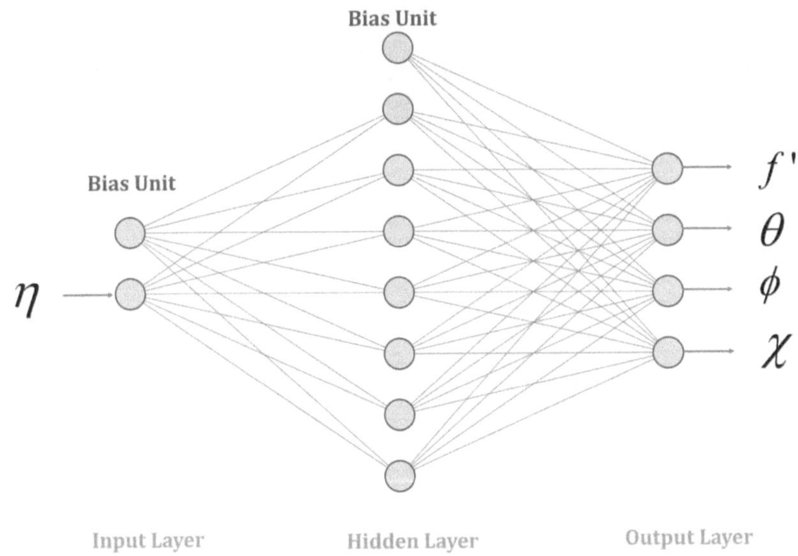

Figure 11.2 Structural layout of LMNN for proposed flow field.

technique, the derivatives of a parameter are expressed as the variations between the values of the variable at different places in space or time. This is done by using the Taylor Series expansion. The recent research that uses FDM to study the flow field (FDM) is shown in [29,30]. The finite difference numerical solutions give us values at discrete points in the domain known as grid points. To solve the set of coupled third-order partial differential equations depicted in Eqs. (11.9)–(11.13), we began by assuming:

$$f' = G \tag{11.16}$$

Then Eqs. (11.9)–(11.13) are converted into

$$G'' + fG' - \left(\frac{2n}{n+1}\right)G^2 + \frac{2}{n+1}\left[\theta - R_b\chi - Nr\phi\right]Ri\cos(\alpha) - \lambda G = 0 \tag{11.17}$$

$$\theta'' + \Pr\left[f\theta' + Nb\phi'\theta' + Nt(\theta')^2\right] = 0 \tag{11.18}$$

$$\phi'' + \frac{Nt}{Nb}\theta'' + Lef\varphi' = 0 \tag{11.19}$$

$$\chi'' + \Pr L_b f\chi' - \mathrm{Pe}\phi'[\chi + \chi'] - \mathrm{Pe}\sigma\phi' = 0 \tag{11.20}$$

Having B.C.

$$\begin{gathered} f(0) = 0, \quad G(0) = 1, \quad \theta(0) = 1, \quad \phi(0) = 1, \quad \chi(0) = 1, \\ \eta \to \infty; \quad G \to 0, \quad \theta \to 0, \quad \phi \to 0, \quad \chi \to 0 \end{gathered} \tag{11.21}$$

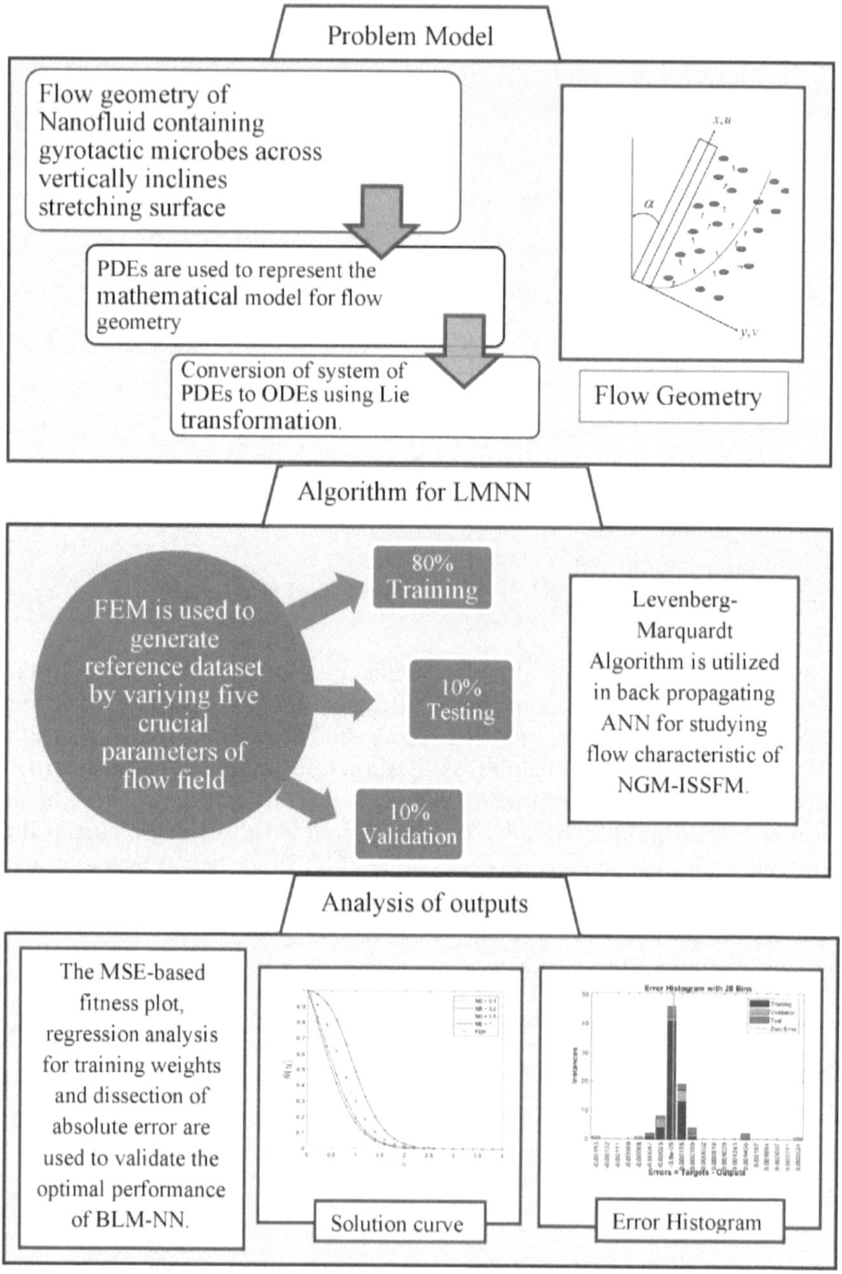

Figure 11.3 Complete work flow of the study.

Now, applying the central-difference formula to a set of Eqs. (11.16) through (11.20), we get:

$$f_2 = f_1 + h_e G_1, \quad f_{i+1} = f_{i-1} + 2h_e G_i, \quad i \geq 2 \tag{11.22}$$

$$\frac{G_{i-1} - 2G_i + G_{i+1}}{h_e^2} + f_i \frac{G_{i+1} - G_{i-1}}{2h_e} - \left(\frac{2n}{n+1}\right) G_i^2$$
$$-\lambda G_i + \frac{2}{n+1} Ri\left[\theta_i - Nr\phi_i - R_b \chi_i\right]\cos(\alpha) = 0 \tag{11.23}$$

$$\frac{\theta_{i-1} - 2\theta_i + \theta_{i+1}}{h_e^2} + \Pr f_i \frac{\theta_{i+1} - \theta_{i-1}}{2h_e}$$
$$+\Pr Nb\left(\frac{\phi_{i+1} - \phi_{i-1}}{2h_e}\right)\left(\frac{\theta_{i+1} - \theta_{i-1}}{2h_e}\right) + \Pr Nt\left(\frac{\theta_{i+1} - \theta_{i-1}}{2h_e}\right)^2 = 0 \tag{11.24}$$

$$\frac{\phi_{i-1} - 2\phi_i + \phi_{i+1}}{h_e^2} + \left(\frac{Nt}{Nb}\right)\left(\frac{\theta_{i-1} - 2\theta_i + \theta_{i+1}}{h_e^2}\right) + Lef_i\left(\frac{\phi_{i+1} - \phi_{i-1}}{2h_e}\right) = 0 \tag{11.25}$$

$$\frac{\chi_{i-1} - 2\chi_i + \chi_{i+1}}{h_e^2} + L_b \Pr f_i\left(\frac{\chi_{i+1} - \chi_{i-1}}{2h_e}\right) - \mathrm{Pe}\left(\frac{\phi_{i+1} - \phi_{i-1}}{2h_e}\right)$$
$$\left[\chi_i + \left(\frac{\chi_{i+1} - \chi_{i-1}}{2h_e}\right)\right] - \sigma \mathrm{Pe}\left(\frac{\phi_{i+1} - \phi_{i-1}}{2h_e}\right) = 0 \tag{11.26}$$

h_e is the step width here. The above-mentioned equations are interconnected and nonlinear, making it impossible to answer them precisely. An iterative approach must be used as a result. Equations being recorded in the following format:

$$x_i = F(K_1, K_2, \ldots\ldots, K_n) \tag{11.27}$$

Each K_i in this instance represents a function of f_i, g_i, θ_i, ϕ_i, χ_i, and x_i can be any of them. Updated iterative results are computed starting from the initial guess values. Repeat this process until the required accuracy is achieved while preserving the lesser absolute error $|x_i - x_{i-1}|$. implying Eq. (11.27) in Eqs. (11.22)–(11.26), we get:

$$f_2 = f_1 + h_e G_1, \quad f_{i+1} = f_{i-1} + 2h_e G_i, \quad i \geq 2 \tag{11.28}$$

$$G_i = \frac{G_{i-1} + G_{i+1}}{2} + f_i h_e \frac{G_{i+1} - G_{i-1}}{4} - \left(\frac{2n}{n+1}\right) h_e^2 G_i^2$$
$$- \frac{h_e^2}{2} \lambda G_i + \frac{h_e^2}{n+1} Ri\left[\theta_i - Nr\phi_i - R_b \chi_i\right] \cos(\alpha) = 0 \tag{11.29}$$

$$\theta_i = \frac{\theta_{i-1} + \theta_{i+1}}{2} + \Pr h_e f_i \frac{\theta_{i+1} - \theta_{i-1}}{4} + \Pr Nb\left(\frac{\phi_{i+1} - \phi_{i-1}}{4}\right)\left(\frac{\theta_{i+1} - \theta_{i-1}}{4}\right)$$
$$+ \frac{\Pr Nt}{2}\left(\frac{\theta_{i+1} - \theta_{i-1}}{2}\right)^2 = 0 \tag{11.30}$$

$$\phi_i = \frac{\phi_{i-1} + \phi_{i+1}}{2} + \left(\frac{Nt}{Nb}\right)\left(\frac{\theta_{i-1} - 2\theta_i + \theta_{i+1}}{h_e^2}\right) + Lef_i h_e\left(\frac{\phi_{i+1} - \phi_{i-1}}{4}\right) = 0 \tag{11.31}$$

$$\chi_i = \frac{\chi_{i-1} + \chi_{i+1}}{2} + L_b \Pr f_i h_e\left(\frac{\chi_{i+1} - \chi_{i-1}}{4}\right) - \text{Pe}\left(\frac{\phi_{i+1} - \phi_{i-1}}{4}\right)$$
$$\left[h_e \chi_i + \left(\frac{\chi_{i+1} - \chi_{i-1}}{2}\right)\right] - \sigma \text{Pe} h_e\left(\frac{\phi_{i+1} - \phi_{i-1}}{4}\right) = 0 \tag{11.32}$$

With boundary conditions:

$$f_1 = 0, \quad G_1 = 0, \quad \theta_1 = 0, \quad \phi_1 = 0, \quad \chi_1 = 0$$
$$G_{21} = 1, \quad \theta_{21} = 1, \quad \phi_{21} = 1, \quad \chi_{21} = 1 \tag{11.33}$$

Eqs. (11.28)–(11.32) have been iteratively solved, and the outcomes at 21 nodes are obtained. MATLAB® is used to run the algorithm's code.

11.3.2 Artificial neural network

A group of machine learning models known as artificial neural networks (ANN) are modelled after the structure and operation of the human brain. They are made up of interconnected nodes known as neurons that are arranged into three levels: an input layer, one or more hidden layers, and an output layer. The interconnections within neurons have weights that measure the robustness of the transmitted signal. Furthermore, each neuron uses an activation function for its input nodes, causing nonlinear changes to the network. For the initial training and evaluation of ANN, the ideal weights were determined using scientific findings. In the training phase, artificial neural networks (ANNs) utilise the back-propagation method to iteratively update their weights, aiming to eliminate the discrepancies between observed outputs and expected outcomes. The schematic representation of the backpropagated neural network is shown in Figure 11.4.

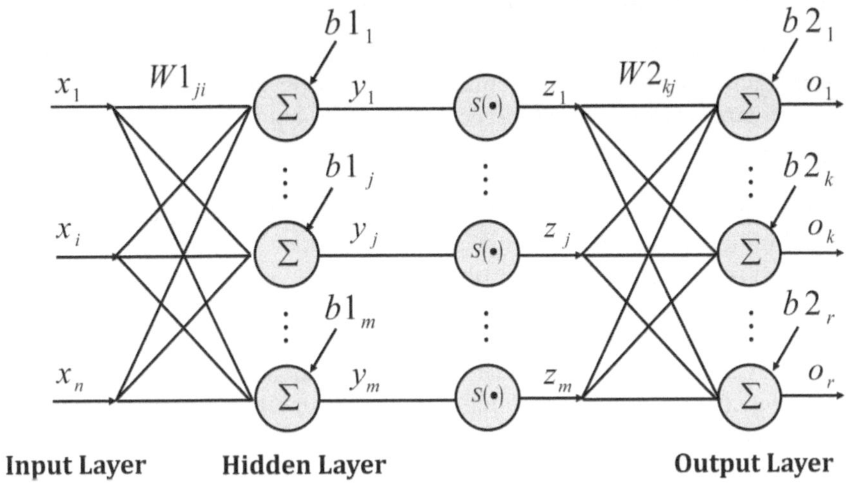

Figure 11.4 General structure for artificial neural network.

This network consists of n input nodes, r output nodes, and a single hidden layer with m nodes. Each link connecting the nodes has a weight that has been assigned to it. The output node has a linear activation function, whereas the hidden node's activation is tansigmoidal. The total input at the jth node of the hidden layer, as shown in Figure 11.4, is as follows:

$$y_j(x) = \sum_{i=1}^{n} x_i w_{ij}^1 + b_j^1$$

w_{ij}^1 represents the weight within ith and jth neuron of the input and hidden layers, here, and the bias at the jth node middle layer is denoted by b_j^1. The output of the jth hidden node is provided by

$$z_j(x) = 1/(1+e^{-y_j(x)})$$

The result $O_k(x)$ offered by the kth output node is stated as for a particular input x.

$$O_k(x) = \sum_{j=1}^{m} w_{jk}^2 z_j + b_k^2$$

w_{kj}^2 represents the weight within jth and kth neuron of the hidden and output layers, here, and the bias at the kth node of outer layer is denoted by b_k^2. The mean square error (MSE) was calculated, and the findings were compared to those of the experiment. The erroneous value was then transmitted backwards using the Levenberg-Marquardt algorithm, and weight modifications were applied at each node throughout the network.

11.3.3 Levenberg-Marquardt algorithm

The Levenberg-Marquardt algorithm, usually referred to as the damped least-squares method, is a popular optimisation approach for nonlinear least squares issues. Kenneth Levenberg and Donald Marquardt invented it in the 1960s. The strategy combines the advantages of the steepest descent technique and the Gauss-Newton algorithm. It solves a linearised least squares problem to update the parameter estimations after each iteration. Combining a Gauss-Newton step and a gradient descent step, the linearised problem is solved. The primary objective of the Levenberg-Marquardt algorithm is to reduce the overall error by adjusting the biases and weights.

The invertible Hessian matrix's approximated expression is $JJ^T + \mu I$. In this particular scenario, the identity matrix is denoted as I, and the combination coefficient is represented by μ. Additionally, the Jacobian matrix, denoted as J, captures the initial gradient of the network error in relation to biases and weights. The Levenberg-Marquardt algorithm makes use of this Hessian matrix's estimation in the following manner [35]:

$$w_{K+1} = w_K - \frac{e_k J_k}{J_K^T J_K + I\mu}$$

w and e are the corresponding weight and error vectors, here. The Quasi-Newton algorithm and the conjugate gradient algorithm are switched during training by the Levenberg-Marquardt algorithm. The behaviour of the Levenberg-Marquardt methodology resembles that of the Quasi-Newton method in the limit when the variable μ approaches zero, as opposed to when μ approaches infinity.

11.4 RESULTS AND DISCUSSION

The ODEs for the nanofluid flow system as shown in Eqs. (11.9)–(11.13) are presented for LMNN numerical experimentation for all cases and scenarios. The four potential scenarios (S1–S4) are explained in order to compute the dataset for $f'(\eta)$, $\theta(\eta)$, $\phi(\eta)$ and $\chi(\eta)$ using LMNN, and the changes are for the Lewis number (Le), Brownian motion parameter (Nb), thermophoresis parameter (Nt), and bioconventional Lewis number (L_b). Table 11.1 provides an overview of the effects of various scenarios and circumstances (C1–C4). Each case of all situations employed 21 input values, which range from 0 to 4, having a 0.1667 step length. From the total dataset, 10% is used for testing, 10% for validation, and the remaining 80% for training the LMNN. The numerical results of various fascinating parameters on interest profiles are displayed in the graphs and tables below.

The performance and transition state of the proposed LMNN for the first instance of each of the four situations may be observed in Figures 11.5a–d

Table 11.1 Scenarios and cases associated with flow fields with variables $Pe = 0.1, \alpha = \frac{\pi}{6}, R_b = 0.1, \sigma = 0.2, Pr = 5, n = 1$ and $Nr = 0.3$, $Pe = \lambda = Ri = 0.1$,

Scenarios	*Cases*	*Le*	*Nt*	*Nb*	L_b
Le	1	3	0.2	0.5	0.5
	2	5			
	3	9			
	4	15			
Nt	1	3	0.1	0.5	0.5
	2		0.2		
	3		0.5		
	4		0.7		
Nb	1	3	0.2	0.1	0.5
	2			0.2	
	3			0.5	
	4			1	
L_b	1	3	0.2	0.5	0.5
	2				0.6
	3				0.7
	4				0.8

and 11.6a–d, respectively. The fitness curves for the initial instance of each scenario are depicted in Figure 11.7a–d. It compares the results obtained from the proposed LMNN technique and the FDM and the corresponding error obtained for training, validation and testing is less than 7.25×10^{-03}, 5.2×10^{-05}, 3.5×10^{-03}, 1.8×10^{-04} and 6.5×10^{-04}, respectively. Figure 11.8a–d shows the error histograms for the first case of each scenario to analyse the errors. The bin value that is close to the zero line for each scenario is 20. Figure 11.9 displays the regression analysis for the first case of all scenarios. The suggested BLM-NN performs admirably in solving the variants of models, and it is suitable for optimal modelling for each of the validation, training, and testing datasets, as seen by the correlation coefficient's consistent value of approximately 1. The solution plots and the AE between the reference solution and the LMNN solutions are shown in Figures 11.10–11.13 Additionally, Table 11.2 shows how the four (C1–C4) of the flow field have converged in terms of MSE, productivity, executed epochs, execution length, and backpropagation metrics. The analysis and findings indicate the accuracy and convergence of the Levenberg-Marquardt-based neural network concept.

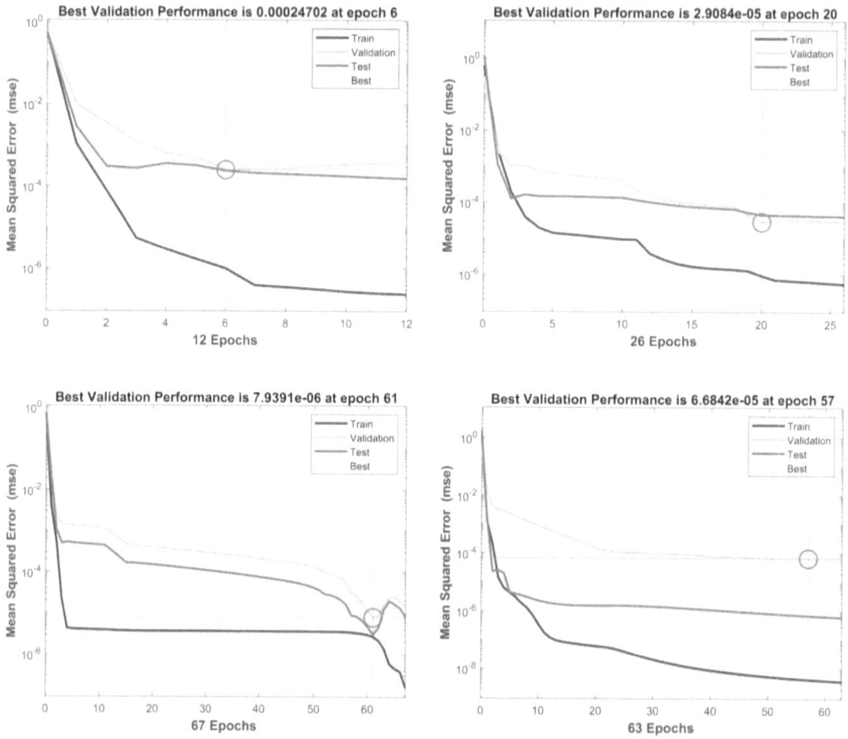

Figure 11.5 Efficiency of the LMNN in case 1 of each scenario for analysing the flow field using mean square error (MSE) (a) S1 (b) S2 (c) S3 (d) S4.

11.4.1 Effects of *Le*, *Nb*, *Nt* and L_b on $f'(\eta)$, $\theta(\eta)$, $\phi(\eta)$ and $\chi(\eta)$ with absolute error estimates

To understand how the Lewis number (Le), Brownian motion parameter (Nb), thermophoresis parameter (Nt), and bioconventional Lewis number (L_b) affect the velocity profile $f'(\eta)$, temperature profile $\theta(\eta)$, nanoparticle concentration profile $\phi(\eta)$, and microorganisms density profile $\chi(\eta)$ the LMNN results are examined utilising MATLAB®. The Lewis number (Le), which has no dimensions, is the ratio of mass diffusivity to heat diffusivity. Figure 11.10a and c show how growing values of the Lewis number Le are responsible for the decreasing trends in temperature and concentration profiles. Their corresponding absolute errors are shown in Figure 11.10b and Figure 11.10d. Higher values of Le indicate active molecular motion, which elevates the temperature of the fluid. A fluid having a lesser Brownian diffusion coefficient, especially one found in a fluid containing a higher

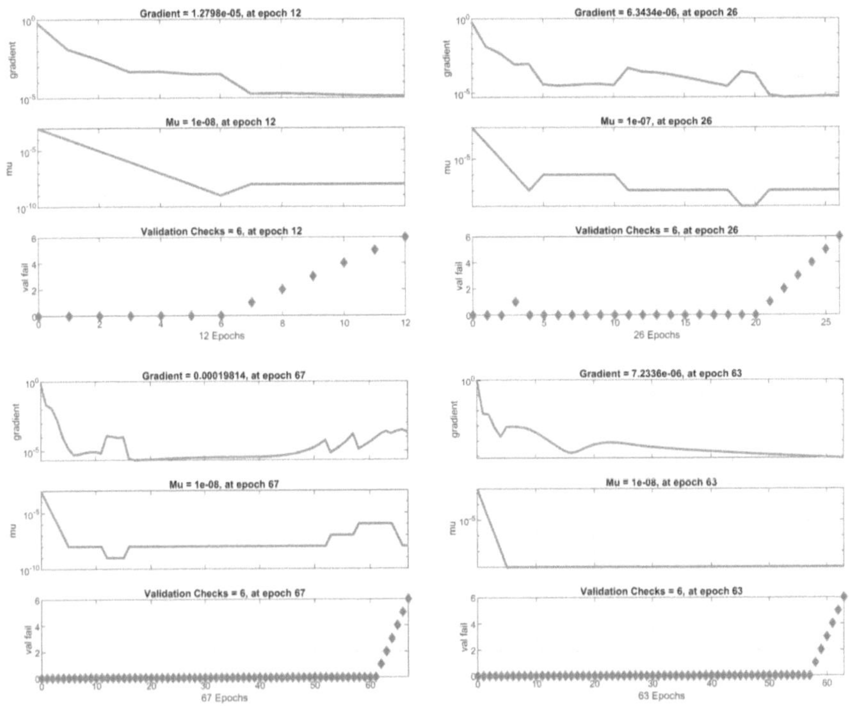

Figure 11.6 Transition state for LMNN for case 1 of each scenario for analysing the flow field (a)S1 (b)S2 (c) S3 (d) S4.

Lewis number, allows particles to diffuse deeply into it. Consequentially, the relationship involving the thermal diffusivity and the microorganism's diffusivity has a shorter depth for higher levels of the Lewis number.

Figure 11.11a and d illustrates the impact of the Brownian motion parameter on temperature profile $\theta(\eta)$ and nanoparticle concentration profile $\phi(\eta)$ and their corresponding absolute error is shown in Figure 11.11a and d. As expected, the structure of the temperature profile is identical to that of the profiles for traditional heat transfer fluids. Brownian motion raises the temperature, but as it accelerates, the volume fraction profile of the nanoparticles decreases.

The effects of the thermophoresis parameter (Nt) on the temperature profile $\theta(\eta)$ and the nanoparticle concentration profile $\phi(\eta)$ are demonstrated in Figure 11.12a and d, and their corresponding absolute errors are shown in Figure 11.12a and d. The temperature and nanoparticle concentration have been shown to increase as the Nt is increased. Physically, increasing

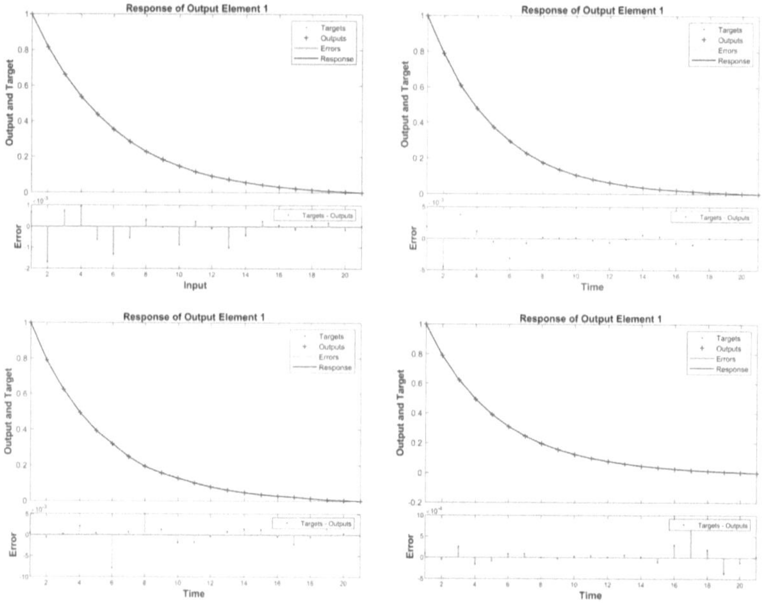

Figure 11.7 Fitness curves for case 1 of each scenario for the LMNN outcomes (a) S1 (b) S2 (c) S3 (d) S4.

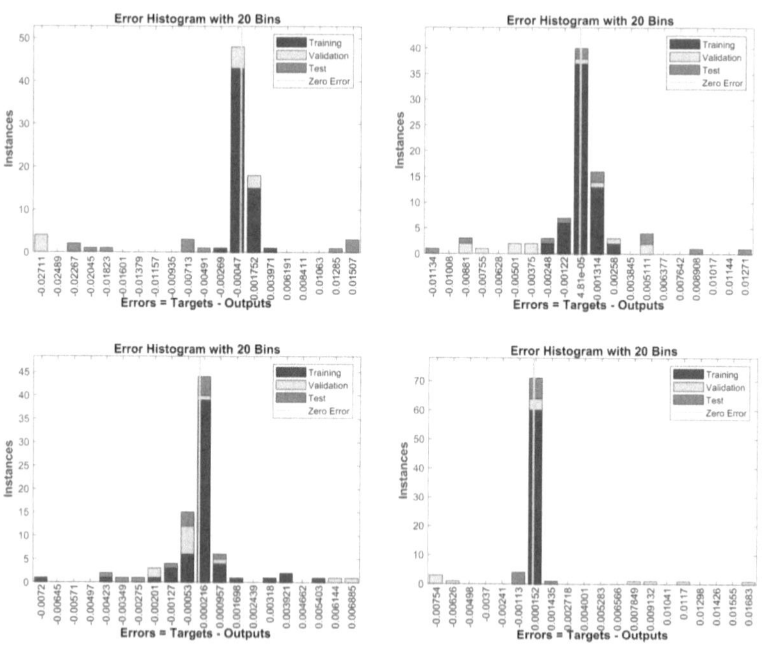

Figure 11.8 Error histograms of the LMNN results for studying error for first case of each scenario (a) S1 (b) S2 (c) S3 (d) S4.

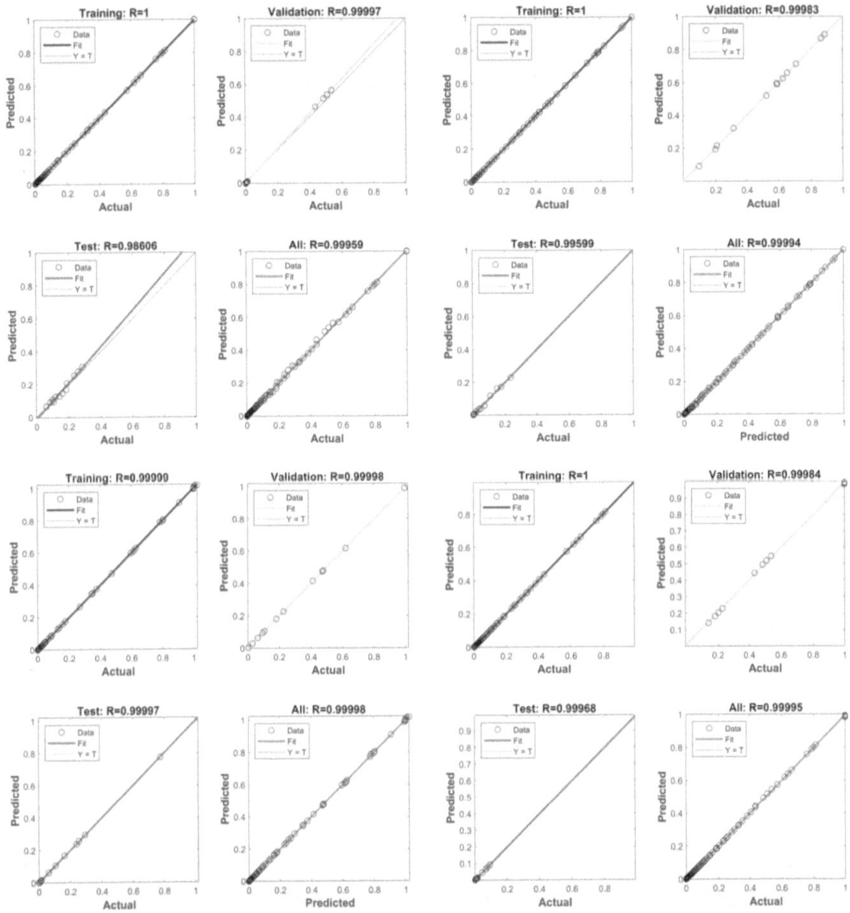

Figure 11.9 Regression analysis of the LMNN results for first case of each scenario (a) S1 (b) S2 (c) S3 (d) S4.

the $\theta(\eta)$ results in an increase in the temperature gradient and, thus, the intermediate force. Figure 11.13 shows that as the bioconvection Lewis number increases, the density of the motile bacteria drops. The L_b opposes the fluid's motion. Physiologically, raising this quantity hinders the germs' ability to diffuse.

The entire Table 11.2 of data shows how consistently reliable LMNN is at resolving diverse scenarios of the fluid flow problem. The results of the LMNN's performance in correctly resolving each instance of the four

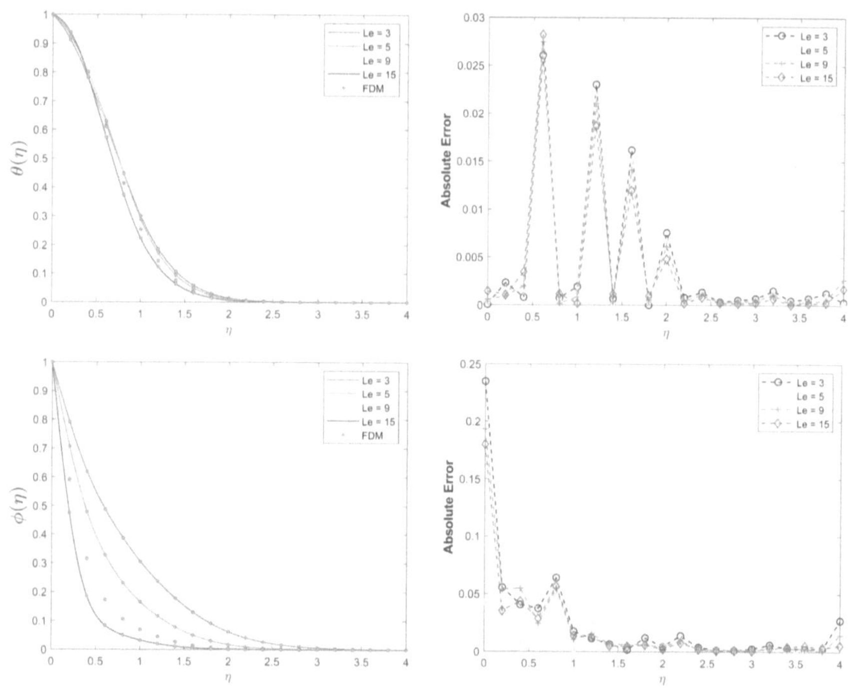

Figure 11.10 Analysis of LMNN results with corresponding absolute errors by varying Lewis number for concentration and microorganisms' density profiles (a) Impact of *Le* on 'θ' (b) Absolute error analysis for 'θ' (c) Impact of *Le* on 'ϕ' (d) Absolute error analysis for 'ϕ'.

circumstances of the proposed flow problem are displayed in Table 11.2. The associated prediction parameters are close to 10^{-07} to 10^{-08}, 10^{-10}, 10^{-11} to 10^{-09}, and 10^{-12} to 10^{-10}, respectively.

11.5 CONCLUSION

The current study uses a computational intelligence heuristic based on backpropagated Levenberg-Marquardt neural networks (LMNN) to analyse the nanofluid flow with gyrotactic microorganisms over a vertically inclined stretching surface. To create the reference dataset for the proposed LMNN, four crucial parameters of the flow problem are varied using the FDM. By using numerical and graphical representations of mean square error

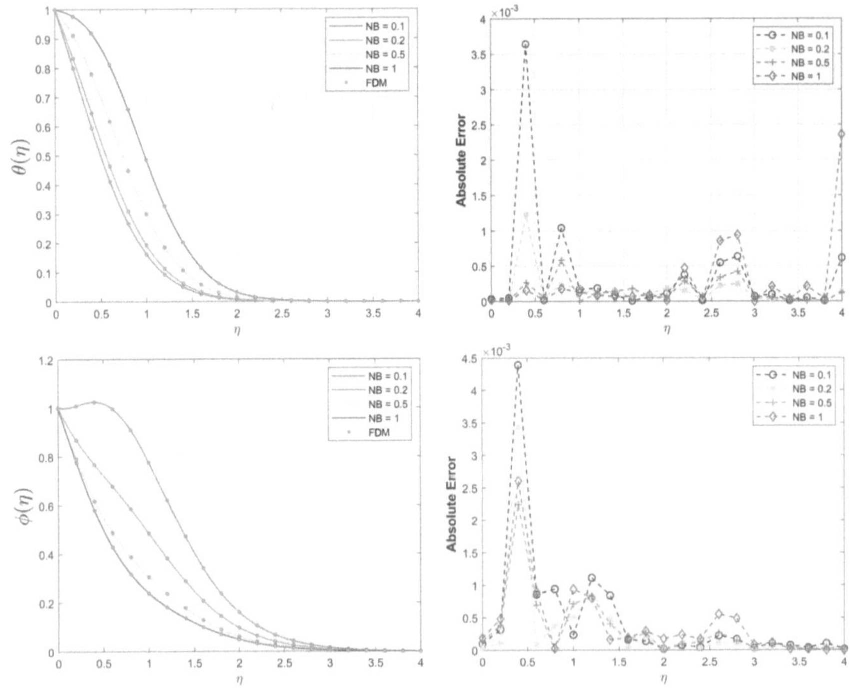

Figure 11.11 Analysis of LMNN results with corresponding absolute errors by varying Brownian motion parameter for temperature and concentration profiles (a) Impact of *Nb* on 'θ' (b) Absolute error analysis for 'θ' (c) Impact of *Nb* on 'ϕ' (d) Absolute error analysis for 'ϕ'.

convergence plots, correlation regression analysis, histogram plots, and the absolute errors between the computed and reference results, we confirm the accuracy of the technique. The suggested methodology offers results as a function of input, in contrast to numerical methods, which only provide sufficiently accurate results for a specified grid of discrete inputs. Therefore, for any continuous input across the whole trained interval, the solution is easily accessible. Additionally, the proposed flow geometry, which utilises the gyrotactic microorganism and nanofluid, has a wide range of industrial and biomedical applications such as aeroplane configurations, powered engineering, nanodrug delivery, pipeline industry, pharmacy, in manufacturing of nano-biopolymer computer graphics, and automotive cooling, which enhance the significance of the proposed study. The key findings of the study are:

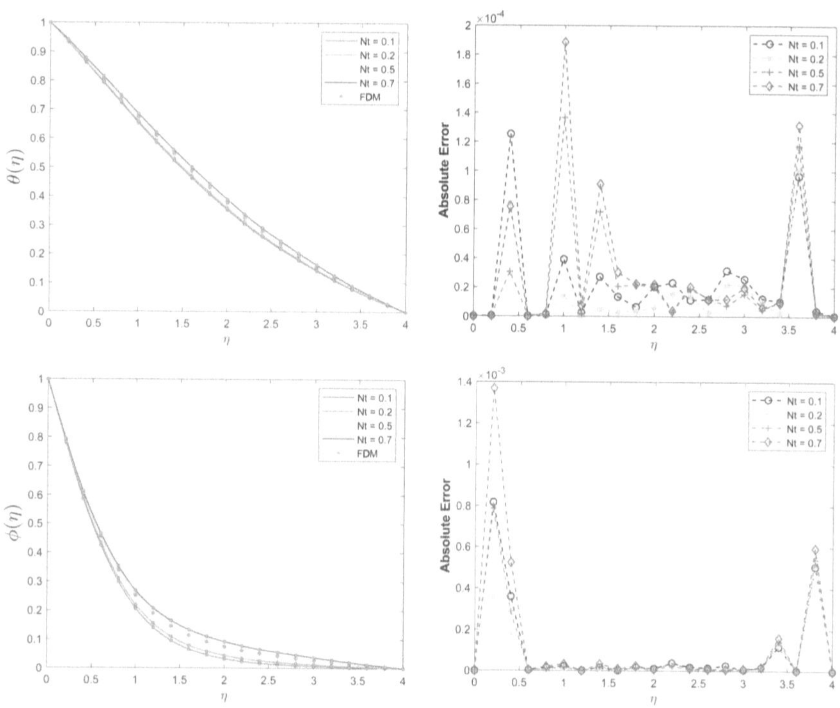

Figure 11.12 Analysis of LMNN results with corresponding absolute errors by varying thermophoresis parameter for temperature and concentration profiles (a) Impact of *Nt* on 'θ' (b) Absolute error analysis for 'θ' (c) Impact of *Nt* on 'ϕ' (d) Absolute error analysis for 'ϕ'.

- As Lewis number increases, the temperature and concentration profile decrease.
- The temperature rises due to Brownian motion, but as it quickens, the nanoparticle volume fraction profile drops.
- As the thermophoresis parameter is increased, the temperature and nanoparticle concentration rise.
- The density of the motile microorganisms decreases as the bioconvection Lewis number rises.

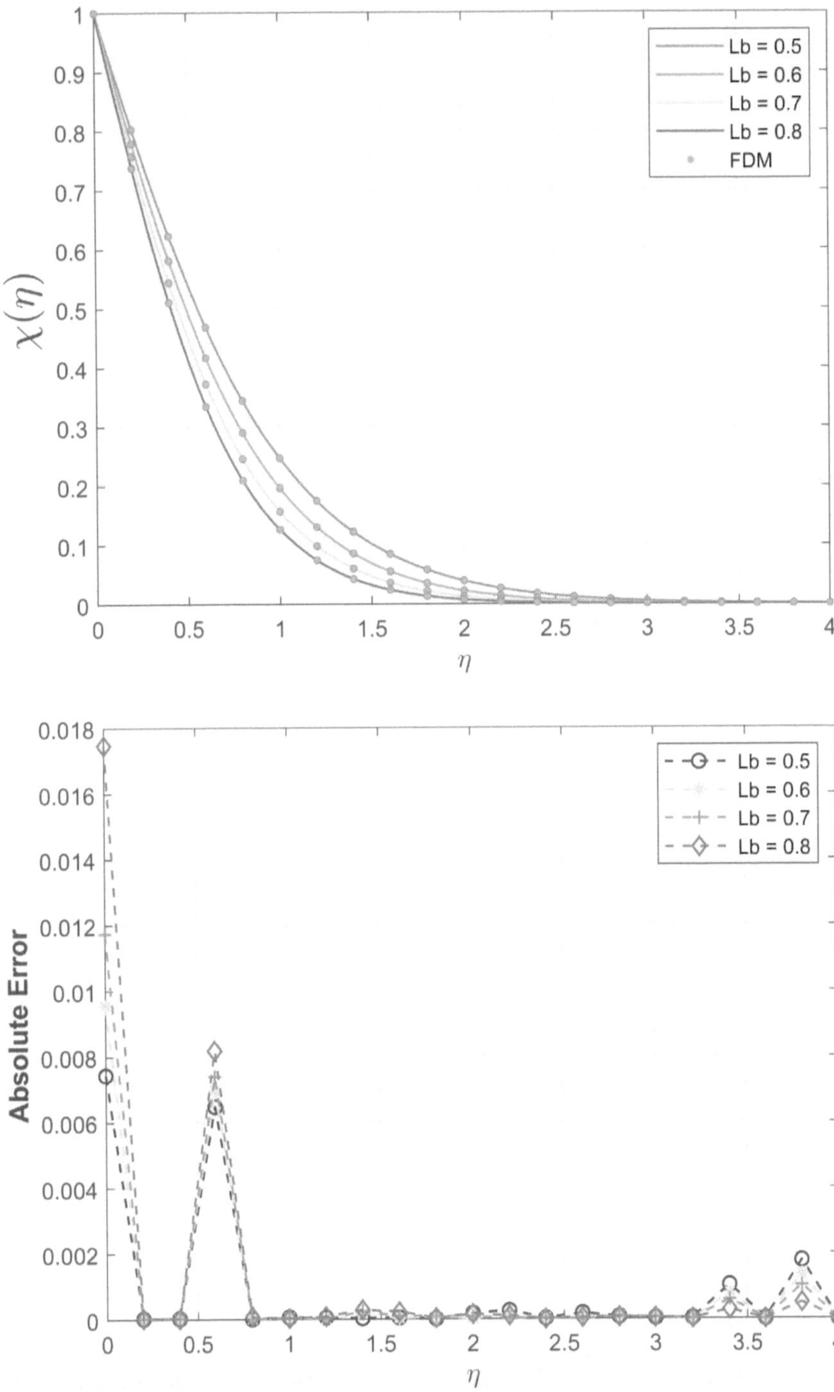

Figure 11.13 Analysis of LMNN results with corresponding absolute errors by varying bioconventional Lewis number for microorganism's density profile (a) Impact of L_b on 'θ' (b) Absolute error analysis for 'θ'.

Table 11.2 Simulation values obtained from LMNN for each scenario and cases of flow field

For scenario 1							
	MSE						
Cases	*Training*	*Validation*	*Testing*	*Performance*	*Mu specifier*	*Epochs*	*Time (s)*
1	6.2E-07	4.4E-06	6.6E-09	5.5E-08	2.6E-07	1.00E-08	298
2	5.2E-06	9.2E-06	6.4E-07	3.5E-06	8.2E-09	1.00E-06	280
3	2.1E-05	8.2E-07	3.2E-06	8.4E-09	9.9E-07	1.00E-06	370
4	3.6E-09	5.9E-07	4.3E-07	2.5E-07	6.4E-04	1.00E-06	304
For scenario 2							
1	1.5E-07	2.9E-06	4.5E-07	4.5E-07	1.9E-05	1.00E-07	70
2	9.5E-07	8.2E-07	9.2E-08	2.1E-08	2.8E-06	1.00E-04	55
3	8.8E-07	2.6E-08	4.1E-08	4.5E-04	5.9E-03	1.00E-05	87
4	9.5E-08	6.5E-07	6.3E-07	9.5E-06	8.6E-06	1.00E-05	67
For scenario 3							
1	9.8E-07	2.9E-05	9.6E-06	5.5E-08	3.9E-04	1.00E-08	66
2	8.8E-06	5.2E-06	7.3E-05	8.5E-05	6.6E-07	1.00E-06	73
3	3.6E-07	5.3E-07	7.2E-05	8.3E-04	7.9E-06	1.00E-09	74
4	6.6E-08	6.9E-08	1.4E-04	7.5E-07	4.5E-03	1.00E-09	56
For scenario 4							
1	2.8E-08	8.6E-07	1.4E-08	4.5E-08	3.8E-06	1.00E-08	104
2	3.4E-09	5.9E-07	2.2E-08	6.5E-05	2.5E-07	1.00E-08	111
3	1.9E-08	2.3E-07	6.6E-07	6.3E-07	3.6E-09	1.00E-09	98
4	8.7E-07	2.4E-07	3.1E-07	2.5E-07	5.7E-03	1.00E-07	86

REFERENCES

[1] Choi, S. U. S., & Eastman, J. (1995). *Enhancing Thermal Conductivity of Fluids with Nanoparticles, Developments and Applications of Non-Newtonian Flows*. ASME, New York.

[2] Buongiorno, J. (2006). Convective transport in nanofluids. *Journal of Heat Transfer*, *128*(3), 240–250. https://doi.org/10.1115/1.2150834

[3] Prabakaran, R., Eswaramoorthi, S., Loganathan, K., & Sarris, I. E. (2022). Investigation on thermally radiative mixed convective flow of carbon nanotubes/Al_2O_3 nanofluid in water past a stretching plate with joule heating and viscous dissipation. *Micromachines*, *13*(9), 1424. https://doi.org/10.3390/mi13091424

[4] Awan, A. U., Ahammad, N. A., Majeed, S., Gamaoun, F., & Ali, B. (2022). Significance of hybrid nanoparticles, Lorentz and Coriolis forces on the dynamics of water based flow. *International Communications in Heat and Mass Transfer*, *135*, 106084. https://doi.org/10.1016/j.icheatmasstransfer.2022.106084

[5] Alam, M. S., Rahman, M. M., & Sattar, M. A. (2007). Similarity solutions for hydromagnetic free convective heat and mass transfer flow along a semi-infinite permeable inclined flat plate with heat generation and thermophoresis. *Nonlinear Analysis: Modelling and Control*, *12*(4), 433–445. https://doi.org/10.15388/na.2007.12.4.14675

[6] Rahman, M. M., Aziz, A., & Al-Lawatia, M. A. (2010). Heat transfer in micropolar fluid along an inclined permeable plate with variable fluid properties. *International Journal of Thermal Sciences*, *49*(6), 993–1002. https://doi.org/10.1016/j.ijthermalsci.2010.01.002

[7] Foroozesh, J., & Kumar, S. (2020). Nanoparticles behaviors in porous media: Application to enhanced oil recovery. *Journal of Molecular Liquids*, *316*(10), 113876. https://doi.org/10.1016/j.molliq.2020.113876

[8] Kuznetsov, A. V., & Avramenko, A. A. (2004). Effect of small particles on this stability of bioconvection in a suspension of gyrotactic microorganisms in a layer of finite depth. *International Communications in Heat and Mass Transfer*, *31*(1), 1–10. https://doi.org/10.1016/S0735-1933(03)00196-9

[9] Uddin, M. J., Alginahi, Y., Bég, O. A., & Kabir, M. N. (2016). Numerical solutions for gyrotactic bioconvection in nanofluid-saturated porous media with Stefan blowing and multiple slip effects. *Computers and Mathematics with Applications*, *72*(10), 2562–2581. https://doi.org/10.1016/j.camwa.2016.09.018

[10] Elbashbeshy, E. M. A., Asker, H. G., & Nagy, B. (2022). The effects of heat generation absorption on boundary layer flow of a nanofluid containing gyrotactic microorganisms over an inclined stretching cylinder. *Ain Shams Engineering Journal*, *13*(5), 101690. https://doi.org/10.1016/J.ASEJ.2022.101690

[11] Rao, M. V. S., Gangadhar, K., Chamkha, A. J., & Surekha, P. (2021). Bioconvection in a convectional nanofluid flow containing gyrotactic microorganisms over an isothermal vertical cone embedded in a porous surface with chemical reactive species. *Arabian Journal for Science and Engineering*, *46*, 2493–2503. https://doi.org/10.1007/s13369-020-05132-y

[12] Chandra, P., & Das, R. (2023b). Impact of gyrotactic microorganisms in nanofluid via porous media along an inclined stretching plate: Finite element analysis. *Numerical Heat Transfer, Part B: Fundamentals*, *83*(6), 367–394. https://doi.org/10.1080/10407790.2022.2163940

[13] Chandra, P., & Das, R. (2023a). Finite-element-based machine-learning algorithm for studying gyrotactic-nanofluid flow via stretching surface. *International Journal of Numerical Methods in Fluids*, *95*, 1–25. https://doi.org/10.1002/fld.5229

[14] Elbashbeshy, E. M. A., & Asker, H. G. (2022). Fluid flow over a vertical stretching surface within a porous medium filled by a nanofluid containing gyrotactic microorganisms. *European Physical Journal Plus*, *137*(5), 541. https://doi.org/10.1140/epjp/s13360-022-02682-y

[15] Rafique, K., Anwar, M. I., Misiran, M., Khan, I., Alharbi, S. O., Thounthong, P., & Nisar, K. S. (2019). Numerical solution of casson nanofluid flow over a non-linear inclined surface with soret and dufour effects by Keller-Box method. *Frontiers in Physics*, *7*, 139. https://doi.org/10.3389/FPHY.2019.00139/XML/NLM

[16] Falodun Bidemi, O., & Ahamed, M. S. S. (2019). Soret and Dufour effects on unsteady Casson magneto-nanofluid flow over an inclined plate embedded in a porous medium. *World Journal of Engineering*, *16*(1), 0144. https://doi.org/10.1108/WJE-04-2018-0144
[17] Sobamowo, M. G., Yinusa, A. A., & Makinde, O. D. (2019). A study on the effects of inclined magnetic field, flow medium porosity and thermal radiation on free convection of casson nanofluid over a vertical. *World Scientific News*, *138*(1), 1–64.
[18] Gholinia, M., Hoseini, M. E., & Gholinia, S. (2019). A numerical investigation of free convection MHD flow of Walters-B nanofluid over an inclined stretching sheet under the impact of Joule heating. *Thermal Science and Engineering Progress*, *11*, 272–282. https://doi.org/10.1016/j.tsep.2019.04.006
[19] Parvin, S., Mohamed Isa, S. S. P., Arifin, N. M., & Ali, F. M. (2021). The inclined factors of magnetic field and shrinking sheet in casson fluid flow, heat and mass transfer. *Symmetry*, *13*(3), 373. https://doi.org/10.3390/SYM13030373
[20] Bukhari, A. H., Raja, M. A. Z., Sulaiman, M., Islam, S., Shoaib, M., & Kumam, P. (2020). Fractional neuro-sequential ARFIMA-LSTM for financial market forecasting. *IEEE Access*, *8*, 71326–71338. https://doi.org/10.1109/ACCESS.2020.2985763
[21] Bukhari, A. H., Sulaiman, M., Raja, M. A. Z., Islam, S., Shoaib, M., & Kumam, P. (2020). Design of a hybrid NAR-RBFs neural network for nonlinear dusty plasma system. *Alexandria Engineering Journal*, *59*(5), 3325–3345. https://doi.org/10.1016/j.aej.2020.04.051
22] Sabir, Z., Umar, M., Guirao, J. L. G., Shoaib, M., & Raja, M. A. Z. (2021). Integrated intelligent computing paradigm for nonlinear multi-singular third-order Emden-Fowler equation. *Neural Computing and Applications*, *33*(8), 3417–3436. https://doi.org/10.1007/s00521-020-05187-w
[23] Ahmad, S. I., Faisal, F., Shoaib, M., & Raja, M. A. Z. (2020). A new heuristic computational solver for nonlinear singular Thomas-Fermi system using evolutionary optim0ed cubic splines. *European Physical Journal Plus*, *135*(1), 55. https://doi.org/10.1140/epjp/s13360-019-00066-3
[24] Umar, M., Sabir, Z., Raja, M. A. Z., Aguilar, J. F. G., Amin, F., & Shoaib, M. (2021). Neuro-swarm intelligent computing paradigm for nonlinear HIV infection model with CD4+ T-cells. *Mathematics and Computers in Simulation*, *188*, 241–253. https://doi.org/10.1016/j.matcom.2021.04.008
[25] Shoaib, M., Raja, M. A. Z., Sabir, M. T., Bukhari, A. H., Alrabaiah, H., Shah, Z., Kumam, P., & Islam, S. (2021). A stochastic numerical analysis based on hybrid NAR-RBFs networks nonlinear SITR model for novel COVID-19 dynamics. *Computer Methods and Programs in Biomedicine*, *202*, 105973. https://doi.org/10.1016/j.cmpb.2021.105973
[26] Ilyas, H., Ahmad, I., Raja, M. A. Z., & Shoaib, M. (2021). A novel design of Gaussian WaveNets for rotational hybrid nanofluidic flow over a stretching sheet involving thermal radiation. *International Communications in Heat and Mass Transfer*, *123*, 105196. https://doi.org/10.1016/j.icheatmasstransfer.2021.105196

[27] Ahmad, I., Ilyas, H., Urooj, A., Aslam, M. S., Shoaib, M., & Raja, M. A. Z. (2019).Novel applications of intelligent computing paradigms for the analysis of nonlinear reactive transport model of the fluid in soft tissues and microvessels. *Neural Computing and Applications*, *31*(12), 9041–9059. https://doi.org/10.1007/s00521-019-04203-y

[28] Manohar, B., & Das, R. (2022a). Artificial neural networks for prediction of COVID-19 in India by using backpropagation. *Expert Systems*, *40*(5), e13105. https://doi.org/10.1111/exsy.13105

[29] Manohar, B., & Das, R. (2022b). Artificial neural networks for the prediction of monkeypox outbreak. *Tropical Medicine and Infectious Disease*, *7*(12), 424. https://doi.org/10.3390/TROPICALMED7120424

[30] Manohar, B., & Das, R. (2023). Comparison of hybrid artificial neural networks with GA, PSO, and RSA in predicting COVID-19 cases: A case study of India. *Multi-Disciplinary Applications of Fog Computing: Responsiveness in Real*. Retrieved August 14, 2023, from https://www.igi-global.com/chapter/comparison-of-hybrid-artificial-neural-networks-with-ga-pso-and-rsa-in-predicting-covid-19-cases/327892

[31] Ahmad, I., Raja, M. A. Z., Ramos, H., Bilal, M., & Shoaib, M. (2021). Integrated neuro-evolution-based computing solver for dynamics of nonlinear corneal shape model numerically. *Neural Computing and Applications*, *33*(11), 5753–5769. https://doi.org/10.1007/s00521-020-05355-y

[32] Chandra, P., & Das, R. (2023). A hybrid machine learning algorithm for studying magnetized nanofluid flow containing gyrotactic microorganisms via a vertically inclined stretching surface. *International Journal for Numerical Methods in Biomedical Engineering*, 40(1), e3780. https://doi.org/10.1002/cnm.3780

[33] Chandra, P., & Das, R. (2023). A hybrid RSA-IPA optimizer for designing an artificial neural network to study the Jeffery-Hamel blood flow with copper nanoparticles: Application to stenotic tapering artery. *Results in Engineering*, 20, 101542. https://doi.org/10.1016/j.rineng.2023.101542

[34] Chandra, P., & Das, R. (2024). Design of a unified physics-informed neural network using interior point algorithm to study the bioconvection nanofluid flow via stretching surface. *Engineering Applications of Artificial Intelligence*, 133(F), 108647. https://doi.org/10.1016/j.engappai.2024.108647

[35] Gavin, H. P. (2019). *The Levenberg-Marquardt Algorithm for Nonlinear Least Squares Curve-Fitting Problems*. Duke University, Durham, NC, 1–19. https://people.duke.edu/~hpgavin/ce281/lm.pdf

Chapter 12

Detection of bearing fault using vibration signal through multiple feature extraction-based machine learning algorithms

Rajeev Kumar, RS Anand, and Nipun Jain

12.1 INTRODUCTION

Induction motors are globally recognised as highly reliable electrical drives extensively utilised in manufacturing industries, owing to their robust design and low maintenance requirements. Therefore, effective condition monitoring of these rotating machines plays an important role in diagnosing and pre-emptively detecting potential machine failures to avoid unplanned shutdowns [1,2]. The induction motors have various integral components, which involve the motor operation. Among these, bearings emerge as important elements responsible for mechanical rotations, assuming a critical role as load-bearing components within the machinery. According to the survey conducted by the electrical agencies, bearing faults are the most dominant issues affecting induction motors with a rate of occurrence of almost 40%. This rate is significantly higher compared to other motor faults such as stator winding fault, rotor bar fault, and rotor eccentricity fault. Stator failure occurs at a rate of 20%, rotor failures at 12%, and other types of failure comprise the remaining 20%. Hence, the machines commonly suffered from the bearing faults [1,3–5].

Bearing faults can be categorised into two types: one is single-point crack fault and the second is multiple-point crack fault. These faults have a significant impact on bearing operation and can lead to various issues, such as misalignment, rough surfaces, waviness, and obstruction of the rolling ball [6]. To ensure smooth and uninterrupted operation, it is essential to detect and diagnose these faults at an early stage. This proactive approach helps prevent substantial economic losses and the need for complete plant shutdown. Hence, it's essential to monitor the health of bearings periodically analysing vibration signals, acoustic analysis, stator current analysis, and temperature analysis during motor operation [6–10]. In the existing literature, numerous vibration monitoring techniques have been reported by researchers to detect and diagnose faults in rotating machinery. The generated vibration signals provide rich information of motor performance and their load conditions without interrupting the motor operations. However, these recorded vibration signals are inherently complex in nature due to environmental noise and interference

 DOI: 10.1201/9781003534761-12

signal components [11]. It is difficult to extract quality fault information from these signals. Therefore, signal pre-processing methods such as averaging, filtering, DC components removal, and noise cancellation are used prior to signal analysis. These pre-processed signals are then used for further analysis using signal processing techniques in time domain, frequency domain, and time-frequency domain [12–15].

In the time domain, various statistical features are discussed including Root Mean Square (RMS), Mean, Peak values, Kurtosis, Standard Deviation, Impulse Factor, and Shape Factor. Researchers also explore the frequency domain techniques such as Fast Fourier Transform (FFT), Power Spectrum, Cepstrum analysis, and Time-frequency domain wavelet transform along with short-time Fourier transform as reported in their studies [11,16–21]. The suggested frequency methods such as Fast Fourier transform, Empirical mode decomposition (EMD), Hilbert Transform, Hilbert-Huang transform, spectral analysis, stochastic resonance, sparse decomposition, and Wigner-vile distribution are discussed to extract the features and classification of faults [22–29]. Advanced fault diagnosis methods incorporate soft computing techniques, such as artificial intelligence, machine learning, and deep learning methods for the classification of bearing faults at different loading conditions [30]. Machine learning-based fault detection techniques are used with various classifier models namely Decision Tree (DT), Random Forest (RF), k-Nearest Neighbour (k-NN), Support Vector Machine (SVM), and Naive Bayes with the features selection techniques [31–37]. These machine learning algorithms enhance the effectiveness of fault detection ability of the self-automated diagnosis systems. The ability of the system can be improved through the dimensional reduction of the feature vectors. Principal component analysis, Relief, Lasso, and mRMR method are reported by the researchers in their literature to easily classify the model accuracy from the high dimensional data [38,39].

This study presents a machine learning-based approach to the investigation and classification of bearing faults. Statistical features, including RMS, Mean, Kurtosis, Skewness, Crest Factor, Variance, Impulse Factor, Shape Factor, Clearance Factor, and Standard Deviation of recorded vibration signals, are utilised to investigate the rolling element bearing fault at different loading conditions. The work is organised as follows: the current section elucidates the significance of induction motor and their condition monitoring schemes under bearing fault conditions. Section 12.1.2 delves into the methodology of the fault investigation. The experimental setup and dataset are discussed in Section 12.1.6. Analysis and results are presented in Section 12.1.7 while the conclusion of the entire work is discussed in Section 12.1.8.

12.2 METHODOLOGY

The proposed methodology for fault classification of ball bearings is shown in Figure 12.1. This scheme is based on vibration signal monitoring (VSM),

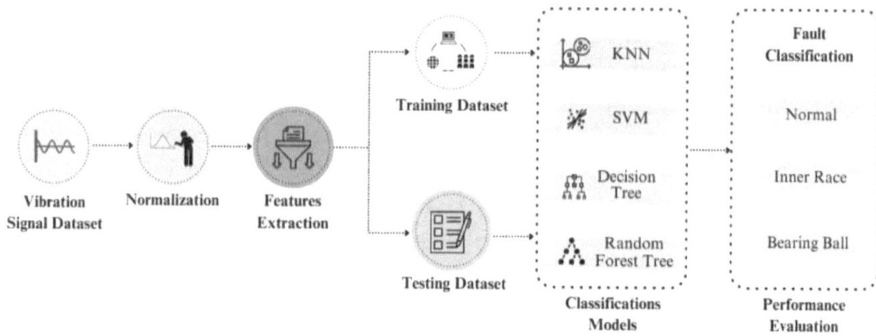

Figure 12.1 Block diagram of proposed algorithms.

a non-invasive technique used to diagnose the bearing health in the induction motors. Vibration signals were acquired from both healthy and faulty bearing for the analysis of bearing health. These signals provide comprehensive insights into the dynamic behaviour of the machine's rotating components. Furthermore, a signal preprocessing technique is applied to normalise the recorded signals, remove the DC bias and unwanted ripple. Subsequently, a feature extraction is performed on these processed signals, computing characteristics features of vibration signals through statistical parameters as discussed in Table 12.1. An optimal dataset is prepared to construct the feature vectors from the extracted information according to the bearing faults and motor operating conditions. Following the data preparation and optimal feature vector construction, various machine learning classification models including DT, Random Forest Tree (RF), SVM, and k-Nearest Neighbors (k-NN) are implemented for the fault classification.

12.2.1 Vibration signal

The vibration signal produced by the machine is a characteristic pattern of oscillation while the machine is in operation. This signal comprises a series of oscillations that arise from various mechanical events occurring within a specific time frame. Such signals help to study health and performance of a machine during its operation. In this study, vibration signals were acquired from the bearings of induction motor when the motor operating under different loading conditions. A seeded fault of 0.007, 0.014, and 0.021 inches was introduced into the inner race, outer race, and rolling ball through electro-discharge machine (EDM). The effects of these introduced faults are reflected in the recorded vibration signals in terms of impulsive noise and fault frequencies. Impulsive noise refers to irregular and sudden spikes of disturbances in the vibration signal while the fault frequencies indicate specific frequency in the signal caused by the presence of faults. Vibration signals in the time domain are shown in Figure 12.2, which is segmented into 800 samples each. These segmented signals can be used for feature

Table 12.1 Signal characteristics features in time domain

Features	*Equations*		*Features*	*Equations*	
Standard Deviation	$X_{\text{std}} = \sqrt{\frac{1}{N}\sum_{i=1}^{N}(x_i - \mu)^2}$	(12.2)	**Skewness**	$X_{\text{skewness}} = \frac{1}{N}\frac{\sum_{i=1}^{N}(x_i - \mu)^3}{\sigma^3}$	(12.7)
RMS	$X_{\text{RMS}} = \frac{\sqrt{\sum_{i=1}^{N} x_i^2}}{N}$	(12.3)	**Crest factor**	$X_f = \frac{X_{\max}}{X_{\min}}$	(12.8)
Mean	$X_{\text{mean}} = \frac{1}{N}\sum_{i=1}^{N} x_i$	(12.4)	**Clearance factor**	$X_{\text{CF}} = \frac{x_p}{\left(\frac{1}{N}\sum_{i=1}^{N}\sqrt{\lvert x_i \rvert}\right)^2}$	(12.9)
Kurtosis	$X_{\text{kurtosis}} = \frac{1}{N}\frac{\sum_{i=1}^{N}(x_i - \mu)^4}{\sigma^4}$	(12.5)	**Shape factor**	$X_{\text{SF}} = \frac{x_{\text{RMS}}}{\frac{1}{N}\sum_{i=1}^{N}\lvert x_i \rvert}$	(12.10)
Variance	$X_{\text{var}} = \frac{\sum_{i=1}^{N}(x_i - \mu)^2}{N-1}$	(12.6)	**Impulse factor**	$X_{\text{IF}} = \frac{x_p}{\frac{1}{N}\sum_{i=1}^{N}\lvert x_i \rvert}$	(12.11)

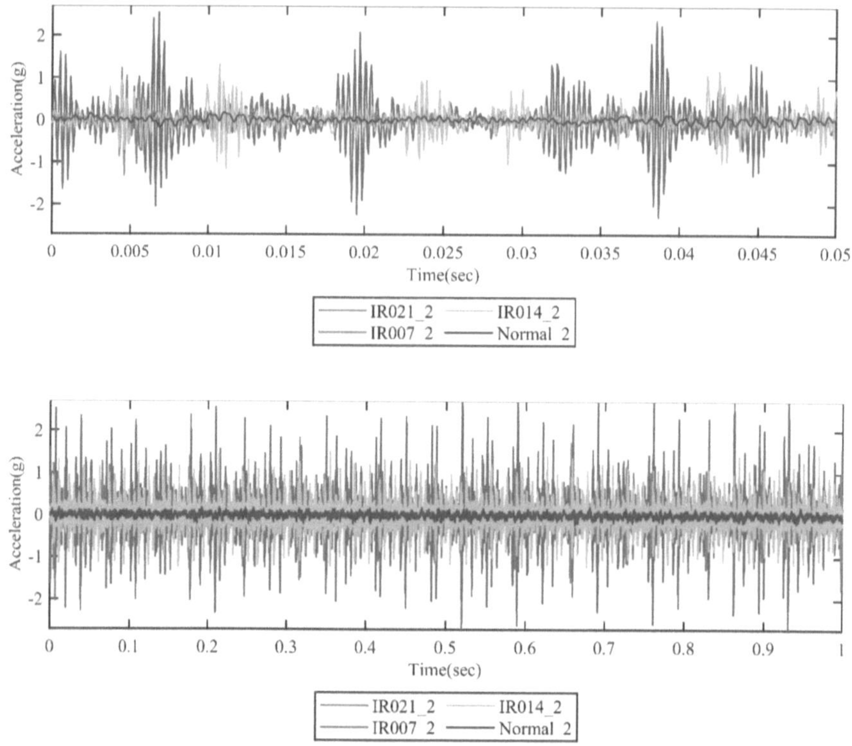

Figure 12.2 Vibration signal of Inner Race fault (a) zoom scaled (b) standard scaled.

extraction, which involves the identification of specific characteristics of the signal that can be used to differentiate the different conditions or state of the machine.

12.2.2 Normalisation

Normalisation is a pre-processing method used to mitigate the effect of amplitude variation and enhances the quality of the vibration signal during feature extraction. Various signal components such as the natural motor frequency, the shaft rotational frequency, the characteristics frequency of incipient faults and their harmonics can be a part of vibration signal. These frequency components describe the machine health state, system structure and degree of component defects. The normalisation mainly focuses on the vibration signal amplitude and its frequency.

$$X_{\text{normalised}} = \frac{X - X_{\min}}{X_{\max} - X_{\min}} \tag{12.1}$$

where X is a signal, X_{max} is the maximum value of signal, X_{min} is the minimum value of signal.

12.2.3 Features extraction

Feature extraction is the most important task in ensuring the successful implementation of machine learning algorithms. However, the outcomes of any classification model mainly depend upon the quality of the selected features. Therefore, in this study, ten statistical features are extracted from the time domain vibration signals for condition assessment of test bearings. These features are computed using the mathematical equations as discussed in Table 12.1, which encompass calculations for mean, rms, standard deviation, skewness, variance, crest factor, clearance factor, impulse factor, and shape factor. These features are subsequently utilised to generate feature vectors, which are then normalised within the range of 0–1 using the given Eq. (12.1).

where x_i= vibration signal, x_p= maximum value of vibration signal, N=number of signal sample.

12.3 CLASSIFICATION MODELS

Classification is a method or procedure used to categorise a given dataset into distinct classes. Alternatively, classification can be described as a method that learns from a provided dataset to make assignments or categorisations into classes. These categories are also referred to as labels, targets, etc. Classification tasks can be applied to both linear and non-linear datasets. In the context of classifications, the training categories can take the form of binary choices, such as Yes or No, 0 or 1. To illustrate the concept of classification, consider an example of bearing defects. Bearing fault dataset is prepared to classify the motor health condition, which is divided into two classes: one indicating the presence of a fault and the other indicating its absence or bearing is in normal state. By associating input data or fault information with the relevant categories and ensuring accurate training of the classifier, it becomes capable of determining whether bearing has any defect or not. In this study, induction motor bearing faults are predicted by four classification models such as (a) SVM, (b) k-NN, (c) DT, and (d) RF.

12.3.1 Support vector machine

The SVM was first introduced by Vapnik in 1995 and used as the supervised machine learning model for data classification and regression analysis. SVM is known for its ability to handle complex datasets and provide robust predictions by maximising the margin between different classes.

SVM operates on the principle of finding a hyperplane that best separates data points belonging to different classes. The objective is to identify the hyperplane that maximises the margin between the two classes while minimising the classification error. The points closest to the hyperplane, called the support vectors, play a crucial role in defining the margin. The margin is the distance between the hyperplane and the nearest data points from each class. SVM aims to find the hyperplane that not only separates the classes but also maximises this margin. In a two-dimensional space, a hyperplane is a line that separates data points of different classes.

In higher-dimensional spaces, a hyperplane becomes a multidimensional surface as shown in Figure 12.3a. The margin is the perpendicular distance between the hyperplane and the closest data points from each class. SVM can handle nonlinear data by transforming the original features space into a higher-dimensional space using a kernel function (e.g., polynomial, or radial basis function). This transformation enables SVM to find a hyperplane that can separate nonlinearly distributed data points. SVM finds applications in a wide range of domains such as text and document classification, image recognition, bioinformatics, finances, healthcare, etc. SVMs offer an effective approach to solving a wide range of real-world problems across diverse domains. However, proper parameter tuning, kernel selection, and consideration of computational resources are essential for maximising their performance.

The two-dimensional space condition is easily explained in Figure 12.3a, where class A represents the semi-circle, class B represents the circle and dotted lines represent the margin. SVM hyperplane is used to separate two different series of points of dataset into separate classes and achieve the optimum hyperplane when the problem is solved.

Hyperplane equations are

$$(w.x)+b=0 \tag{12.12}$$

w is a normal vector of the hyperplane b is distance from the centre and x is the input vector. Class A and Class B equations are given below:

$$(w.x)+b=1 \tag{12.13}$$

$$(w.x)+b=-1 \tag{12.14}$$

The data point limits correspond to the support vector for a particular class and the decision functions can be defined for the entire class A and class B as

$$F(x)=\text{Sign}((w.x)+b) \tag{12.15}$$

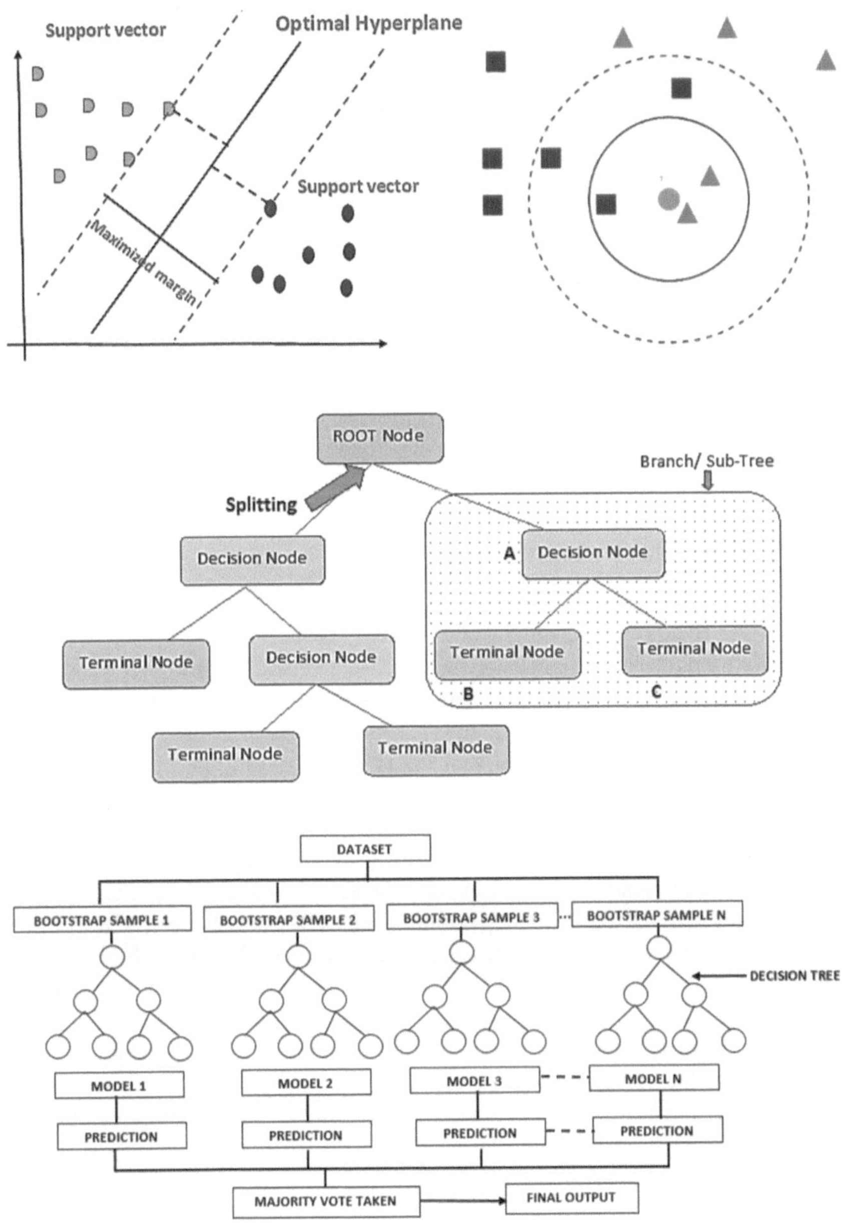

Figure 12.3 (a) support vector machine (b) K nearest neighbour (c) Decision tree (d) Random forest.

Optimisation of hyperplane may be achieved by minimising the

$$=\frac{1}{2}\|w\|^2 \tag{12.16}$$

$$y_i\left(w_i x_i + b\right) \geq 0 \tag{12.17}$$

SVM training is performed by using the following equations.

$$L(\propto) = \sum_{i=1}^{k} \propto_1 - \frac{1}{2}\sum_{i,j=0}^{k} \propto_i \propto_j y_i y_j k\left(x_i x_j\right) \tag{12.18}$$

$$\sum_{i=1}^{k} y_i \propto_i = 0 \tag{12.19}$$

12.3.2 K-nearest neighbour

The k-NN was first introduced by Evelyn Fix and Joseph Hodges in 1951 known as a fundamental machine learning algorithm that operates on the principles of pattern recognition and instance-based learning. It's a simple yet versatile algorithm used for both classification and regression tasks. k-NN is particularly appealing due to its ease of implementation, intuitive nature, and adaptability to various types of data. The k-NN classifier makes predictions based on the majority class of the k nearest data points to a given data point. The algorithm requires a dataset with labelled examples. The input features of these examples are used to calculate distances between data points. The first step is to select the value of k, which represents the number of nearest neighbours that will contribute to the prediction. A smaller k can result in noise affecting the prediction, while a larger k can lead to over-smoothing. For a new input data point, the distances to all other data points in the leading dataset are computed using metrics like Euclidean distance, Manhattan distance, or other suitable distance measures. The k data points with the shortest distances to the input point are selected. The majority class among the k neighbour is determined, and the input point is assigned to that class.

The Euclidean distance is the distance between the two classes p and q. The Euclidean distance between two points can be calculated by p and q from Eq. (12.10):

$$d_E = \sqrt{\sum_{i=1}^{n}\left(p_i - q_i\right)^2} \tag{12.20}$$

In Figure 12.3b, solid circle may be assigned to the first class (triangle) or the second class (square) based on the value of K or the number of samples inside the circle. If $K=3$, it may be assigned to the first class (triangle) because there are two triangles and one square inside the circle. If $K=5$, it will be assigned to the second class (square) because three squares and two triangles are present inside the dashed circle. Similarly, for $K=11$, the result will be square or triangle, respectively. The optimum value of K is usually determined from the Euclidean distance between the testing and training samples.

k-NN is one of the simplest algorithms to understand and implement. It doesn't make strong assumptions about the data distribution, making it suitable for a wide range of scenarios. Unlike many other algorithms, k-NN doesn't have a training phase. It simply stores the training dataset and performs predictions directly. k-NN has its limitations; it is sensitive to the scale of features. Features with larger scales can dominate the distance calculations. Normalising or standardising the data is recommended. k-NN is utilised in various domains, including image classification, medical diagnostics, recommendation systems, anomaly detection, text analysis, etc.

12.3.3 Decision tree

The DT was first introduced by British researcher William Belson in 1959 and is used as a supervised machine learning model for data classification and regression analysis. It is a graphical representation of a series of decisions and their potential outcomes. The DT works by recursively splitting the dataset into subsets based on the values of input features, ultimately assigning a class label to each instance.

The tree structure resembles a flowchart as shown in Figure 12.3c, where internal nodes represent decisions or tests on specific features, and branches originating from these nodes depict the possible outcomes of these tests. At the leaf nodes, the final predictions or class labels are assigned. The construction of a DT involves selecting the best feature to split the data at each step, aiming to maximise the purity or homogeneity of the resulting subsets. This is usually achieved through metrics like Gini impurity or Information gain. DTs are advantageous due to their interpretability, which allows for clear visualisation of the decision-making process. They can handle both numerical and categorical data, making them versatile for various applications. However, DTs can be prone to overfitting, especially when the tree becomes too deep and complex. To mitigate this, techniques like pruning, which involves removing branches that do not contribute much to overall accuracy, can be applied. Moreover, ensemble methods like RFs and Gradient Boosting use multiple decision tress to improve generalisation and accuracy.

In summary, the DT Classifier is a fundamental algorithm in machine learning known for its simplicity, interpretability, and ability to handle diverse data types. While it has advantages such as ease of interpretation and handling non-linear relationships, precautions must be taken to prevent overfitting and improve generalisation performance and its limitations, its straightforward structure, combined with advanced techniques, makes it a powerful tool for solving classification problems across various domains.

12.3.4 RF classifier

RF was first introduced by Tim Kam Ho in 1995, which is a supervised learning method used for classification and regression problems. It belongs to the ensemble learning family of algorithms (known as bagging and bootstrap aggregation), which combine the predictions of multiple base models to improve overall performance and robustness. RF is particularly effective in handling complex and high-dimensional datasets, making it a popular choice across various domains, including finance, healthcare, and natural language processing.

At its core, RF algorithm constructs a multitude of decision tress during training and combines their outputs to make predictions. Each DT is trained on a random subset of the training data, and the final prediction is obtained by aggregating the predictions of individual tress. This process helps to reduce the risk of overfitting, a common issue in single decision tress models, by introducing randomness and diversity in the learning process. The benefits of using a RF classifier include robustness, accuracy, variable importance, non-linearity handling, outlier robustness, etc. However, there are also some considerations and limitations associated with RF, which includes interpretability, computational sources and hyperparameters, etc. RF has several hyperparameters that need to be tuned, such as the number of tresses, the maximum depth of each tree, and the size of feature subsets. However, like any programming, proper hyperparameter tuning and thoughtful analysis of results are essential for maximising its effectiveness.

12.4 VALIDATION METHODS

12.4.1 K-fold for cross-validation

The K-fold is a validation approach that involves dividing the data into K-folds for validation purpose. To validate the holding method, each subset of k is utilised as a test, while the remaining (k–1) subsets serve as the training set. To determine the accuracy of trained model, the average of all k iterations is computed. This approach is beneficial in mitigating both underfitting and overfitting of the data, as it ensures the utilisation of each segment within the validation process.

12.4.2 Stratified K-fold cross-validation

Stratified K-fold is an extension of K-fold validation approach primarily used for classification problems. This technique involves randomly partitioning the entire dataset into K-folds while ensuring that the distribution of the target classes remains consistent across each fold. In other words, the proportions of different target classes are maintained at relatively constant level in each fold. This approach is particularly useful for maintaining representative class distributions within the folds and enhancing the reliability of validation process in classification tasks.

12.5 PERFORMANCE EVALUATION MATRIX

The evaluation of classification models relies on four common measures such as accuracy, precision, recall, and $F1$. Accuracy assesses the correct predictions made by the model in relation to the total number of predictions. Precision determines the ratio of true positive (TP) to all instances predicted as positives by the model. Recall calculates the accuracy of correctly predicted TP instances out of all actual positive instances. The $F1$ score computes the harmonic mean of precision and recall, providing a balanced assessment of the model's performance.

$$\text{Accuracy} = \frac{|TP + TN|}{|TP + FP + TN + FN|} \tag{12.21}$$

$$\text{Precision} = \frac{|TP|}{|TP + FP|} \tag{12.22}$$

$$\text{Recall} = \frac{|TP|}{|TP + FN|} \tag{12.23}$$

$$F1 = 2 \times \frac{\text{Precision Recall}}{\text{Precision} + \text{Recall}} \tag{12.24}$$

where TP – True Positive Rate, TN – True Negative Rate, FP – False Positive Rate, and FN – False Negative Rate.

12.5.1 Confusion matrix

The confusion matrices are used as a pivotal tool in assessing the performance of the classification models. It provides a comprehensive overview of how accurately the predicted classes align with the actual classes as shown in Figure 12.4. The evaluation process involves computing various

	ACTUAL CLASS	
PREDICTED CLASS	True Positive (TP)	False Positive (FP)
	False Negative (FN)	True Negative (TN)

Figure 12.4 Confusion matrix.

metrics derived from the confusion matrix, including true positive, false positive (FP), false negative (FN), and true negative (TN). TP represents the instances that are correctly predicted as positive by the model, matching the actual positive instances. FP, on the other hand, corresponds to cases where the model inaccurately predicted a positive outcome when the actual class is negative. FN denotes instances where the model wrongly classified a negative outcome when the actual class is positive. TN signifies instances that are accurately classified as negative by the model, in accordance with the actual negative instances.

By meticulously analysing the values within metrics, the confusion matrix provides a comprehensive understanding of a classification model's strengths and weaknesses, shedding light on its performance accuracy and predictive capabilities across different class outcomes.

12.6 EXPERIMENT SETUP AND DATASETS

The experiment setup of a bearing test stand of the Case Western Reserve University (CWRU) Bearing Data Center, USA [40] is shown in Figure 12.5. This setup comprises various components including 2 hp electric motor, drive and fan end bearing, encoder, dynamometer, and torque transducer. The SKF 6205-2RS JEM and SKF 6203-2RS JEM bearings were placed at both drive and fan end to support the motor shaft. In access to bearing performance, a single point fault was introduced on the test bearing with the fault diameter of 0.007, 0.014, and 0.021 inches using the electro-discharge machine. The depth of the introduced fault is the same as 0.11 inches for all the fault and load conditions. The details of the bearings are given below:

- Bearing types – 6205-2RS JEM SKF
- Inside diameter – 0.9843 in.
- Outside diameter – 2.0472 in.
- Ball diameter – 0.3216 in.
- Pitch diameter – 1.537 in.
- Roller number -09

This experiment was conducted for the healthy and faulty bearings, during the experimentation, the motor was running at a constant speed of approximately 1,797–1,720 rpm for the motor load of 0–3 hp. Vibration data were recorded by using an accelerometer at 12 kHz sample rate when

Figure 12.5 Bearing test rig experiment setup and defected bearings at CWRU Bearing Data Center, USA.

both healthy and faulty bearings were reinstalled into the motor. Here, the data was recorded for a duration of 10 seconds (equivalent to 120.000 samples) for each fault case. Subsequently, this recorded data was being processed within MATLAB® software environment and then data was stored in (.mat) file format, denoted by the nomenclature "XA_B". In this naming format, X denotes the bearing component, A represents the fault class and B indicates the connected load. As an illustrative example, in case of an inner fault with diameter of 0.007 inches and no load, the corresponding datafile was saved with the name IR007_0.

12.7 RESULTS AND DISCUSSION

In this present study, vibration signatures of damaged bearings were obtained from an experimental setup designed by the CWRU laboratory. Before the feature extraction procedure, the recorded vibration signals are normalised to achieve the required accuracy from the classification models. The classification models with proposed methods are simulated on the Jupyter-Notebook environment with Python programming.

Table 12.2 Computed statistical features from vibration signal

	STD	RMS	MEAN	KURTOSIS	VARRIANCE	CREST FACTOR	SKEWNESS	Clearance Factor	Shape Factor	Impulse Factor	Class
2726	0.092629	0.464515	0.455198	3.473341	0.008580	1.579956	-0.062403	1.630228	1.020469	1.612296	B021
852	0.054210	0.522198	0.519380	30.642565	0.001825	1.891249	-0.010821	1.907696	1.005426	1.901510	IR014
9	0.120009	0.510439	0.496149	3.024380	0.014402	1.749927	-0.037912	1.828819	1.028802	1.028802	Normal
1363	0.038841	0.496416	0.494896	3.857392	0.001509	1.278845	0.071760	1.284757	1.003071	1.282772	B014
1460	0.034280	0.491981	0.490787	2.968986	0.001175	1.209774	-0.020499	1.214204	1.002433	1.212718	B021
2042	0.064863	0.476064	0.471630	7.569774	0.004207	1.771108	0.289041	1.796464	1.009401	1.787758	IR021

The performance outcomes of the proposed techniques are discussed for healthy and three bearing faults classes for three individual scenarios. The details of the extracted features are discussed in Table 12.2. The process of fault classification is carried out separately for each scenario, which is computed using the classification models including DT, RF, SVM, and k-NN. The assessment of these classification models is evaluated for the different fault classes using the both K-fold and stratified fold approach, with a specified number of cross-validation iterations (cv=5).

12.7.1 Scenario 1

In scenario 1, the proposed method is used to evaluate the healthy (H), inner race (IR), and ball bearing (B) fault through multiclass machine learning models. These classes are categorised as shown in Table 12.3 where Class1-H has 600 instances, Class2-IR and Class3-B have 1,800 instances each. Each of these instances is constructed utilising a features vector extracted from vibration signal, considering 800 samples in one segment.

The performance evaluation of the classification models is discussed in Table 12.4 with k fold and stratified fold approach for scenario 1. A five-fold cross-validation is executed on training dataset, which obtained the remarkable results. The average accuracy is achieved at 100% for both RF and DT, 96% for k-NN, and 94% average accuracy in SVM. Furthermore, the accuracy of these classification models experienced an approximate improvement of 3.76% when the stratified fold approach was applied instead of the k-fold approach, especially in case of k-NN. The confusion matrices, shown in Figure 12.6, provide a visual representation of the accuracy of bearing fault prediction for each individual class.

12.7.2 Scenario 2

In scenario 2, the proposed methodology is used to evaluate the bearing fault with fault diameters of 0.007, 01014, and 0.021 inches in inner race (IR) and ball bearing (B) utilising through multiclass machine learning models. The classes are classified as follows: Class1-H signifies Healthy bearings and 'ClassX-AB' associated with a fault in the bearings, where X represents the class number, A represents the bearing component and B represents the

Table 12.3 CWRU dataset classes for scenario 1

New class	*Original class*	*Instances*
Class 1 - H	H_0 to H_3	4 × 150=600
Class 2 - IR	IR_0 to IR_3	4 × 3 × 150=1,800
Class 3 - B	B_0 to B_3	4 × 3 × 150=1,800

Table 12.4 Performance evaluation of classification models with K-fold and stratified fold cross-validation method

	RF		*k-NN*		*SVM*		*DT*	
Fold	*K-fold*	*SK-fold*	*K-fold*	*SK-fold*	*K-fold*	*SK-fold*	*K-fold*	*SK-fold*
1	99.52	100	95.23	99.76	93.80	94.76	99.76	99.76
2	100	100	95.71	99.52	93.80	92.85	100	99.76
3	99.76	99.76	95.47	99.52	93.57	91.19	99.76	99.52
4	100	99.76	95.95	99.76	95.00	94.52	99.76	99.76
5	100	100	96.66	99.52	95.47	94.04	100	99.76
Average	**100**	**99.90**	**96.00**	**99.61**	**94.00**	**93.47**	**100**	**99.71**

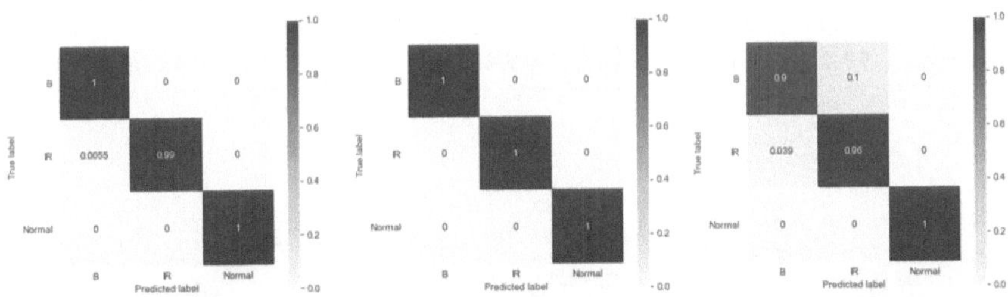

Figure 12.6 Confusion matrixes for scenario 1.

fault diameter. The details of the fault classes are shown in Table 12.5. Each class comprises 600 instances that are computed from vibration signal, considering the 800 samples in each segment. These fault classes are evaluated across different operating conditions, including no load, half load, full load and overload conditions.

The performance analysis of the classification models for scenario 2 is discussed in Table 12.6, using with k-fold and stratified fold cross-validation approach. The computational approach for the classification models is similar to the scenario 1. Therefore, the accuracy of the classification models has experienced an improvement of approximately 5.9% for k-NN model and 2.09% for the SVM model utilising the stratified fold approach. The confusion matrices are shown in Figure 12.7, which provides the visual representation of bearing fault prediction accuracy for seven classes.

Table 12.5 CWRU dataset classes for scenario 2

New class	*Original class*	*Instances*
Class 1 – H	H_0 to H_3	4 × 150=600
Class 2 – IR007	IR_0 to IR_3	4 × 150=600
Class 3 – IR0014	IR_0 to IR_3	4 × 150=600
Class 4 – IR0021	IR_0 to IR_3	4 × 150=600
Class 5 – B007	B_0 to B_3	4 × 150=600
Class 6 – B014	B_0 to B_3	4 × 150=600
Class 7 – B021	B_0 to B_3	4 × 150=600

Table 12.6 Performance evaluation of classification models with K-fold stratified fold cross-validation method

	RF		*k-NN*		*SVM*		*DT*	
Fold	*K-fold*	*SK-fold*	*k-fold*	*SK-fold*	*K-fold*	*SK-fold*	*K-fold*	*SK-fold*
1	98.80	99.20	93.80	98.09	91.90	94.10	98.09	98.40
2	99.40	98.80	92.02	97.30	92.38	92.60	98.57	97.90
3	98.80	98.40	91.19	96.90	88.69	93.60	97.97	98.09
4	98.92	99.04	92.97	98.09	92.02	94.50	98.57	97.70
5	98.80	98.50	91.77	97.02	91.06	93.80	98.33	98.40
Average	**99.00**	**98.8**	**92.00**	**97.48**	**91.00**	**93.7**	**98.00**	**98.10**

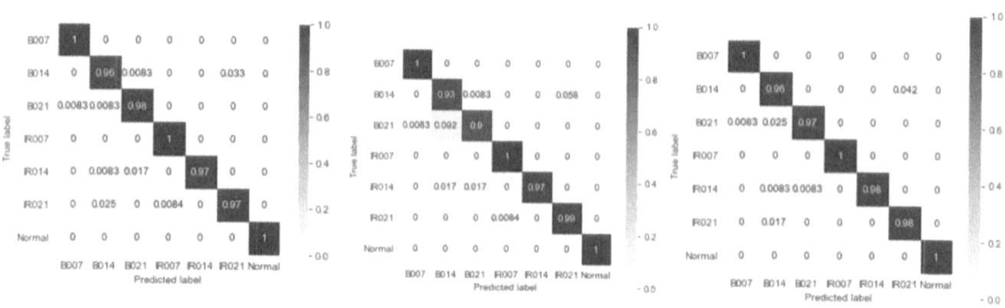

Figure 12.7 Confusion matrixes for scenario 2.

Normal and B007 class bearing faults have been predicted with 100% accuracy by all the classification models than the other fault classes.

12.7.3 Scenario 3

In scenario 3, the proposed method is used to evaluate with the same bearing faults such as 0.007, 0.014, and 0.021 inches in inner race (IR) and ball bearing (B) through multiclass machine learning models. This analysis is conducted only for two loading conditions such as half load and full load. The healthy bearing represents Class1-H and faulty bearing labelled as ClassX-AB as mentioned in scenario 2. These fault categories and their corresponding details are summarised in Table 12.7. Each class also comprises 600 instances that are computed from vibration signal, considering the 800 samples in each segment.

The performance analysis of the classification models for scenario 3 is discussed in Table 12.8, using with k-fold and stratified fold cross-validation approach. The computational approach for the classification models is similar to the scenarios 1 and 2. Therefore, the accuracy of the classification models experienced an improvement of approximately 3.3% for the k-NN model and decreased by about 0.56% for SVM model utilising the stratified fold approach. The confusion matrices are shown in Figure 12.8, which provides the visual representation of bearing fault prediction accuracy for seven classes. Normal, IR007 and B007 class bearing faults have been predicted with 100% accuracy by all the classification models than the other fault classes.

Table 12.7 CWRU dataset classes for scenario 3

New class	*Original class*	*Instances*
Class 1 – H	H_1 to H_2	4 × 150=600
Class 2 – IR007	IR_1 to IR_2	2 × 150=300
Class 3 – IR0014	IR_1 to IR_2	2 × 150=300
Class 4 – IR0021	IR_1 to IR_2	2 × 150=300
Class 5 – B007	B_1 to B_2	2 × 150=300
Class 6 – B014	B_1 to B_2	2 × 150=300
Class 7 – B021	B_1 to B_2	2 × 150=300

Table 12.8 Performance evaluation of classification models with K-fold stratified fold cross-validation method

	RF		*k-NN*		*SVM*		*DT*	
Fold	*K-fold*	*SK-fold*	*K-fold*	*SK-fold*	*K-fold*	*SK-fold*	*K-fold*	*SK-fold*
1	99.52	100	95.23	99.76	93.80	94.76	99.76	99.76
2	100	100	95.71	99.52	93.80	92.85	100	99.76
3	99.76	99.76	95.47	99.52	93.57	91.19	99.76	99.52
4	100	99.76	95.95	99.76	95.00	94.52	99.76	99.76
5	100	100	96.66	99.52	95.47	94.04	100	99.76
Average	**100**	**98.8**	**96.00**	**99.61**	**94.00**	**93.47**	**100**	**99.71**

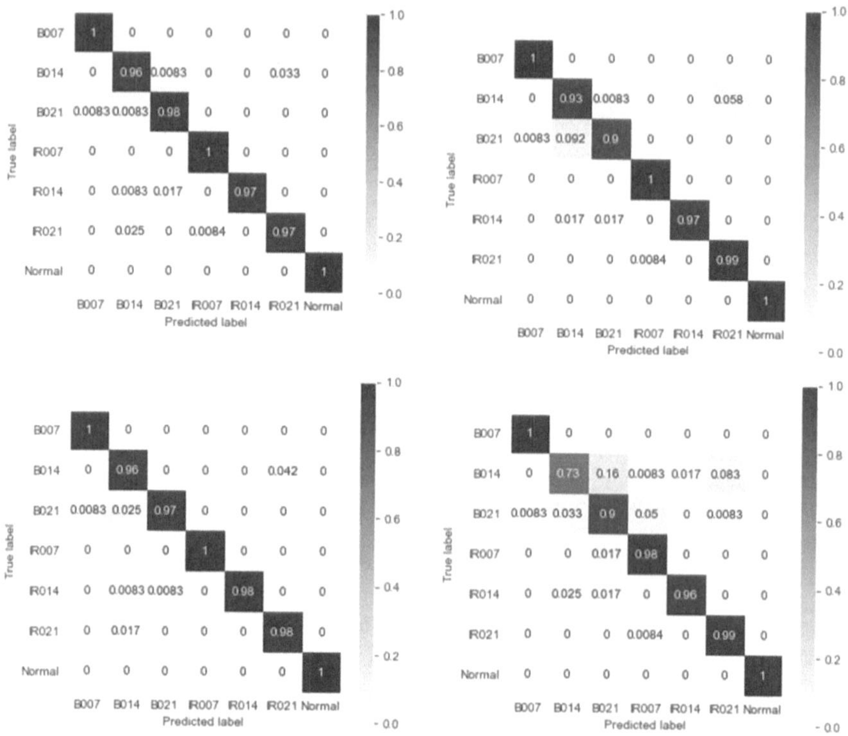

Figure 12.8 Confusion matrixes for scenario 3.

12.8 CONCLUSION

This work presents a comprehensive analysis of bearing fault classification through a combination of soft computing techniques involving statistical feature extraction and machine learning classification methods. Vibration signals were acquired from artificially defective bearings in the CWRU laboratory, USA under different loading conditions. There are ten statistical time domain features that are extracted from the vibration signal in the time domain after passing through the normalisation process. These computed statistical features are being used with machine learning models such as SVM, K-Near Neighbour (k-NN), DT, and RF for the purpose of classification of the bearing faults. The outcomes of fault classification are presented using both the k-fold and stratified fold approaches. It is found that the accuracy of the RF method is relatively higher (approximately 100%) compared to other classification methods namely the SVM, k-Near Neighbour (k-NN), RF, and DT for scenario 1, scenario 2, and scenario 3. On the other hand, the SVM classification model exhibited the lowest fault classification accuracy, with 94% in scenario 1, 91% in scenario 2, and 94% in scenario 3 using the k-fold approach. However, with the stratified fold method, accuracy has been improved across all four classification learning models. Therefore, k-NN and DT have moderate fault classification accuracy with 99% across all considered scenarios. Thus, the RF classifier emerged as a highly accurate and powerful method when compared to the other classification models such as SVM, k-NN, and DT.

Furthermore, the frequency domain analysis is not discussed in this study. The performance of the classification methods can be improved by using the frequency domain features extraction and the feature selection methods. This approach will be the future objective of bearing fault classification with machine learning models.

REFERENCES

[1] Faiz, J., & Moosavi, S. M. M. (2016). Eccentricity fault detection - From induction machines to DFIG - A review. *Renewable and Sustainable Energy Reviews*, *55*, 169–179. https://doi.org/10.1016/j.rser.2015.10.113

[2] Bhavsar, R. C. (2013). Various techniques for condition monitoring of three phase induction motor- A review. *International Journal of Engineering Inventions*, *3*(4), 22–26. https://www.ijeijournal.com/papers/v3i4/D0342226.pdf

[3] Kumar, N., & Satapathy, R. (2023). Bearings in aerospace, application, distress, and life: A review. *Journal of Failure Analysis and Prevention*, *23*(3), 915–947. https://doi.org/10.1007/s11668-023-01658-z

[4] Nandi, S., Toliyat, H. A., & Li, X. (2005). Condition monitoring and fault diagnosis of electrical motors - A review. *IEEE Transactions on Energy Conversion*, *20*(4), 719–729. https://doi.org/10.1109/TEC.2005.847955

[5] Choudhary, A., Goyal, D., Shimi, S. L., & Akula, A. (2019). Condition monitoring and fault diagnosis of induction motors: A review. *Archives of Computational Methods in Engineering*, *26*(4), 1221–1238. https://doi.org/10.1007/s11831-018-9286-z

[6] Kumar, S., Goyal, D., & Dhami, S. S. (2018). Statistical and frequency analysis of acoustic signals for condition monitoring of ball bearing. *Materials Today: Proceedings*, *5*(2), 5186–5194. https://doi.org/10.1016/j.matpr.2017.12.100

[7] Bauer, M., Balaratnam, N., Weidenauer, J., Wagner, F., & Kley, M. (2022). Comparison of envelope demodulation methods in the analysis of rolling bearing damage. *JVC/Journal of Vibration and Control*, *29*, 21–22. https://doi.org/10.1177/10775463221129155

[8] Xi, F., Sun, Q., & Krishnappa, G. (2000). Bearing diagnostics based on pattern recognition of statistical parameters. *Journal of Vibration and Control*, *6*(3), 375–392. https://doi.org/10.1177/107754630000600303

[9] Jain, P. H., & Bhosle, S. P. (2021). A review on vibration signal analysis techniques used for detection of rolling element bearing defects. *International Journal of Mechanical Engineering*, *8*(1), 14–29. https://doi.org/10.14445/23488360/ijme-v8i1p103

[10] Yilmaz, M. S., & Ayaz, E. (2009). Adaptive neuro-fuzzy inference system for bearing fault detection in induction motors using temperature, current, vibration data. *IEEE EUROCON, 2009*, 1140–1145. https://doi.org/10.1109/EURCON.2009.5167779

[11] Lee, D. H., Hong, C., Jeong, W. B., & Ahn, S. (2021). Time-frequency envelope analysis for fault detection of rotating machinery signals with impulsive noise. *Applied Sciences (Switzerland)*, *11*(12), 11125373. https://doi.org/10.3390/app11125373

[12] Patel, R., Gupta, S. P., & Kumar, V. (2014). Bearing fault severity estimation using time-based descriptors for rotating electric machines. *International Journal of Power and Energy Conversion*, *5*(2), 197–209. https://doi.org/10.1504/IJPEC.2014.061764

[13] Nayana, B. R., & Geethanjali, P. (2017). Analysis of statistical time-domain features effectiveness in identification of bearing faults from vibration signal. *IEEE Sensors Journal*, *17*(17), 5618–5625. https://doi.org/10.1109/JSEN.2017.2727638

[14] Singh, P., & Harsha, S. P. (2019). Statistical and frequency analysis of vibrations signals of roller bearings using empirical mode decomposition. *Proceedings of the Institution of Mechanical Engineers, Part K: Journal of Multi-Body Dynamics*, *233*(4), 856–870. https://doi.org/10.1177/1464419319847921

[15] Helmi, H., & Forouzantabar, A. (2019). Rolling bearing fault detection of electric motor using time domain and frequency domain features extraction and ANFIS. *IET Electric Power Applications*, *13*(5), 662–669. https://doi.org/10.1049/iet-epa.2018.5274

[16] Bellini, A., Filippetti, F., Franceschini, G., Tassoni, C., & Kliman, G. B. (2001). Quantitative evaluation of induction motor broken bars by means of electrical signature analysis. *IEEE Transactions on Industry Applications*, *37*(5), 1248–1255. https://doi.org/10.1109/28.952499

[17] Fu, S., Liu, K., Xu, Y., & Liu, Y. (2016). Rolling bearing diagnosing method based on time domain analysis and adaptive fuzzy C -means clustering. *Shock and Vibration*, *2016*, 9412787. https://doi.org/10.1155/2016/9412787

[18] Deng, X. W., Yang, P., Ren, J. S., & Yang, Y. W. (2014). *Rolling bearings time* and frequency domain fault diagnosis method based on Kurtosis analysis. *Asia-Pacific Power and Energy Engineering Conference, APPEEC, 2015-March*, Kowloon, Hong Kong. https://doi.org/10.1109/APPEEC.2014.7066018

[19] Yang, Y. (2017). A signal theoretic approach for envelope analysis of real-valued signals. *IEEE Access*, *5*, 5623–5630. https://doi.org/10.1109/ACCESS.2017.2688467

[20] Li, D. Z., Wang, W., Member, S., & Ismail, F. (2015). A spectrum synch technique for induction motor health condition monitoring. *IEEE Transactions on Energy Conversion*, *30*(4), 1348–1355. https://doi.org/10.1109/TEC.2015.2454440

[21] Šoda, J., Vujović, I., & Kulenović, Z. (n.d.). Analysis of the vibration signal using time-frequency methods. *Transactions of FAMENA*, *39*(3), 23–34.

[22] Romero-Troncoso, R. D. J. (2017). Multirate signal processing to improve FFT-based analysis for detecting faults in induction motors. *IEEE Transactions on Industrial Informatics*, *13*(3), 1291–1300. https://doi.org/10.1109/TII.2016.2603968

[23] Toma, R. N., Prosvirin, A. E., & Kim, J. M. (2020). Bearing fault diagnosis of induction motors using a genetic algorithm and machine learning classifiers. *Sensors (Switzerland)*, *20*(7), 20071884. https://doi.org/10.3390/s20071884
[24] Rai, V. K., & Mohanty, A. R. (2007). Bearing fault diagnosis using FFT of intrinsic mode functions in Hilbert-Huang transform. *Mechanical Systems and Signal Processing*, *21*(6), 2607–2615. https://doi.org/10.1016/j.ymssp.2006.12.004
[25] Amar, M., Gondal, I., & Wilson, C. (2015). Vibration spectrum imaging: A novel bearing fault classification approach. *IEEE Transactions on Industrial Electronics*, *62*(1), 494–502. https://doi.org/10.1109/TIE.2014.2327555
[26] Li, D. Z., Wang, W., & Ismail, F. (2017). An intelligent harmonic synthesis technique for air-gap eccentricity fault diagnosis in induction motors. *Chinese Journal of Mechanical Engineering (English Edition)*, *30*(6), 1296–1304. https://doi.org/10.1007/s10033-017-0192-3
[27] Tian, K., Zhang, T., Ai, Y., & Zhang, W. (2018). Induction motors dynamic eccentricity fault diagnosis based on the combined use of WPD and EMD-simulation study. *Applied Sciences (Switzerland)*, *8*(10), 8101709. https://doi.org/10.3390/app8101709
[28] Chen, H. C., & Pu, H. Y. (2016). Fault analysis of induction motor based on discrete fractional Fourier transform. *Proceedings -2016 IEEE International Symposium on Computer, Consumer and Control, IS3C 2016*, *2*, 69–72. https://doi.org/10.1109/IS3C.2016.28
[29] Xie, S., Wang, X., Qu, C., Wang, X., & Guo, J. (2013). Impacts of different wind speed simulation methods on conditional reliability indices. *International Transactions on Electrical Energy Systems*, *20*, 1–6. https://doi.org/10.1002/etep
[30] Jose, G., & Jose, V. (2014). Fault diagnosis in induction motor using soft computing techniques. *ICACCS 2013- Proceedings of the 2013 International Conference on Advanced Computing and Communication Systems: Bringing to the Table, Futuristic Technologies from Around the Globe, December 2013*, Novi Sad, Serbia. https://doi.org/10.1109/ICACCS.2013.6938693
[31] Jain, N., & Kumar, R. (2022). A review on machine learning & it's algorithms. *International Journal of Soft Computing and Engineering*, *12*(5), 1–5. https://doi.org/10.35940/ijsce.E3583.1112522
[32] Shah, K., Patel, H., Sanghvi, D., & Shah, M. (2020). A comparative analysis of logistic regression, random forest and KNN models for the text classification. *Augmented Human Research*, *5*(1), 12. https://doi.org/10.1007/s41133-020-00032-0
[33] Matić, D., & Kulić, F. (2012). SVM broken bar detection based on analysis of phase current. *15th International Power Electronics and Motion Control Conference and Exposition, EPE-PEMC 2012 ECCE Europe*, Novi Sad, Serbia, 31–34. https://doi.org/10.1109/EPEPEMC.2012.6397456
[34] Deng, L., Zhang, A., & Zhao, R. (2022). Intelligent identification of incipient rolling bearing faults based on VMD and PCA-SVM. *Advances in Mechanical Engineering*, *14*(1), 16878140211072990. https://doi.org/10.1177/16878140211072990

[35] Hasan, A. I. (2021). Vibration classification of power plant equipment using decision tree algorithm approach. *ICT-PEP 2021- International Conference on Technology and Policy in Energy and Electric Power: Emerging Energy Sustainability, Smart Grid, and Microgrid Technologies for Future Power System, Proceedings*, 97–101, Yogyakarta, Indonesia. https://doi.org/10.1109/ICT-PEP53949.2021.9601062

[36] Somvanshi, M., & Chavan, P. (n.d.). A review of machine learning techniques using decision tree and support vector machine. *2016 International Conference on Computing Communication Control and automation (ICCUBEA)*, Pune, India.

[37] Dos Santos, T., Ferreira, F. J. T. E., Pires, J. M., & Damasio, C. (2017). Stator winding short-circuit fault diagnosis in induction motors using random forest. *2017 IEEE International Electric Machines and Drives Conference, IEMDC 2017*, Miami, Florida, USA. https://doi.org/10.1109/IEMDC.2017.8002350

[38] Urbanowicz, R. J., Meeker, M., La Cava, W., Olson, R. S., & Moore, J. H. (2018). Relief-based feature selection: Introduction and review. *Journal of Biomedical Informatics*, *85*, 189–203. https://doi.org/10.1016/j.jbi.2018.07.014

[39] Zhang, Y., Ray, S., & Guo, W. (2016). On the consistency of feature selection with lasso for non-linear targets. *Proceedings of The 33rd International Conference on Machine Learning*, *48*, 183–191.

[40] Kenneth Loparo. (n.d.). *Bearings vibration test data, The Case Western Reserve University Bearing Data Center.* Retrieved June 3, 2022, from https://engineering.case.edu/bearingdatacenter/welcome

Chapter 13

Artificial neural network-based stock price prediction using Levenberg-Marquardt algorithm

Balakrishnama Manohar, Raja Das, Lakshmi M, Jaykumar Ishvarbhai Prajapati, and K. Madhusudhan Reddy

13.1 INTRODUCTION

The nature of the stock market is ever-changing, uncertain, and non-sequential. Forecasting stock prices are challenging due to their reliance on numerous factors, including political circumstances, the global economy, corporate financial statements, performance, and more. As a result, to optimize gains and decrease losses, techniques that foresee stock values in advance by examining trends over the past few years might prove exceedingly valuable for stock market fluctuations [1]. Throughout history, two main approaches have been put forth for projecting a company's stock price: To anticipate a stock's future price, the technical analysis approach employs historical stock prices such as closing and opening figures, trading volume, and neighbouring closing values [2]. The subsequent kind of examination is qualitative and is performed by an economic expert [3] grounded in external elements such as corporate profile, market condition, political and economic influences, and textual content in the structure of financial news pieces, social media updates, and even online journals. In the current landscape, sophisticated intelligent techniques rooted in either technical or fundamental analysis are harnessed for stock price prediction. This is particularly significant when evaluating the stock market, where the volume of information is extensive and intricate. To handle this array of information, an effective model capable of recognizing hidden patterns and intricate connections within this extensive dataset is necessary. In this domain, machine learning methodologies have demonstrated a capacity to elevate efficiency by 60%–86% in comparison to conventional approaches [2,4].

When it comes to projecting stock prices, the bulk of previous endeavours in this domain has relied on classical algorithms such as linear regression [5], moving average convergence/divergence (MACD) [6], alongside linear models like autoregressive moving average (ARMA) and autoregressive integrated moving average (ARIMA) [7,8]. Recent studies suggest that machine learning has the potential to enhance the prognosis of the stock market. Techniques such as support vector machine (SVM) and

DOI: 10.1201/9781003534761-13

Random Forest (RF) have surfaced [9,10]. Various techniques founded on neural networks, including artificial neural network (ANN), convolutional neural network (CNN) [11], recurrent neural network (RNN) [12,13], and advanced architectures such as long short-term memory (LSTM) [14], have exhibited encouraging results [15,16].

Artificial neural networks (ANNs) have gained widespread utilization in addressing numerous challenges due to their adaptable characteristics. ANNs possess the ability to unveil latent attributes through a self-learning process. These versatile approximators are adept at establishing connections between inputs and outputs within intricate and extensive datasets. As a result, ANNs are a viable choice for approximating the valuation of a corporation's stocks. Selvin et al. projected the stock prices of businesses listed on the NSE by conducting a relative analysis of diverse deep learning algorithms [13]. Hamzaebi et al. explored multi-periodic forecasting in the stock market by employing iterative and guided methodologies such as the ANN paradigm [14]. Yodele et al. and Y. Kara et al. delved into stock price index trends using dual models based on ANNs and SVMs [10,17]. They examined how well both models worked and found that, on average, the ANN model achieved quite better than the SVM mode. They evaluated the efficacy of both models and ascertained that, on average, the ANN model outperformed the SVM approach. Qi & Zhang investigated optimal techniques for implementing a neural network to depict temporal trends [18]. For modelling distinct trend patterns, they employed four varying approaches: unprocessed data, unprocessed data with a temporal index, detrending, and differencing. Their findings concluded that neural networks yielded superior outcomes. In 2003, H.K. Cigizoglu developed the ANN methodology for predicting, estimating, and extrapolating daily river flow data in Turkey's East Mediterranean region [19,20]. Their research demonstrated that ANN yielded a more accurate fit to the data compared to other methodologies. ANNs can be regarded as computational or mathematical models, mirroring the operational principles of biological neural networks. The architectural design of these neural networks allows them to identify patterns within noisy datasets. ANNs commence by training a system with an extensive dataset, known as the "training phase." Subsequently, during the "validation or prediction phase," the network encounters data that was not part of the training phase, solely for predictive purposes. Bishop et al. introduced the notion that ANN's capacity to learn and forecast stems from the adaptive nature of the human brain [21]. Consequently, ANNs have been harnessed across a diverse array of applications and have demonstrated competence in numerous fields [22,23].

The novelty of the current study is that the opening price of a company was predicted using two classic algorithms: ANN with LM and ANN with GDX. The models incorporate a series of novel factors generated by incorporating High, Low, Open, Close, Adjacent Close, and Volume data into a company's financial information. These new factors will be very important for making the models better at predicting the opening price of a company the next day. Three performance measures, *RMSE*, R^2, *and MAPE*, are used to test how well the models work.

13.2 METHODS AND MATERIALS

13.2.1 Artificial neural networks

The architecture of ANNs drew inspiration from the human brain design. In ANNs, nodes, connections, and weights emulate the roles of neurons, synapses, and the electrical signal strengths found in the brain's nervous system, respectively [24,25]. ANNs have the flexibility to adopt a wide array of configurations, permitting the incorporation of diverse patterns involving links, loops, and connections. However, the scope of this work exclusively involves the utilization of a specific neural network type: the feed-forward network.

All information originating from the input nodes is transmitted to each node within the hidden units and subsequently to every node located in the output units within an ANN. Each tier has the flexibility to incorporate varying quantities of nodes, often involving multiple concealed layers that must be traversed before reaching the ultimate output layer [26]. When in the process of constructing a neural network, the selection of the appropriate number of nodes and layers holds utmost significance. The neural network employed for this study is a feed-forward network, and its depiction can be observed in Figure 13.1.

Neurons form connections through a shared mechanism, facilitated by a set of designated weights. The Multilayer Perceptron (MLP) stands as a prevalent technique in addressing regression-oriented problems. This network configuration encompasses three layers: the input layer, output layer, and hidden layer [17,19]. Neurons introduce a bias to the cumulative values of input factors after weighting, followed by the application of a transfer function to determine output values [20]. The input layer's neuron count aligns with the number of input factors [21,22]. As depicted in Figure 13.1, the conventional MLP structure is outlined. The sigmoid function was employed as the activation function. According to Figure 13.1, the net input to the jth neuron in the hidden layer is expressed as in Eq. (13.1):

$$y_j(x) = \sum_{i=1}^{n} v_{ji} x_i + v0_j \tag{13.1}$$

where the weight linking the *i*th node in the input layer to the *j*th node in the hidden layer is denoted as v_{ji}, and v_{0j} represents the bias associated with the *j*th node in the hidden layer. The output of the *j*th node in the hidden layer is determined in Eq. (13.2):

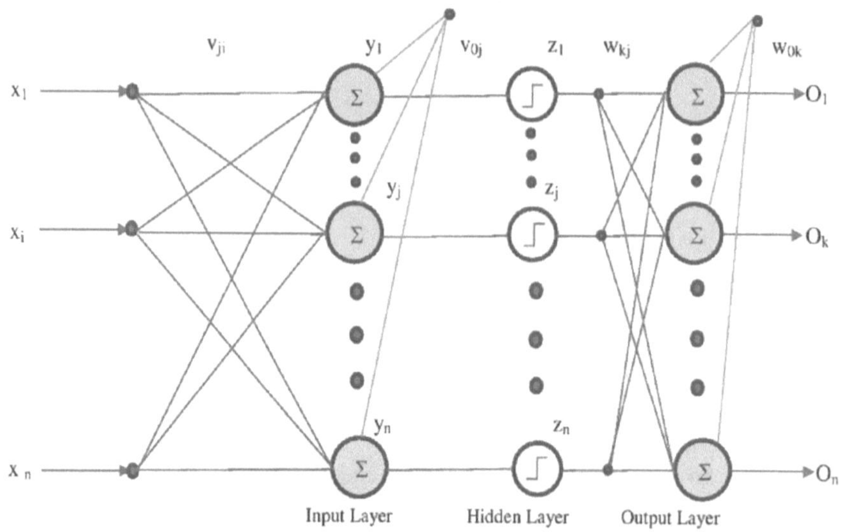

Figure 13.1 A schematic diagram of artificial neural network.

$$z_j(x) = \frac{1}{\left(1 + e^{-y_j(x)}\right)} \tag{13.2}$$

When provided with an input vector x, the resulting value $O_k(x)$ of the kth node in the output layer is obtained by summing the products of the outcomes from the hidden layer nodes, each multiplied by their corresponding weights, along with the bias specific to the kth node in the output layer. This is mathematically expressed as in Eq. (13.3):

$$O_k = \sum_{j=1}^{m} w_{kj} z_j + w0_k \tag{13.3}$$

In this context, where w_{kj} represents the weight connecting the jth node in the hidden layer to the kth node in the output layer, and w_{0k} denotes the bias term associated with the kth node in the output layer.

13.2.2 ANNs training algorithms

In this study, the backpropagation method was used to train feed-forward ANNs [27]. Various learning methods, such as Levenberg-Marquardt (LM), gradient descent with momentum and adaptive learning rate backpropagation (GDX), are used to train and improve the built models.

where N represents the complete count of training datasets; and E_α stands for the error related to the Nth training data, the approximation for E_α is calculated using the formula provided in Eq. (13.4):

$$E_\alpha = \frac{1}{2}\sum_{q=1}^{r}\left(O_q - P_q\right)^2 \tag{13.4}$$

In this context, where r signifies the overall count of outcome nodes, O_q denotes the observed result within the qth outcome node, and P_q stands for the anticipated result at the same qth outcome node.

In each training algorithm, a challenge has been made to reduce the overall error by adjusting the weights and biases.

13.2.3 Levenberg-Marquardt

Nonlinear least squares difficulties are resolved using the LM method. The key benefit of adopting the LM technique is that Hessian matrix (H) computation is not necessary [22]. In these kinds of issues, the Jacobian matrix (J) is used to approximate the Hessian matrix as in Eq. (13.5):

$$H = J^T J \tag{13.5}$$

The LM method interpolates the findings using the Gauss-Newton and gradient descent techniques. The LM algorithm iteration might be represented by the Eq. (13.6):

$$V_{k+1} = v_k - \left(J^T J + \mu I\right)^{-1} J^T e \tag{13.6}$$

In this context, the calculated error at the output is denoted as e, the identity matrix is represented by I, and μ stands as a scalar. When μ equals 0, Eq. (13.6) entirely encapsulates the Gauss-Newton approach. LM method functions similarly to the gradient descent approach for $\mu > 0$. While the Gauss-Newton method tends toward local convergence, the gradient descent approach exhibits global convergence. The LM method uses larger μ values initially for a global convergence and gradient descent algorithm later on for faster local convergence [28]. The work flow chart is shown in Figure 13.2.

13.2.4 Gradient descent with momentum (GDX) and adaptable learning rate backpropagation

Gradient descent utilizing momentum and incorporating backpropagation of an adaptable learning rate. A conventional backpropagation approach often maintains a consistent learning rate throughout the training duration.

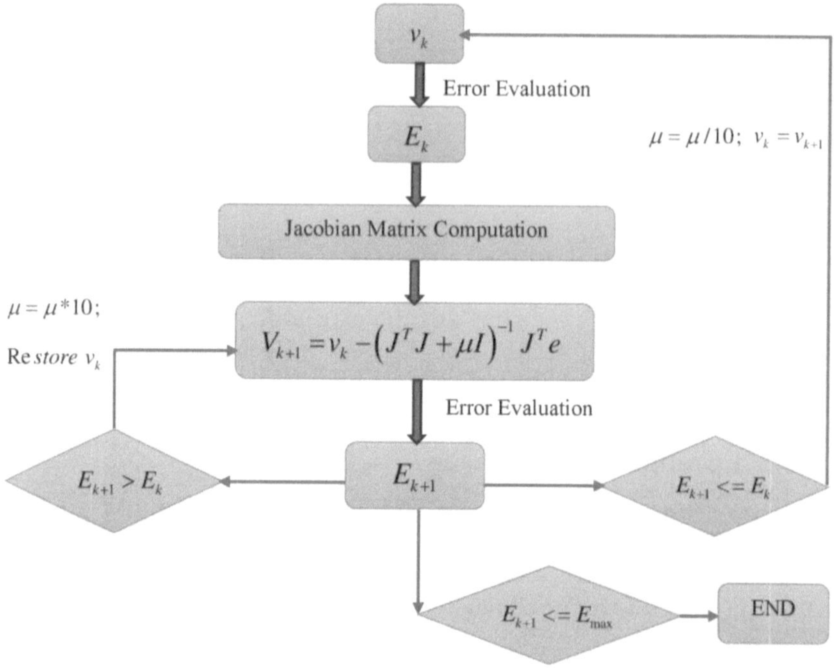

Figure 13.2 A flow chart of LM algorithm.

The appropriate adjustment of the learning rate significantly impacts the algorithm's effectiveness. Setting the learning rate excessively high can lead to algorithmic instability and fluctuations, whereas setting it overly low can protract convergence. To tackle this challenge, the gradient descent with momentum and adaptive learning rate backpropagation (GDX) technique amalgamates momentum training with adaptive learning rate mechanisms [29]. The goal of an adaptive learning rate is to maintain learning stability while maintaining a large learning step size. According to the gradient descent with momentum, each variable is changed. Momentum permits the network to disregard minor characteristics on the error surface by acting as a low-pass filter. One of the simplest and most popular methods for training a network is using this training algorithm [23].

13.2.5 Implementation (construction, training, and testing)

The stock market is a chaotic system, which makes predictions impossible. The stock market can be impacted by a wide range of variables, either directly or indirectly. There is no discernible connection between the factors and pricing. The variables do not have a mathematical connection that we can determine. Inferring a stock's value from such data does not seem to be possible using any formulae. Due to the fact that we do not need to know the solution, the ANN approach is appropriate for this kind of chaotic system. This is a fairly important benefit of ANN approaches. We just need to use the ANN to provide the appropriate results for the supplied inputs [27]. The system will be able to do the function after adequate training. An additional advantage of ANN is that, while learning, it develops the ability to ignore input that does not contribute to the output. The backpropagation technique is also employed for the training phase of our designed system, which is where the parameters, referred to as weights, are given. To employ these weights in the prediction stage, the same formulae used in the training phase are used. MATLAB® was used to develop and run the suggested ANN model for our investigation. The design, training, testing, and validation of the ANN were all included in the execution. The following steps are taken in order to carry out the aforementioned procedure [24].

13.2.6 Description of data

The information was collected from the Yahoo/Finance platform [30]. The data comprises daily records for TCS and Reliance Index stocks. The parameters employed as signals encompass the stock's opening price at the start of the day (open), the maximum price attained by the stock throughout the day (high), the minimum price observed during the day (low), the closing price at the end of the day (close), and the volume of stock shares traded within the same day (volume). The next day's opening price of the stock is utilized as output. Another financial indicator is employed in this research at the same time, but there is no difference in the anticipated following day opening value [31]. The dataset takes place over the years January 2022 to January 2023. There are a total of 366 records in the collection. 329 records are utilized to train and validate the network, whereas 37 records are used to test it. Table 13.1 Displays statistical information regarding the dataset utilized for both training and testing purposes.

Table 13.1 Statistics of the dataset for both training and testing purposes

	Total datasets	*Train data*	*Test data*
Time period	11/01/2022–12/01/2023	11/01/2022–17/12/2022	18/12/2022–12/01/2023

13.2.7 Evaluation criteria and performance analysis

A variety of metric sets can be employed and compared to gauge the efficacy of the ANN models proposed [32]. These encompass regression analysis, error histograms, as well as assessments of root mean square error (RMSE) and mean absolute percentage error (MAPE) for accuracy evaluation [33].

Regression Analysis: Regression analysis constitutes a collection of statistical methodologies employed to predict the connection between variables, encompassing both independent and dependent factors [34]. An independent variable serves as a manipulated input, hypothesis, or stimulus used to examine its potential influence on a dependent variable [35].

Error Histogram: The representation of discrepancies between projected values and actual forecasted values after training a feed-forward neural network (FFNN) is referred to as the error histogram [36]. These error values can encompass negatives, indicating the degree to which predicted values diverge from the desired outcomes [37].

To assess the precision of the model, this study employs statistical equations for RMSE, coefficient of estimation (R^2) and MAPE, as indicated by Eqs. (13.7–13.9):

$$RMSE = \sqrt{\frac{\sum_{i=1}^{n}(O_i - P_i)^2}{N}} \tag{13.7}$$

$$R^2 = 1 - \frac{\sum_{i=1}^{n}(O_i - P_i)^2}{\sum_{i=1}^{n}(O_i - P_i)^2 + \sum_{i=1}^{n}\left(P_i - \hat{O}\right)^2} \tag{13.8}$$

$$\text{MAPE} = \frac{100}{n}\sum_{i=1}^{n}\left|\frac{O_i - P_i}{O_i}\right| \tag{13.9}$$

In this context, O_i signifies the observed values, P_i represents the predicted values, $\hat{O}_i$ symbolizes the mean of observed values, and n stands for the overall count of observations. Evidently, R^2 a greater value would indicate enhanced predictive capability. Conversely, a reduced RMSE value would ensure the resilience of the constructed models [38,39].

13.3 RESULTS

The present investigation focuses on forecasting next day morning stock market prices. We employed two algorithms for training the network and estimating next day morning prices. This section presents the results obtained from these two algorithms, followed by a succinct discourse. Initially, a discussion of the particular results of each algorithm is conducted. Each is presented with its regression value, error histogram, accuracy performance testing for test data. We have conducted experiments on countless configurations for each algorithm. At the conclusion of this section, a comprehensive comparison of the accuracy performance of these two algorithms is presented and elucidated.

13.3.1 Performance plot

The performance plot serves as a tool to pinpoint the specific number of iterations (epochs) at which the mean squared error (MSE) performance function attains its minimum value or stabilizes. It's important to note that the number of epochs doesn't directly correlate with the time required to train the network. Within the plot, the blue line corresponds to the training dataset, constituting approximately 70% of the entire data. The green line represents the validation dataset, encompassing around 15% of the data, while the red line portrays the test outcomes, comprising the remaining 15%. In Figures 13.3 and 13.4, the performance plot results for the LM algorithm across both TCS and Reliance datasets are displayed. Remarkably, the LM algorithm achieves optimal outcomes in just 12 (4) epochs for both datasets. In contrast, Figures 13.5 and 13.6 illustrate the performance plot outcomes for the GDX algorithm in both datasets. In this context, the GDX algorithm demonstrates superior performance after 144 (255) epochs in both TCS and Reliance datasets. However, it's crucial to acknowledge that epochs don't directly signify the time taken by the model for training, testing, and validation purposes. Overall, it's evident that the LM algorithm converges rapidly (in terms of epochs) compared to GDX in both datasets, resulting in the least mean squared error across both TCS and Reliance datasets.

13.3.2 Regression plots

Regression plots are employed to validate the network's performance and illustrate the correlation between the outcomes of the proposed system and the target values. Hence, these plots showcase the network's outputs in relation to the targets for training, validation, testing, and overall datasets. The dashed line within each graph represents the ideal outcome, while the solid line signifies the optimal-fitting linear regression line linking outputs

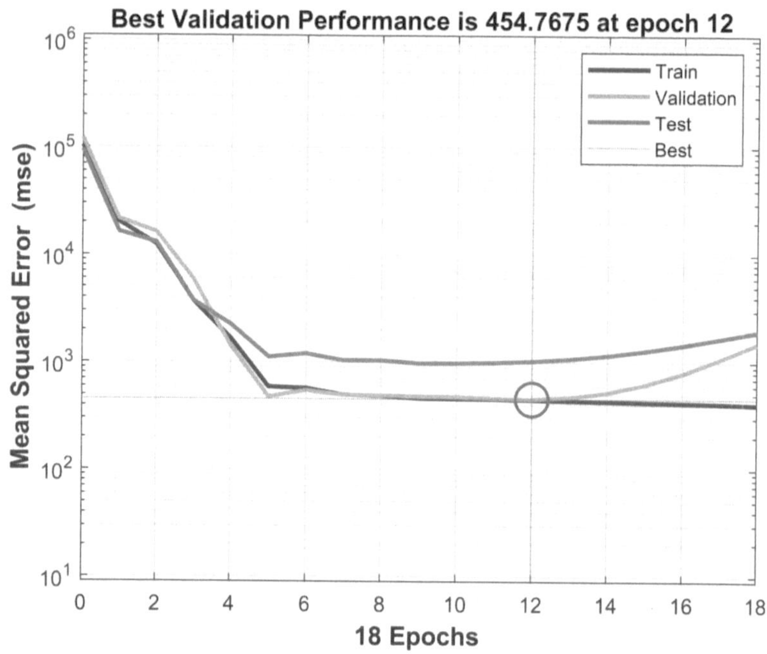

Figure 13.3 Plot of validation performance of ANN-LM for TCS.

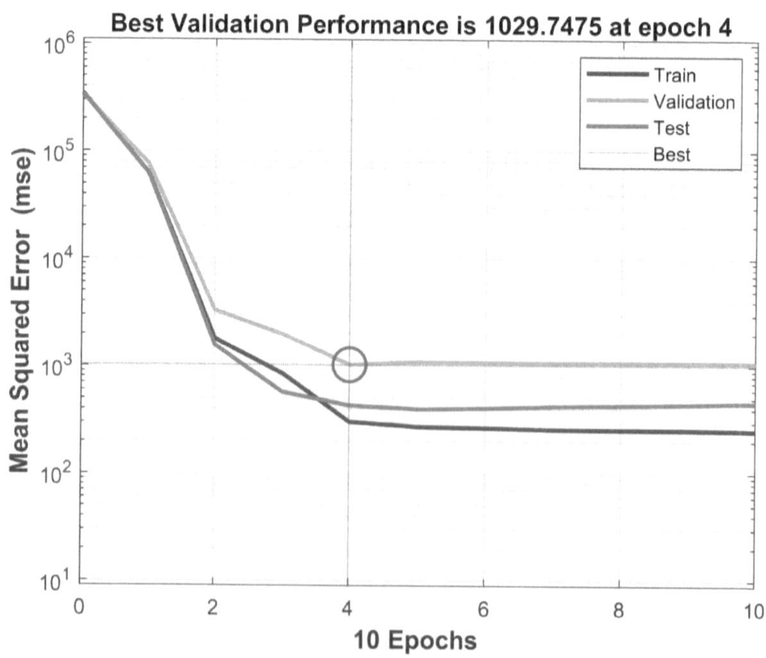

Figure 13.4 Plot of validation performance of ANN-LM for reliance.

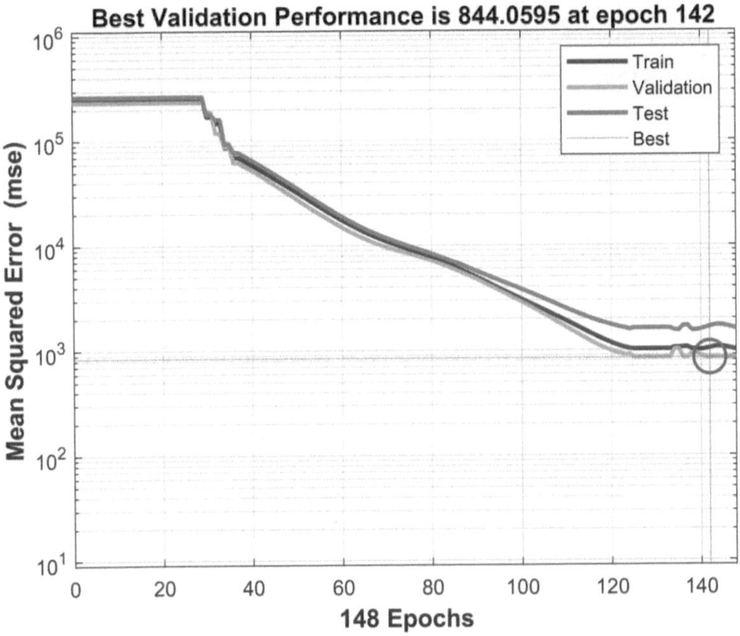

Figure 13.5 Validation performance plot of ANN-GDX for TCS.

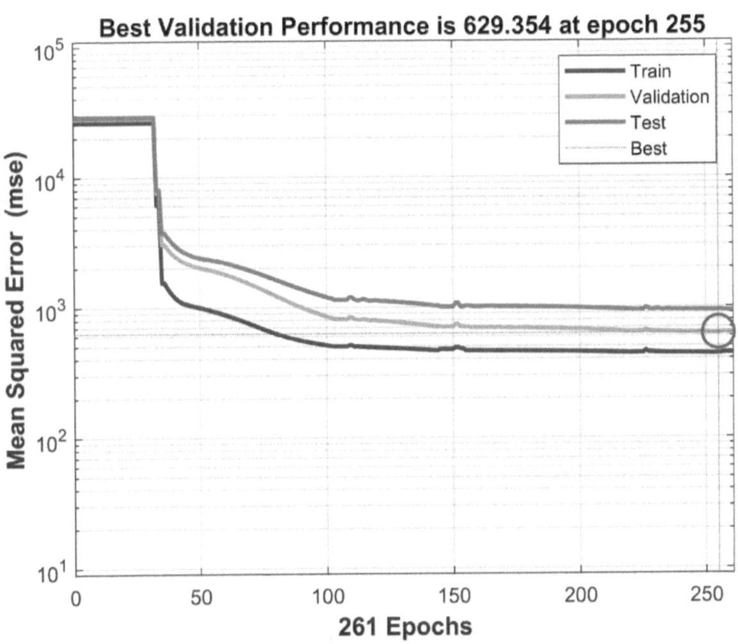

Figure 13.6 Validation performance plot of ANN-GDX for reliance.

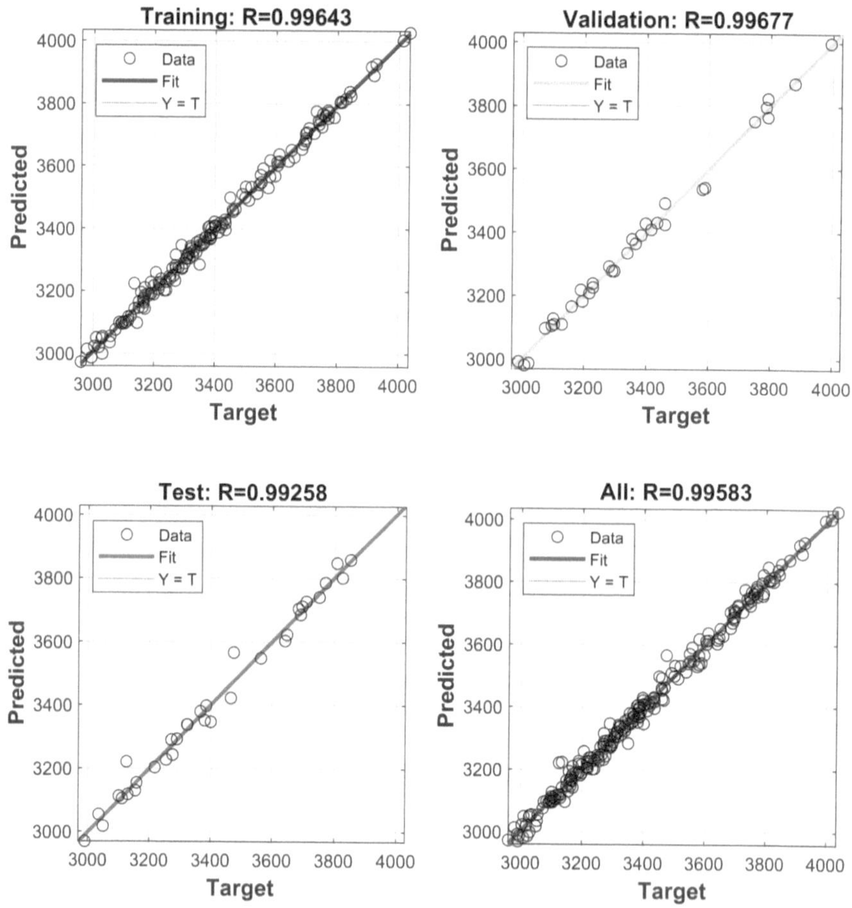

Figure 13.7 Regression plots of ANN-LM during training, testing, validation, and overall, for TCS.

and targets. An R-value of 1 indicates a perfect linear relationship between outputs and targets, whereas an R-value close to zero suggests the absence of a linear relationship.

Remarkably, both ANN models exhibit a very strong fit in the two datasets, as evidenced by *R*-values exceeding 0.98. In Figures 13.7 and 13.8, the LM algorithm displays 99.64% (99.08%) accuracy in training, 99.25% (98.56%) accuracy in testing, 99.67% (96.44%) accuracy in validation, and 99.85% (98.64%) accuracy overall for the TCS (Reliance) datasets. Conversely, in Figures 13.9 and 13.10, the GDX algorithm demonstrates 99.21% (98.48%) accuracy in training, 98.95% (97.62%) accuracy in testing, 99.26% (98.13%) accuracy in validation, and 99.16% (98.25%) accuracy overall for the TCS (Reliance) datasets. Overall, the LM algorithm

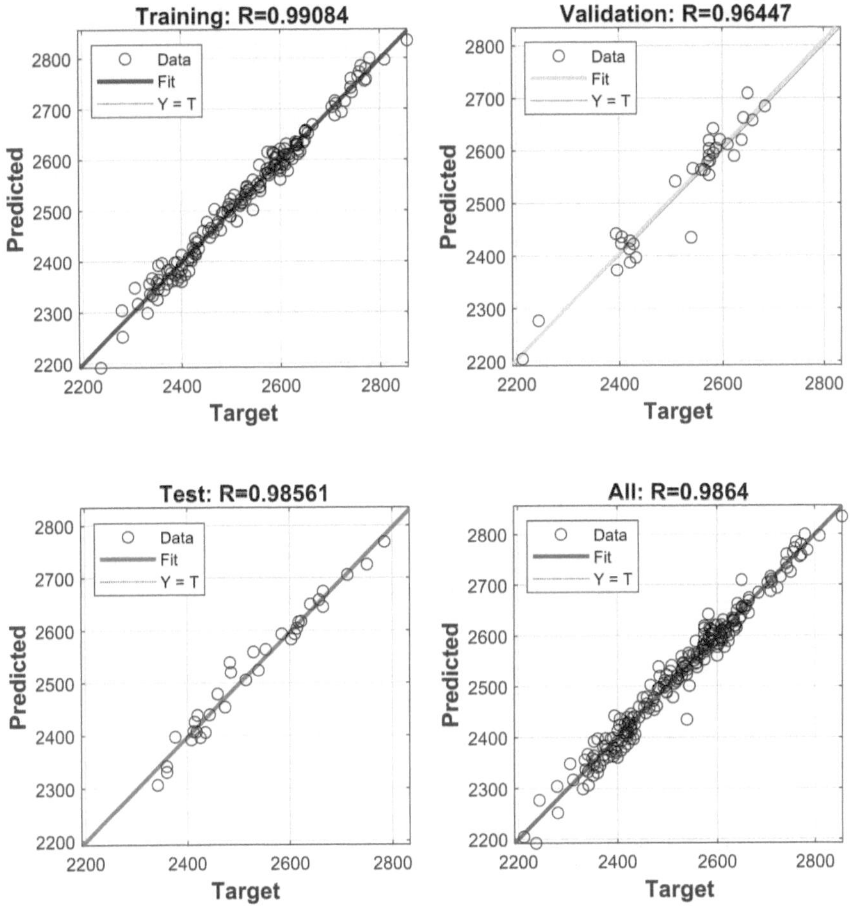

Figure 13.8 Regression plots of ANN-LM during training, testing, validation, and overall, for reliance.

yields superior *R*-values (around 99%) compared to the GDX algorithm across both datasets. Hence, the authors' conclusion is that the LM algorithm outperforms the GDX algorithm in predicting the Indian stock market trends for TCS and Reliance stocks.

Figures 13.11–13.14 illustrate blue bars corresponding to the training dataset, red bars representing the testing dataset, and green bars denoting the validation dataset. The error range, encompassing the maximum positive and maximum negative values, is divided into 20 bins. This error histogram is particularly valuable for identifying outliers within the dataset, which are data points exhibiting significantly poorer fit compared to others. Additionally, the error histogram centres around a singular performance

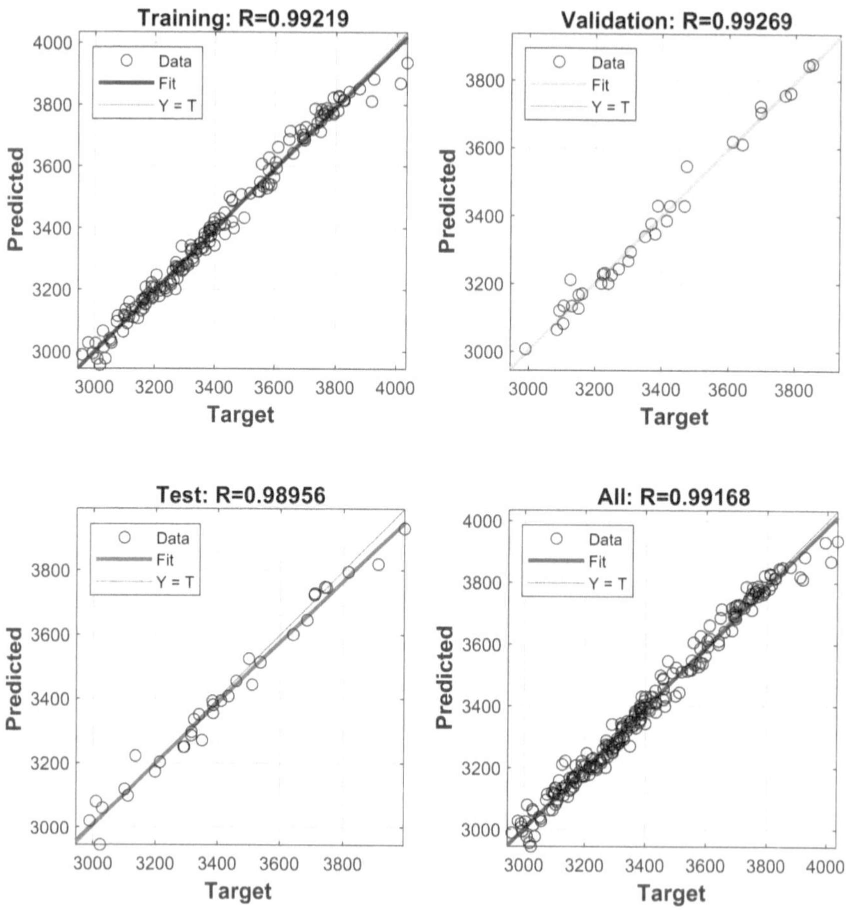

Figure 13.9 Regression plots of ANN-GDX during training, testing, validation, and overall, for TCS.

metric, the MSE. Figures 13.11 and Figure 13.12 display the error histogram graphs for data trained using the LM and Figures 13.13 and 13.14 display the error histogram plot for data trained using GDX algorithms, respectively, for both datasets. If the majority of instances exhibit lower error values, the algorithm's training performance can be considered successful. Figures 13.11 and 13.12 illustrate that an ANN trained with the LM algorithm showcases the least error for most instances in both datasets (TCS and Reliance).

Based on the error histogram graphs, it can be inferred that the LM algorithm outperforms GDX for both datasets. Moving to Figures 13.15 and 13.16 depict the Predicted vs Actual (Expected) opening stock prices for the next 25 days using the ANN-LM and Figures 13.17 and 13.18 depict the

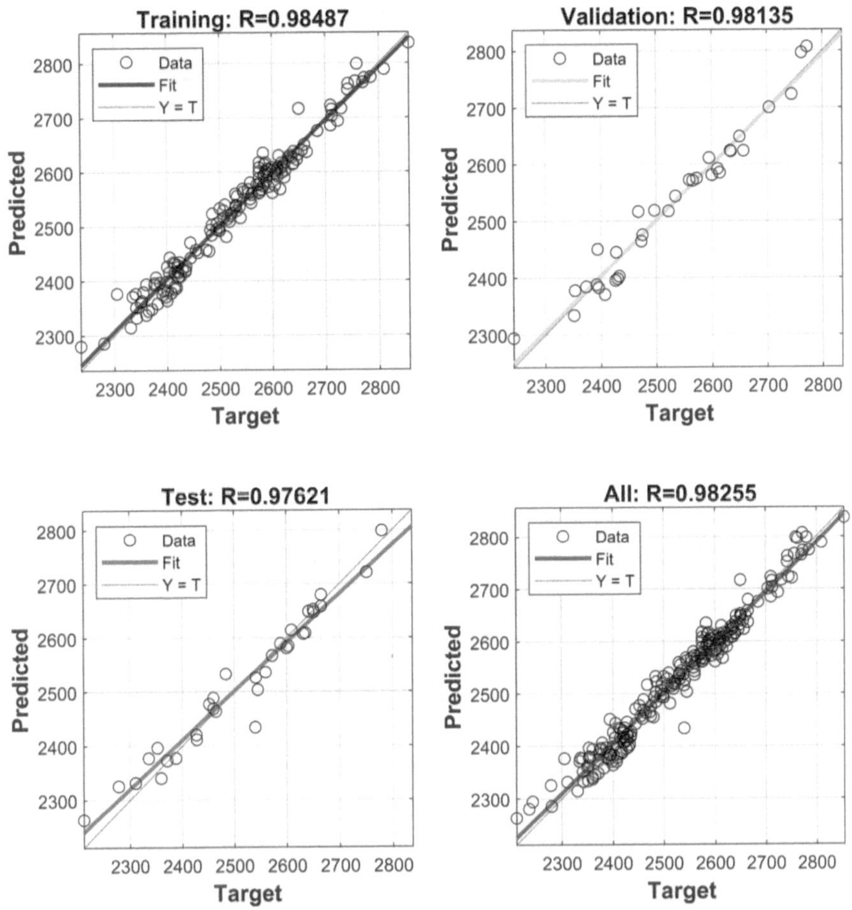

Figure 13.10 Regression plots of ANN-GDX during training, testing, validation, and overall, for reliance

Predicted vs Actual (Expected) opening stock prices for the next 25 days using the ANN-GDX methods for both dataset (TCS and Reliance) respectively. Furthermore, Table 13.2 presents the original opening prices of stocks juxtaposed with the predicted opening prices utilizing ANN models for two distinct companies (TCS and Reliance).

In conclusion, the evidence from the error histogram graphs and other visualizations supports the authors' assertion that the LM algorithm's performance is superior to GDX for both datasets.

According to Table 13.3, ANN-LM performs better than compared to ANN-GDX forecasting of daily closing prices because the error in ANN-LM is much less than the ANN-GDX. The experiments demonstrate that the predictability of daily stock returns varies. According to the

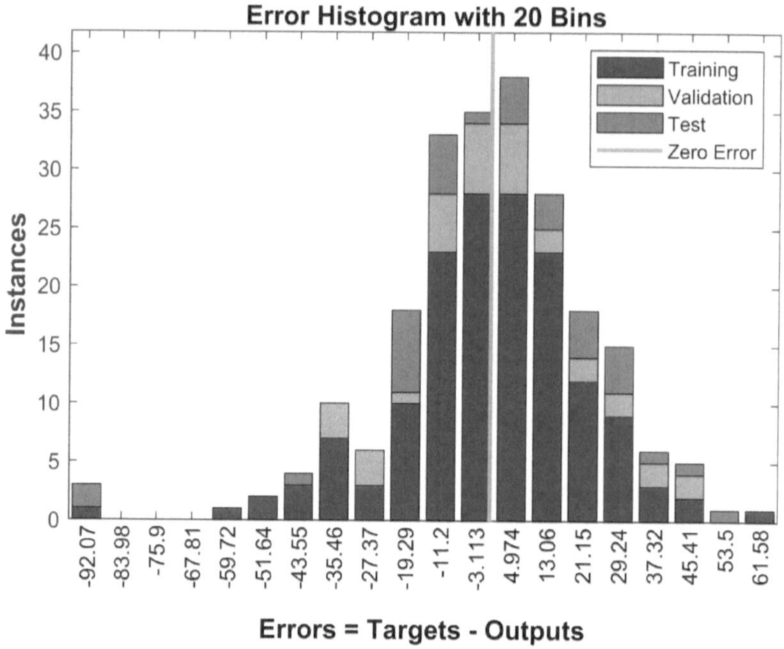

Figure 13.11 Error histogram in the ANN-LM training process for TCS.

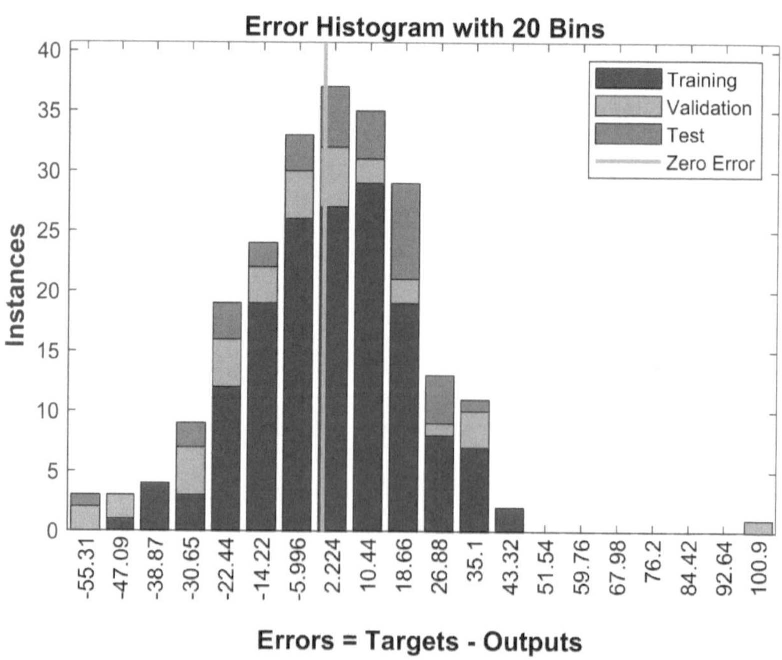

Figure 13.12 Error histogram in the ANN-LM training process for reliance.

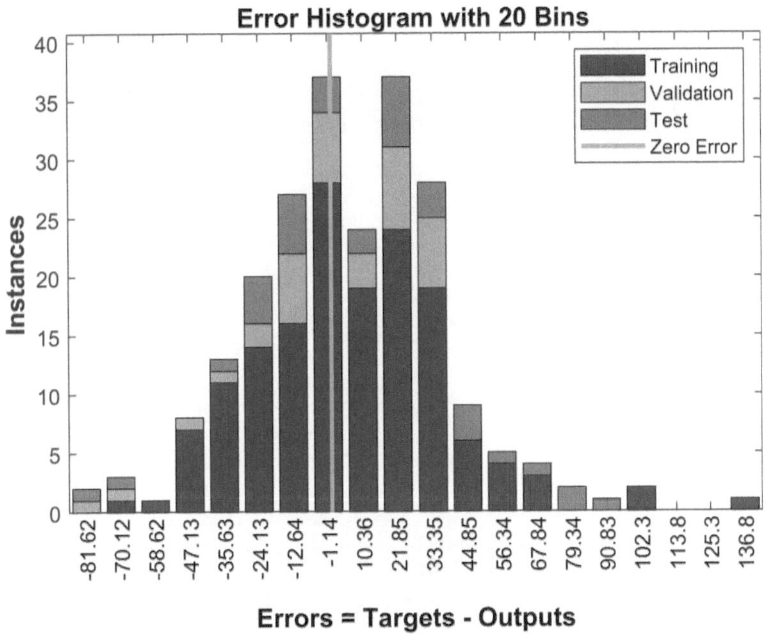

Figure 13.13 Error histogram in the ANN-GDX training process for TCS.

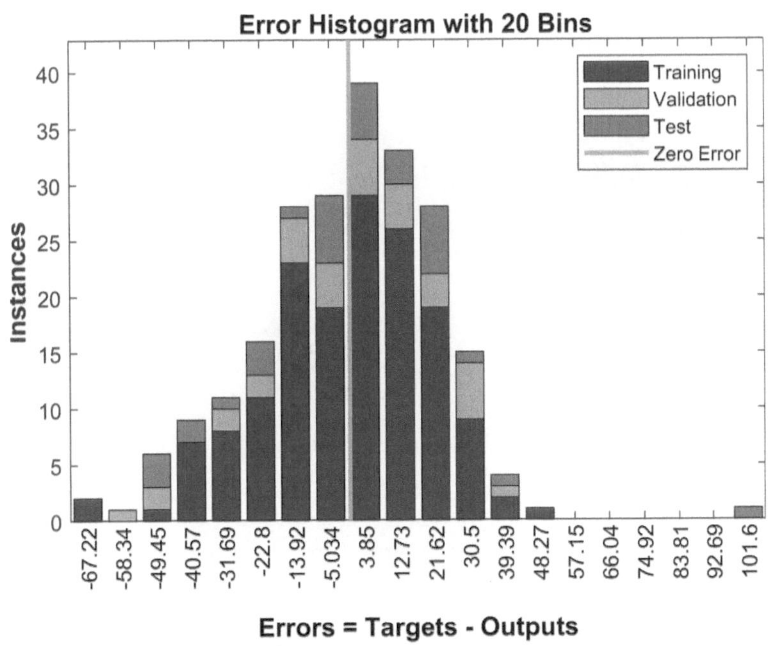

Figure 13.14 Error histogram in the ANN-GDX training process for reliance.

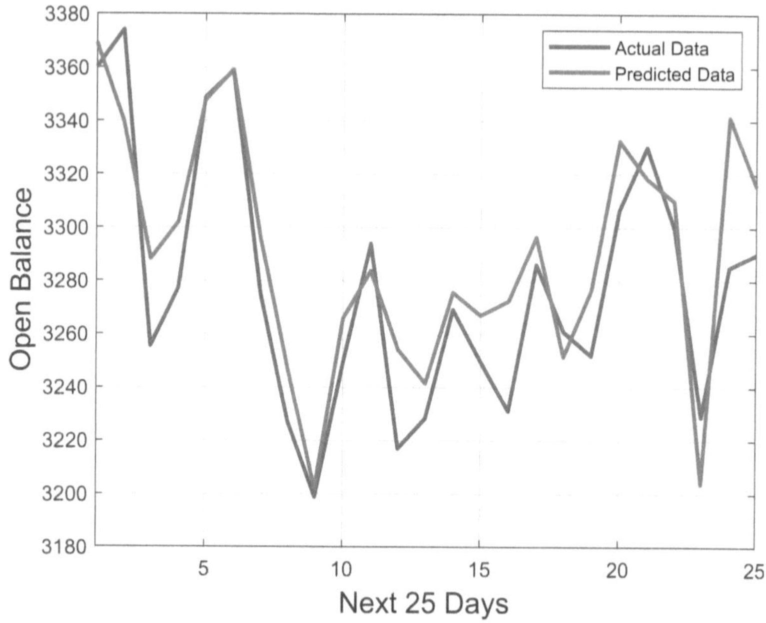

Figure 13.15 Stock market prediction (actual vs predicted) of next day opening price for 25 days using ANN-LM in TCS.

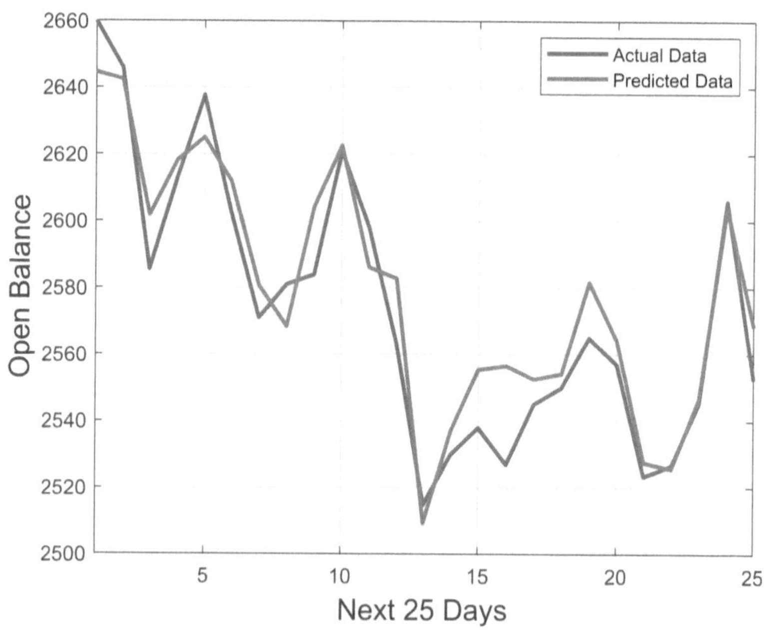

Figure 13.16 Stock market prediction (actual vs predicted) of next day opening price for 25 days using ANN-LM in reliance.

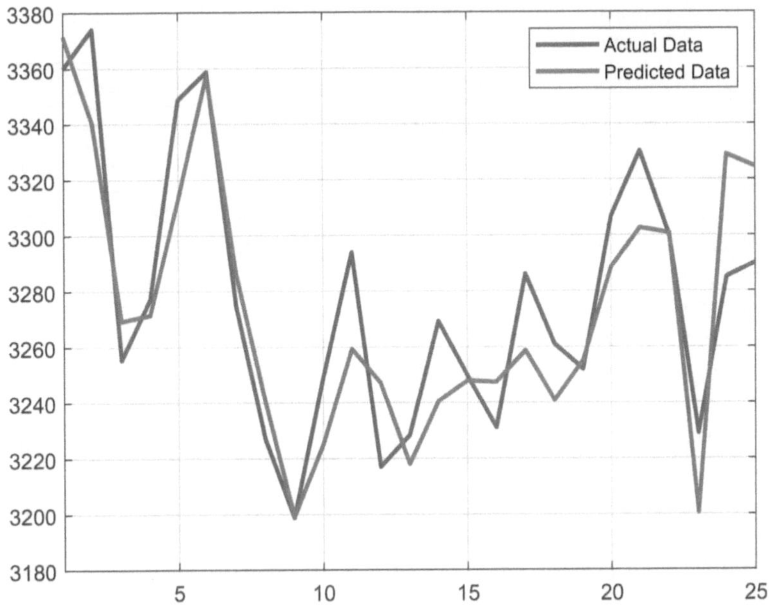

Figure 13.17 Stock market prediction (actual vs predicted) of next day opening price for 25 days using ANN-GDX in TCS.

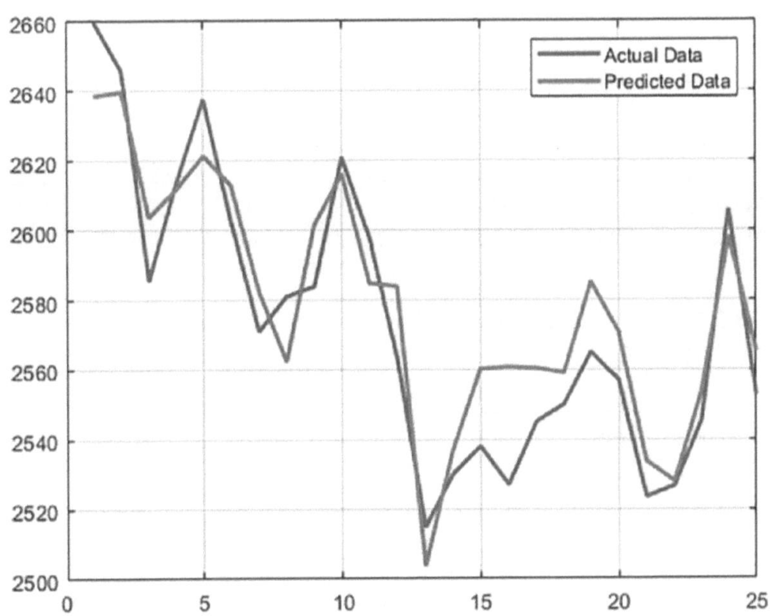

Figure 13.18 Stock market prediction (actual vs predicted) of next day opening price for 25 days using ANN-GDX in reliance.

Table 13.2 Comparison between actual and predicted prices for both datasets using ANN models

	TCS			Reliance		
	Actual	Predicted		Actual	Predicted	
Date		ANN-LM	ANN-GDX		ANN-LM	ANN-GDX
19-12-2022	3,360	3,669	3,372	2,660	2,645	2,638
20-12-2022	3,374	3,345	3,342	2,646	2,623	2,640
21-12-2022	3,256	3,254	3,249	2,586	2,605	2,610
22-12-2022	3,277	3,278	3,280	2,613	2,607	2,611
23-12-2022	3,349	3,350	3,265	2,638	2,623	2,620
24-12-2022	3,359	3,360	3,367	2,603	2,605	2,617
25-12-2022	3,275	3,282	3,287	2,571	2,584	2,580
26-12-2022	3,227	3,233	3,236	2,581	2,576	2,564
27-12-2022	3,199	3,203	3,209	2,584	2,591	2,589
28-12-2022	3,249	3,252	3,256	2,621	2,623	2,618
29-12-2022	3,294	3,287	3,258	2,598	2,594	2,586
30-12-2022	3,217	3,241	3,247	2,563	2,569	2,556
31-12-2022	3,228	3,239	3,214	2,515	2,509	2,510
01-01-2023	3,269	3,274	3,252	2,530	2,558	2,538
02-01-2023	3,250	3,262	3,248	2,538	2,545	2,560
03-01-2023	3,231	3,265	3,229	2,527	2,549	2,554
04-01-2023	3,286	3,293	3,281	2,545	2,556	2,561
05-01-2023	3,261	3,256	3,245	2,550	2,558	2,560
06-01-2023	3,252	3,260	3,241	2,565	2,582	2,586
07-01-2023	3,307	3,326	3,286	2,557	2,562	2,567
08-01-2023	3,330	3,324	3,319	2,524	2,529	2,536
09-01-2023	3,300	3,316	3290	2,527	2,526	2,528
10-01-2023	3,229	3,219	3,216	2,545	2,546	2,547
11-01-2023	3,285	3,340	3,325	2,606	2,604	2,598

Table 13.3 Performance metrics of ANN models during the testing period (TCS and Reliance)

	TCS			Reliance		
Model/company	*RMSE*	*MAPE*	R^2	*RMSE*	*MAPE*	R^2
ANN-LM	17.7144	0.369987	0.953384	12.04326	0.369624	0.988917
ANN-GDX	25.05833	0.534513	0.926132	13.81738	0.455274	0.960198

values of RMSE, R^2, MAPE, and other statistics, for example. The above comparison of ANN-LM and ANN-GDX proves that the prediction error of ANN-LM is significantly lower than that of ANN-GDX; therefore, the ANN-LM prediction is superior to that of ANN-GDX.

13.4 CONCLUSION

Predicting stock market returns is a complex task, primarily due to the volatility of stock prices influenced by a multitude of factors, resulting in intricate patterns. Historical data, which are retrievable from the company's website, encompass a limited set of variables including high, low, open, close, adjusted close, stock price value, trading volume, among others. For forecasting the opening stock price, the following next day morning, the ANN-LM model is employed. Furthermore, the ANN-GDX model is employed for the purpose of comparative analysis. The best values (actual and prediction) were obtained by the ANN-LM model, as shown by the results. The comparative analysis of graphs, i.e., performance plot, histogram plot, and regression plot, shows that ANN-LM provides a more accurate prediction of stock prices than ANN-GDX. In the face of intense competition, any solution must undergo continual improvement to remain relevant. As such, there exists the potential for future advancements in our work. Subsequent research may explore novel ANN prediction algorithms to ascertain whether they yield improved accuracy. Moreover, the inclusion of fresh data from companies could be considered to further enhance prediction capabilities.

REFERENCES

[1] Das, R., & Jaiswal, J. K. (2018). Improvement of ANN models via data envelopment analysis for stock prices forecasting. *International Journal of Engineering & Technology*, 7(4.10), 282–289.

[2] Jaiswal, J. K., & Das, R. (2018). Artificial neural network algorithms based nonlinear data analysis for forecasting in the finance sector. *International Journal of Engineering & Technology*, 7(4.10), 169–176.

[3] Murkute, A., & Sarode, T. (2015). Forecasting market price of stock using artificial neural network. *International Journal of Computer Applications*, *124*(12), 11–15.

[4] Jaiswal, J. K., & Das, R. (2017). Application of artificial neural networks with backpropagation technique in the financial data. In *IOP Conference Series: Materials Science and Engineering* (Vol. 263, No. 4, p. 042139). IOP Publishing, Bristol, UK.

[5] Montgomery, D. C., Peck, E. A., & Vining, G. G. (2021). *Introduction to Linear Regression Analysis*. John Wiley & Sons, Hoboken, NJ.

[6] Chong, T. T. L., & Ng, W. K. (2008). Technical analysis and the London stock exchange: Testing the MACD and RSI rules using the FT30. *Applied Economics Letters*, *15*(14), 1111–1114.

[7] Zhang, G. P. (2003). Time series forecasting using a hybrid ARIMA and neural network model. *Neurocomputing*, *50*, 159–175.

[8] Dey, P. P., Nahar, N., & Hossain, B. M. (2020). Forecasting stock market trend using machine learning algorithms with technical indicators. *International Journal of Information Technology and Computer Science*, 12(3), 32–38.

[9] Liaw, A., & Wiener, M. (2002). Classification and regression by randomforest. *R News*, *2*(3), 18–22.
[10] Adebiyi, A. A., Ayo, C. K., Adebiyi, M., & Otokiti, S. O. (2012). Stock price prediction using neural network with hybridized market indicators. *Journal of Emerging Trends in Computing and Information Sciences*, *3*, 1.
[11] Hiransha, M., Gopalakrishnan, E. A., Menon, V. K., & Soman, K. P. (2018). NSE stock market prediction using deep-learning models. *Procedia Computer Science*, *132*, 1351–1362.
[12] Gandhmal, D. P., & Kumar, K. (2019). Systematic analysis and review of stock market prediction techniques. *Computer Science Review*, *34*, 100190.
[13] Yang, F., Chen, J., & Liu, Y. (2023). Improved and optimized recurrent neural network based on PSO and its application in stock price prediction. *Soft Computing*, *27*(6), 3461–3476.
[14] Moghar, A., & Hamiche, M. (2020). Stock market prediction using LSTM recurrent neural network. *Procedia Computer Science*, *170*, 1168–1173.
[15] Li, L., Wu, Y., Ou, Y., Li, Q., Zhou, Y., & Chen, D. (2017). Research on machine learning algorithms and feature extraction for time series. In *2017 IEEE 28th Annual International Symposium on Personal, Indoor, and Mobile Radio Communications (PIMRC)* (pp. 1–5). IEEE, New York.
[16] Oyeyemi, E. O., McKinnell, L. A., & Poole, A. W. (2007). Neural network-based prediction techniques for global modeling of M(3000)F2 ionospheric parameter. *Advances in Space Research*, *39*(5), 643–650.
[17] Kara, Y., Boyacioglu, M. A., & Baykan, Ö. K. (2011). Predicting direction of stock price index movement using artificial neural networks and support vector machines: The sample of the Istanbul stock exchange. *Expert Systems with Applications*, *38*(5), 5311–5319.
[18] Qi, M., & Zhang, G. P. (2008). Trend time-series modeling and forecasting with neural networks. *IEEE Transactions on Neural Networks*, *19*(5), 808–816.
[19] Cigizoglu, H. K. (2003). Estimation, forecasting and extrapolation of river flows by artificial neural networks. *Hydrological Sciences Journal*, *48*(3), 349–361.
[20] Gopi Kiran, M., Das, R., Behera, S. K., Pakshirajan, K., & Das, G. (2021). Modelling a rotating biological contactor treating heavy metal contaminated wastewater using artificial neural network. *Water Supply*, 21(5), 1895–1912.
[21] Bishop, C. M. (1995). *Neural Networks for Pattern Recognition*. Oxford University Press, Oxford, UK.
[22] Manohar, B., & Das, R. (2022). Artificial neural networks for the prediction of monkeypox outbreak. *Tropical Medicine and Infectious Disease*, *7*(12), 424.
[23] Selvamuthu, D., Kumar, V., & Mishra, A. (2019). Indian stock market prediction using artificial neural networks on tick data. *Financial Innovation*, *5*(1), 1–12.
[24] Manohar, B., & Das, R. (2023). Comparison of hybrid artificial neural networks with GA, PSO, and RSA in predicting COVID-19 cases: A case study of India. In *Multi-Disciplinary Applications of Fog Computing: Responsiveness in Real-Time* (pp. 207–244). IGI Global, Hershey, PA.

[25] Chandra, P., & Das, R. (2023). Impact of gyrotactic microorganisms in nanofluid via porous media along an inclined stretching plate: Finite element analysis. *Numerical Heat Transfer, Part B: Fundamentals*, *83*(6), 367–394.

[26] Chandra, P., & Das, R. (2023). Finite-element-based machine-learning algorithm for studying gyrotactic-nanofluid flow via stretching surface. *International Journal for Numerical Methods in Fluids*, *95*(12), 1888–1912.

[27] Manohar, B., & Das, R. (2023). Artificial neural networks for prediction of COVID-19 in India by using backpropagation. *Expert Systems*, *40*(5), e13105.

[28] Ali, M. S., Ayaz, M., & Mansoor, T. (2022). Prediction of discharge through a sharp-crested triangular weir using ANN model trained with Levenberg-Marquardt algorithm. *Modeling Earth Systems and Environment*, *8*(1), 1405–1417.

[29] Aiken, M. W., & Bsat, M. (1999). Forecasting market trends with neural networks. *Information Systems Managers*, *16*(4), 1–7.

[30] "Yahoo Finance - Stock Market Live, Quotes, Business & Finance News" [Online]. Available: https://finance.yahoo.com/. [Accessed: 26-Jun-2023].

[31] Vijh, M., Chandola, D., Tikkiwal, V. A., & Kumar, A. (2020). Stock closing price prediction using machine learning techniques. *Procedia Computer Science*, *167*, 599–606.

[32] Balamurugan, M., Narendiran, S., Sahoo, S. K., Das, R., & Sahoo, A. K. (2016). Application of particle swarm optimization for maximum power point tracking in PV system. In *2016 3rd International Conference on Electrical Energy Systems (ICEES)* (pp. 35–38). IEEE, New York.

[33] Rahman, M. M., Shakeri, M., Tiong, S. K., Khatun, F., Amin, N., Pasupuleti, J., & Hasan, M. K. (2021). Prospective methodologies in hybrid renewable energy systems for energy prediction using artificial neural networks. *Sustainability*, *13*(4), 2393.

[34] Narendiran, S., Sahoo, S. K., & Das, R. (2016). Control and analysis of MPPT techniques for maximizing power extraction and eliminating oscillations in PV system. *International Energy Journal*, *16*, 3.

[35] Hewage, P., Trovati, M., Pereira, E., & Behera, A. (2021). Deep learning-based effective fine-grained weather forecasting model. *Pattern Analysis and Applications*, *24*(1), 343–366.

[36] Jali, B. R., Behura, R., Barik, S. R., Parveen, S., Mohanty, S. P., & Das, R. (2018). A brief review: biological implications of naphthoquinone derivatives. *Research Journal of Pharmacy and Technology*, *11*(8), 3698–3702.

[37] Palanivel, K., & Surianarayanan, C. (2019). An approach for prediction of crop yield using machine learning and big data techniques. *International Journal of Computer Engineering and Technology*, *10*(3), 110–118.

[38] Das, R., & Pradhan, M. K. (2014). General regression neural network and back propagation neural network modeling for predicting radial overcut in EDM: A comparative study. *World Academy of Science, Engineering and Te chnology, International Journal of Mechanical, Aerospace, Industrial, Mech atronic and Manufacturing Engineering*, *8*(4), 799–805.

[39] Behm, C., Nolting, L., & Praktiknjo, A. (2020). How to model European electricity load profiles using artificial neural networks. *Applied Energy*, *277*, 115564.

Chapter 14

Technological transformation in learning patterns in media and entertainment

Priya Shome, Sumedha Chaudhury, Sudhir Kumar, and Sumit Kumar Rohit

14.1 INTRODUCTION

Ever since COVID-19 crept into the lives of people's dependency on over-the-top (OTT) platforms, AI and other such technologies for entertainment have increased to the greatest extent. The new media platforms have also influenced the target audiences in various ways. The new technologies can be utilised to easily know about the preferences of its target audience and show them content as per their likings and preferences. Machine learning and deep learning can be utilised to improve content creation, recommendation systems and audience engagement, which can simplify the task of content creator and, in turn, provide a great experience to the target audience as well.

If we talk about deep learning, it is a type of machine learning called deep learning, which is based on artificial neural networks, and uses numerous layers of processing to extract increasingly more complex properties from data. It is utilised to detect and investigate the cross platform spread of disinformation, the prototype will combine deep learning, natural language processing, and dynamic network analysis.

Machine learning focuses on using algorithms and statistical models to analyse data patterns and deduce conclusions from them, machine learning is the use and development of computer systems that can learn and adapt without being given explicit instructions.

Artificial Intelligence has been used in numerous ways to improve customer experience and streamline business operations. This will continue even in future where media companies will continue working alongside AIs for personalised content recommendations and effective advertisements.

Using AI, and data based on user behaviour patterns, media companies can personalise what ads to show to garner user interest. While the scope of this was for larger groups in the past, this will zero in for more specific buyer personas, and eventually be targeted for individuals [1].

Every other day we hear fake news in some form or the other. Social media platforms have become a hotspot for these people for spreading fake news. It affects society at large. People tend to believe more in such

 DOI: 10.1201/9781003534761-14

misinformation than following print media that provides genuine and verified information. Content available in the media moulds the opinion of the society. Sometimes it brings a change in how we behave with others and how we judge them. People need to figure out the correct source of information. Copyright issues are also increasing nowadays, all thanks to malicious websites. At times we hear that movies get leaked on these sites before their official release. It demotivates the artists working hard to create something and bringing it to society.

In these ways, we allow technology to govern our behaviour, be it emotional, social, societal, or personal. But we must aim to use it effectively to improve productivity in our daily routine in the long run. It must act as a supplement to our lives instead of being the controller.

This research paper also highlights the risks associated with over utilisation of Artificial Intelligence. Media usage is related to some adverse consequences, especially in the most vulnerable people. The health emergency had a strong impact on the mental and psychological health of adolescents causing changes in their routine and daily activities. Forced isolation increased anxiety and stress especially in the most fragile individuals, such as children and adolescents, leading to a change in habitual lifestyles. The greatest risk was that of taking refuge in excessive use of smartphones, electronic devices, and social networks, running into a "digital overdose. Chatbots can solve customer problems in second [2].

But leads to greater corrupt practices among youngsters of not using their own efforts, talents and brain to solve certain problems. In media industry, the introduction of AI news anchors looks innovative and interesting but the point to ponder about is that this can be an immense source of creating further unemployment for the pool of talented youth who will be soon replaced, and the media industry will no longer require real news anchors. The power of AI is immense in all sense, we cannot deny the fact that the future belongs to AI but we humans will be left with some scope to earn or explore our talents for job or any other purpose. AI is manmade but if it replaces all the industries or in this case let's take the film and media industry, we won't have real actors even in the best of films, natural scenic beauty will be replaced with the technology-based ambience. The threat is not limited to the media or film industry, but it lingers around all the other industries. It is a mere fact that the content creators task gets simplified, but it is also a harsh reality that the next generation after generation will be aloof with all the natural film making process, a film will be generated sitting in front of a laptop or desktop. The hard work and pride of a journalist to create a news story are well known to all but imagine if a media channel is totally dependent on AI for news gathering and content generation, the employment of new anchors will be hampered by corrupt practices, which will make the budding journalists lethargic and totally without utilisation of brain and hard work. This represents a major source of brain drain. This research is not against the utilization

Figure 14.1 Representing the benefits provided by machine learning and deep learning to target audience.

of AI, deep learning, or machine learning but it focuses on the apt use of such technologies so that there remains a balance and the media and film industry is totally not driven away only by AI. The bliss of AI for the media industry also cannot be denied, as it easily detects the fake news. Figure 14.1 represents the benefits provided by machine learning and deep learning to target audience.

Artificial intelligence can keep the target audience on cloud nine with the kind of experience they get by the usage of AI, OTT platform and other aspects. Personalised content can lead to gratification amongst the target audience as they will find it useful as per their purpose. Besides, the AI also acts as a guide for proper decision-making, as the task gets simplified. So, this research has tried to act as a mirror and tried to bring about the reality of media and film industry's dependence on AI, machine learning and deep learning.

The objectives of the research were to find out how technology contributes and impacts films as well as observe the relation between technology and media overall.

A sample size of as many as 59 residents was taken based on mixed research approach focusing on both qualitative and quantitative research approaches and probability sampling across most of the states of India aged between 18 years and 45 years irrespective of gender wherein they were approached through a questionnaire as a tool including various close and open-ended questions through survey as a research method (part of applied research) with the help of Google form for a month in 2023. Survey is one of the oldest and common research methods in communication research. It is to acquaint with the common idea, perspective, and attitude of the population. In the type of survey, longitudinal survey was chosen wherein Cohort Studies was adapted. It is a study in which samples taken have a common cohort. Cohort in research means any group of individuals who have experienced the same significant life event within the given period. The research tried to connect individuals as a source, followed by message analysis, channel analysis and audience analysis.

14.2 RELATED WORK

The authors have suggested that digital technologies invaded all phases of print media production. Electronic media and multimedia are also explained and positional. The technologies and production systems covered are state-of-the-art and represent a multitude of products from well-known manufacturers worldwide [3].

Authors have proposed in their research that while multitasking is not a new concept, it has received increasing attention in recent years with the development of new media and technologies [4]. Contemporary trends appear to suggest that multitasking is on the rise among the younger

In terms of revenue, it has been a mixed year with stronger titles pulling ahead while others falter. The Reuters Institute Digital News Report showed a 7% increase in subscription in the US last year led by the New York Times and Washington Post. The Guardian also defied its critics by reporting 800,000 paying customers, including half a million subscribers or members, and 300,000 one-off donations [5].

Examples of cognitive tools include databases, spreadsheets, semantic networks, expert systems, communications software such as teleconferencing programs, online collaborative knowledge construction environments, multimedia/hypermedia construction software, and computer programming languages. In the cognitive tools approach, media and technology are given directly to learners to use for representing and expressing what they know. Learners themselves function as designers using media and technology as tools for analysing the world, accessing, and interpreting information, organising their knowledge, and representing what they know to others [6]

Current developments in communication technologies require all of us to become intelligent participants in technological decision-making. Where in this dawning of the Information Age, does one draw the line regarding types of technologies alone? If computers are communication technologies, so too are robotics and genetic engineering. It no longer makes sense to think about radio, television, telephone, and computers as communication technologies, and assembly-line production and heavy industry as industrial technologies [7].

Notably, Media and society are closely related to each other. The extensive impact of media on society can easily be seen these days. Media reflects our society, how it works, and what it constitutes. With the advancement in the technological area, our society has also observed the expansion in the thoughts and ideas of people. Every invention from the printing press to the latest smartphones our society has accepted. Earlier people used to communicate things with the help of sketch and print forms but as time passed the medium became more advanced [8].

14.3 SURVEY STUDY

59 responses were received from across India. Respondents were from different professions like, businessmen, housewives, students, IT professionals, filmmakers and cinematographers. Gender wise there were 23 Female Respondents and 36 male respondents respectively as represented in Figures 14.2. Figure 14.3 represents the technology used in entertainment and media industry.

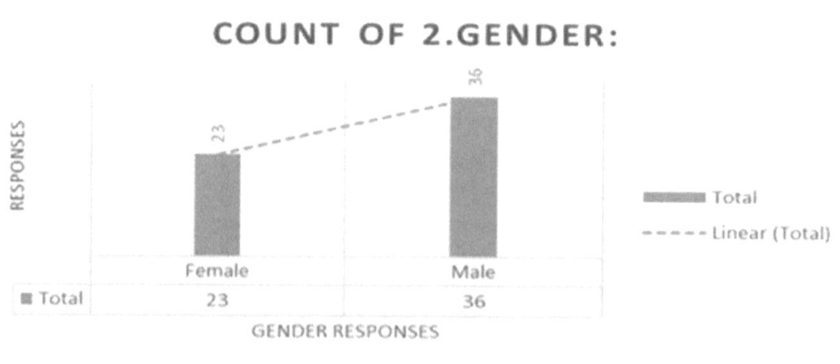

Figure 14.2 Male and female count of responses.

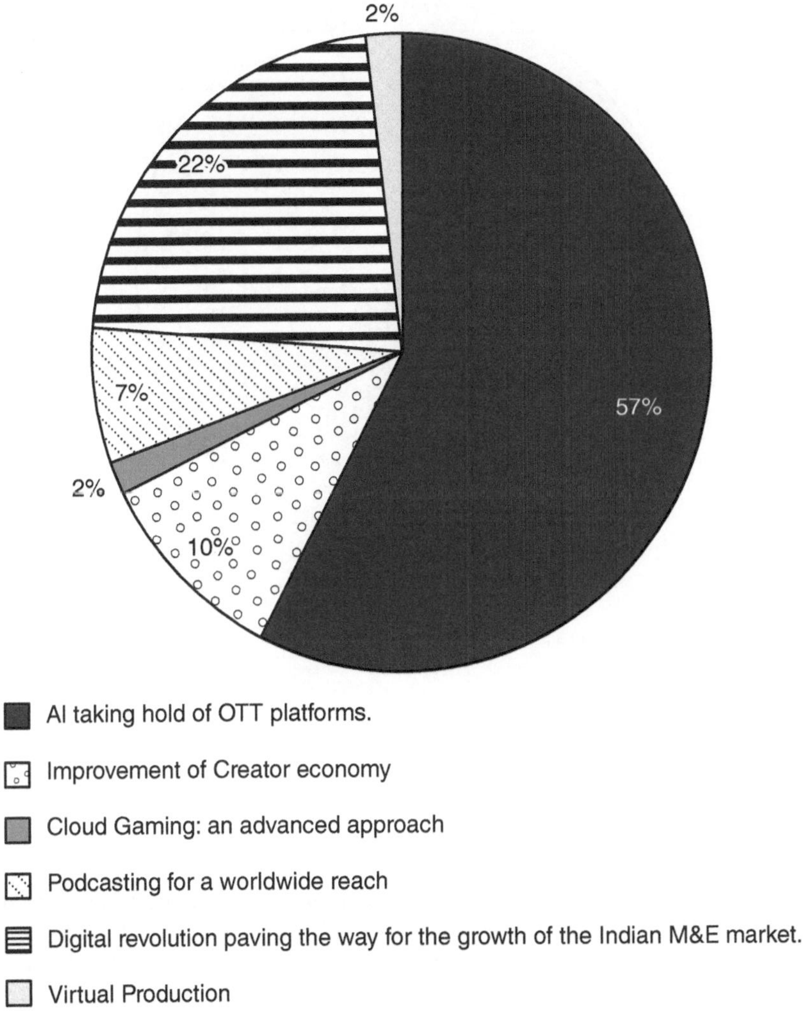

Figure 14.3 Technology used in entertainment and media industry.

From the survey, it was found that 54.4% of respondents said that AI taking hold of OTT platforms.

AI is positioned to alter the entertainment business in ways we can only begin to imagine, from transforming the way we consume content to producing totally new kinds of entertainment. The personalisation of content is one of the most significant ways AI is anticipated to affect the entertainment sector.

Virtual reality (VR) and augmented reality (AR) are redefining how consumers consume information and find entertainment. The way individuals interact with content is changing because of these two immersive technologies.

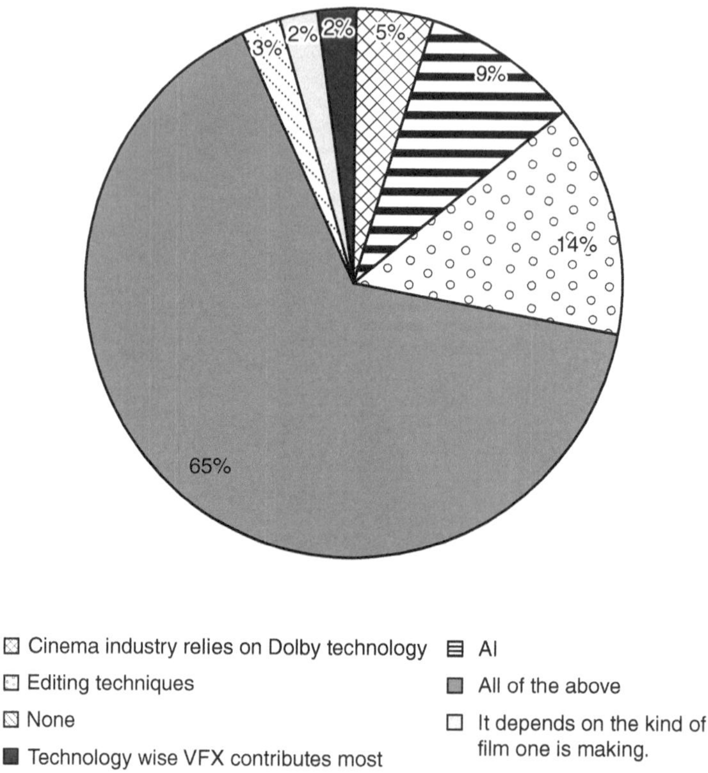

Figure 14.4 Survey outcome for technology contributing in film industry.

Future media and entertainment sector themes include utilising data for insights, AI for content development and administration, and automation for optimising workflows. Among the applications of digital technologies in the M&E sector is trend analysis-based content optimization. Figure 14.4 represents the survey outcome for technology contributing in film industry.

In essence, this technology allows an audience to witness superimposed images of real objects.

Technology developments like VR, AR, and 3D are set to alter how films are produced and enjoyed. Movie experiences could become more interactive and immersive because of filmmakers' experiments with novel forms of storytelling and immersion. Figure 14.5 represents the survey outcome for the advantage of digital film.

Filmmaking has been significantly simplified and cost-cutting thanks to digital cameras, which have replaced expensive film with digital storage techniques that are more durable, secure, and affordable. Figure 14.6 represents the survey outcome for technology used to improve film.

The use of technology has enhanced the quality of filmmaking, as cutting-edge technologies ensure captivating and clear visuals, improved

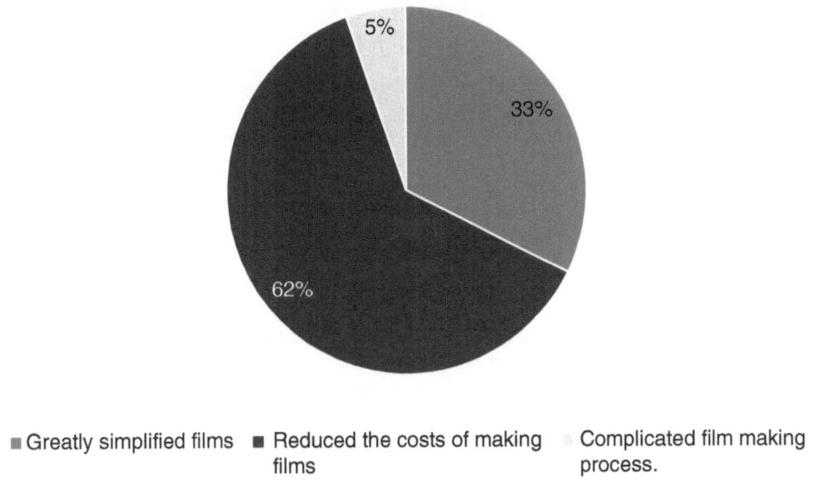

Figure 14.5 Survey outcome for the advantage of digital film.

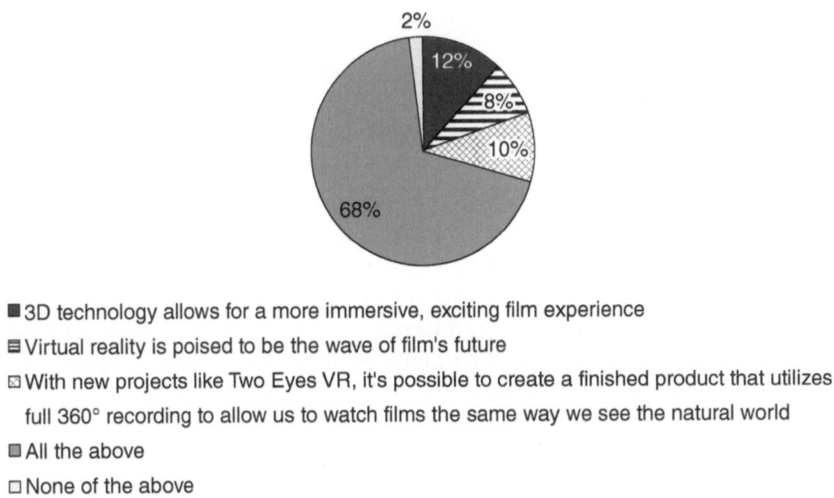

Figure 14.6 Survey outcome for technology used to improve film.

sound effects, simpler editing tools, and digital film distribution, helping to provide spectators with a more satisfying viewing experience.

The global VR and AR industry is anticipated to increase from $37 billion in 2019 to $1.3 trillion by 2030, according to market research firm Research and Markets. Several experts believe that new technologies will enable consumers to engage fully with products prior to purchase, assisting in the conversion of advertising budgets into actual consumer purchases. Additionally, by combining print and digital media, these technologies may offer customers compelling, individualised experiences.

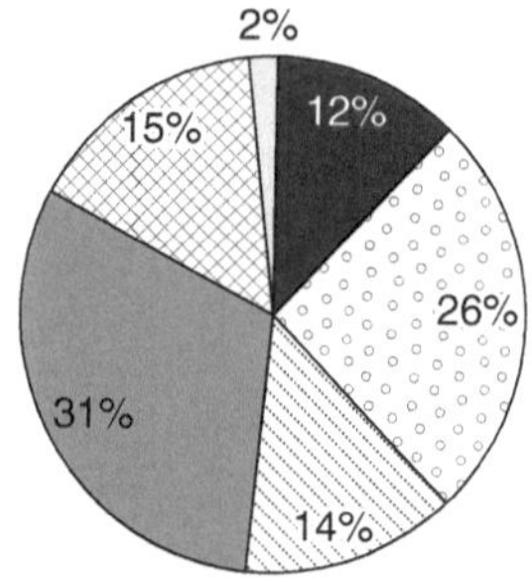

- Algorithmic Video Editing spoils the natural base of the film for audience
- Audience getting aloof of the basic realities of the world
- Adaptation of 3D printing for movies is proof that new technology in the film industry rarely starts in the film industry
- 3D Previsualization revolutionizes the pre-production process by conjuring fully immersive, digital replicas of the physical sets or locations to be used on a live-action production
- Internet of things, revolutionizing the average film set
- Technology is useful as long as we have the control button.

Figure 14.7 Survey outcome of latest technologies used to pose to the target audiences.

Technology developments like VR, AR, and 3D are set to alter how films are produced and enjoyed. Movie experiences could become more interactive and immersive because of filmmakers' experiment with novel forms of storytelling and immersion. Figure 14.7 represents the survey outcome of latest technologies used to pose to the target audiences.

The prominence of industry and the rate at which it overtook economies were accelerated by technological advancements. The ability to create and distribute excellent content has been made possible by advancements in cell phones, cameras, rendering software, computing processors, smart Televisions, earphones/headphones, improved sound systems, and wireless. Media companies and advertisers now have excellent options to interact with people thanks to new technology. For information, people used to rely on newspapers and television. People can now consume material through a variety of venues, including websites, video platforms, and social media.

14.4 DISCUSSION

The outcomes of the survey were

i. Nowadays OTT companies are slowly overtaking the industry, which is good and bad. Due to OTT companies' competition must rise immensely and more competition means more opportunity. In short, nowadays earning money, or working in the film /media industry has become very easy which was not possible in the early days. Small markets generate monopoly businesses the same as the increase in market gives equal opportunities. Due to OTT and digital platforms, the audience has an immense variety of content and education rapidly because of that today's audience is more knowledgeable about movies and media than an older audience.

ii. Nowadays content providers like OTT are more connected with people, as they have strong story plots, and the characters don't exaggerate as compared to the existing commercial cinemas.
iii. Nowadays people seek convenience and no one will take out time to visit theatres for movies so OTT should be accepted by all as changes are necessary and it targets all segments of audiences as well.
iv. There are different genres of OTT platforms serving the needs of users instantly. The OTT market is challenging the STB industry.
v. Netflix is the most successful OTT platform, and it can be easily seen that it has covered different genres for its target audience with impressive quality and technology-wise, its user interface is audience friendly and it has a successful journey altogether.
vi. There is no doubt that OTT is the future that will eventually diminish the market capture of television. People are relying more on modern technology for entertainment such as YouTube, Netflix, and Amazon Prime Video. These platforms have come up with a great volume of content encompassing quality content as well which is somehow outshining the previous technologies for entertainment like TV channels, Theatres CDs and even radio. It can be assumed that AI (artificial intelligence) aided film technology will come in the future and set a new dimension to the film industry completely overturning the previous set of beliefs about filmmaking.
vii. People find Netflix to be the most relevant for the media industry and target audience. Netflix has successfully positioned itself as a leading OTT company and has a vast library of diverse content that caters to a wide range of audience preferences. Its user-friendly interface, personalised recommendations, and original programming have enabled it to meet customers' demands effectively. With its global reach and continuous investment in content creation, Netflix has thrived by adapting to evolving consumer expectations and shaping the future of the media industry.

In today's digital world, media and entertainment companies leverage the power of digital technologies in engaging with customers and provide them with numerous channels for consuming information and entertainment.

One cannot deny the fact that information reaches us very fast nowadays due to the advancement in digital technology and this is a great advantage in many circumstances. But one cannot deny the fact that the spread of fake news is also a big problem today and this also affects the thinking process of the common mass. Even deliberate editing of facts with malicious intentions has been observed. Ways and means to control these aspects need to be developed at the earliest. By leveraging technology, Endless entertainment is available for consumers.

Today information is everything, the one who is more informed seems to be more powerful, so a customer would always feel good in consuming all diverse fields of information.

There is a vast of information available on the internet, but an individual consumes it when they are not working. They do have to look for information on the internet or media and entertainment company websites or apps.

Indeed, while Google provides almost every kind of information under the sky, the OTT platform has revolutionised the domain of entertainment. It's like a dream come true. It's clearly a market with different competitors having a complex connection with their consumers. Of course, it has the advantage of global reach.

Things are changing, we previously consume entertainment from traditional media like radio, Newspapers, magazines, and radio but new technologies are replacing them slowly in the entertainment sector. Digital technologies are more user-friendly. They generate AI data on the content you like and accordingly show the content in your news feed. In this way, they know what you want more than you yourself know what you want. This has pros and cons as you get the desired content but may get addicted to those content and waste valuable time.

Easy access to the world of art and entertainment is a huge help. They provide convenience, allowing users to access information and entertainment anytime, anywhere, and on various devices. We have the option to choose from a large variety of content giving one the opportunity to choose his/her favourite one.

14.5 RATIONAL VIEWPOINTS

The entertainment sector has benefited greatly from digital transformation, including better accessibility, a wider audience reach, audience engagement, data-driven insights, cost-effectiveness, and a new revenue model.

Automation has become a very popular buzzword in the media and entertainment sector lately, and for good reason: media and entertainment businesses employ it to drive growth while maintaining client satisfaction. By enhancing personalisation, cutting expenses, enhancing customer service, and enhancing reputations.

The use of technology has improved the quality of filmmaking because cutting-edge technologies guarantee captivating and clear visuals, improved sound effects, simpler editing tools, and digital film distribution, all of which contribute to giving audiences a more enjoyable movie experience.

Consumers may now access material that is specifically catered to their unique preferences thanks to the usage of AI in more immersive and personalised experiences. The capacity to customise material is one of the biggest advancements AI is bringing to the entertainment sector.

AI will undoubtedly affect how films are made and experienced as society progresses towards a more automated future. Faster production times are made possible by AI, which also improves viewer and filmmaker interaction, making it simpler than ever to produce excellent visual content.

Digital technologies like Big Data Analytics, Mobile Apps, AI, and User Experience, which are assisting media and entertainment enterprises in reinventing themselves and offering a wide range of media products for consumers to examine and consume, are at the centre of this disruption

The following are the main disruptive technologies that are changing how media and entertainment companies impact citizens:

i. Machine learning - Algorithms for machine learning are used to comprehend consumer consumption trends and evaluate the effects of shifting trends in consumption patterns across demographic groupings. Based on this information, curated content can be promoted across digital channels. Today, movie plots are based on data from machine learning algorithms that assist studios in forecasting consumer preferences and have a higher likelihood of generating income and delivering hits.
ii. AI and Big Data Analytics - AI and Big Data Analytics are essential differentiators for media and entertainment organisations to evaluate the effects of new business models, measure the efficacy of advertising spending across numerous channels, and provide customised content across multiple channels.
iii. Mobile apps- are being utilised more and more to provide content to mobile devices. All content providers, whether they supply content for business (LinkedIn), entertainment (Netflix), or social media (Instagram, Facebook), now have mobile apps due to the spread of smartphones and the rise in consumer device usage. Smartphone apps are used to consume a lot of content, which generates income from subscriptions.

Following are the advantages of AI in media and entertainment

i. Cost reduction - Digital media channels have reduced the price of advertising. Increased spending on advertising on digital channels has lowered the cost of advertising relative to traditional channels because of expanding tendencies towards consuming more digital material. Media firms can provide carefully selected adverts to their target audiences thanks to Big Data Analytics and machine learning, which results in more effective use of advertising.

ii. Transparency and Customer Experience - Media and entertainment companies can give customers a more transparent and tailored experience by using digital technology and improved customer service capabilities. The customer experience is made richer and the hunger for more material is increased by placing emphasis on user experience design and selected content.
iii. Revenue Models - With a focus on the customer, media and entertainment companies strive to produce curated content and offer unique goods and services for consumers. New revenue streams have been created because of the growth of digital subscriptions and multi-product packages. The capacity to combine numerous data sources and produce insights opens new prospects and revenue-generating methods.

Since it offers a competitive advantage in addition to being a differentiator, digital transformation in the media and entertainment sector is a growing trend globally. Several areas of the industry are still facing difficulties, such as the news sector, which is in decline as fewer people are choosing to pay for online content. For the media and entertainment sector, personalisation and curation are essential to success and survival.

Thus, the impact of AI in media and entertainment industry is huge. Not only media and entertainment, many other industries are using the advantages including health sector, security, agriculture, etc. [9–13]

14.6 CONCLUSION

Based on the objectives, it was discovered that AI controls more than 50% of OTT platforms. AI has the potential to transform the entertainment industry in ways we can only begin to imagine, from changing the way we consume content to creating entirely new types of entertainment. One of the most significant ways AI is expected to impact the entertainment industry is through content personalisation. The study showed that technological advancements accelerated the prominence of industry. The ability to create and distribute excellent content has been made possible by advancements in cell phones, cameras, rendering software, computing processors, smart Televisions, earphones/headphones, improved sound systems, and wireless. Media companies and advertisers now have started counting on the latest technology advancements.

REFERENCES

[1] Butler, R. W. (1990). The influence of the media in shaping international tourist patterns. *Tourism Recreation Research*, 15(2), 46–53.
[2] Holmes, D. (2005). *Communication Theory: Media, Technology, and Society*. Sage, NewYork.

[3] Hooper, S., & Rieber, L. P. (1995). Teaching with technology. *Teaching: Theory into Practice*, 2013, 154–170.
[4] Kipphan, H. (Ed.). (2001). *Handbook of Print Media: Technologies and Production Methods*. Springer Science & Business Media, Berlin.
[5] Lee, J., Lin, L., & Robertson, T. (2012). The impact of media multitasking on learning. *Learning, Media, and Technology*, 37(1), 94–104.
[6] Meron, E. (1992). Pattern formation in excitable media. *Physics Reports*, 218(1), 1–66.
[7] Ohm, J. (2003). *Multimedia Communication Technology: Representation, Transmission, and Identification of Multimedia Signals*. Springer Science & Business Media, Berlin.
[8] Khadidos, A. O., Alyoubi, K. H., Mahato, S., Khadidos, A. O., & Mohanty, S. N. (2023). Computer aided detection of major depressive disorder (MDD) Using Electroencephalogram Signals,. *IEEE Access*, 11, 41133–41141. https://doi.org/10.1109/ACCESS.2023.3262930
[9] Mahato, S., Paul, S., Goyal, N., Mohanty, S. N., & Jain, S. (2023). 3EDANFIS: Three channel EEG-based depression detection technique with hybrid adaptive neuro fuzzy inference system. *Recent Patents on Engineering*, 17(6), 32–48.
[10] Andrew, J., Deva Priya, I., Martin Sagayam, K., Bhushan, B., Sei, Y., & Eunice, J. (2023). Blockchain for healthcare systems: Architecture, security challenges, trends and future directions. *Journal of Network and Computer Applications*, 215, 103633. https://doi.org/10.1016/j.jnca.2023.103633
[11] Bhushan, B., Kumar, A., Agarwal, A. K., Kumar, A., Bhattacharya, P., & Kumar, A. (2023). Towards a secure and sustainable internet of medical things (IoMT): Requirements, design challenges, security techniques, and future trends. *Sustainability*, 15, 6177. https://doi.org/10.3390/su15076177
[12] Reeves, T. C. (1998). The impact of media and technology in schools. *Journal of The Journal of Art and Design Education*, 2, 58–63.
[13] Newman, N. (2018). *Journalism, Media and Technology Trends and Predictions 2018*. Reuters Institute for the Study of Journalism, Oxford, UK.

Chapter 15

Intelligent heart disease prediction using machine learning

Dev Narula, Shalini Mahato, Bam Bahadur Sinha, and R. Badlishah Ahmad

15.1 INTRODUCTION

Heart Diseases are one of the leading causes of death worldwide, a total of 17.9 million deaths occur each year due to cardiovascular diseases (CVDs) accounting for a total of 32% (one-third) of total deaths per year. CVD/heart diseases are a broad group of disorders related to the heart, blood vessels, and circulatory system. The most important reason for such a high number of CVD deaths is unhealthy lifestyle, physical inactivity, consumption of detrimental substances (e.g., tobacco), and obesity [1]. Lifestyles play a major role in determining the health of the heart. Due to our busy lifestyle, health becomes a secondary concern which leads to poor physical health and vulnerability to CVDs. CVDs need to be diagnosed and treated early so that one can recover.

Machine learning (ML) is a discipline that gives computers the ability to produce output without being explicitly programmed. ML aims to more efficiently and effectively emulate human abilities [2]. Computers are faster and more accurate than humans and ML relies on it to make accurate predictions from a given data by using past experiences such as data from events. Success of ML in other sectors such as marketing has led to its widespread use in other areas also.

Today a large number of lethal diseases are identified at very late stages when most of the harm has been done leading to the demise of the patient. Hence it has become important to predict the risk of CVD's at an early stage. A lot of CVDs require long-term (and expensive) treatment, even after being treated one can be left to live a restricted lifestyle and have difficulty in daily life activities such as lifting weights or bending. Heart diseases also introduce the risk of developing Chronic Angina (chest pain) which can cause lifetime inconvenience among other complications such as those of medical side-effects. The average age of CVD in India is 53 and 70 in the USA and Europe [3].

This chapter discusses briefly CVDs, ML, and comparative analysis of various ML algorithms and their effectiveness in predicting the risk of

 DOI: 10.1201/9781003534761-15

various CVDs. These techniques help in disease prediction as well as saving enormous time and money as well as most importantly the life of an individual.

15.2 CARDIOVASCULAR DISEASES

The heart is one of the most important organs in our body. It is the central part of the circulatory system and its main job is to pump blood throughout our body. CVDs are a group of diseases that affect the functioning of the heart, vascular tissues of the brain and blood vessels leading to ineffective circulation, which can cause health deterioration and various other ailments and even lead to death if the heart stops pumping blood/fails [4]. CVD is the world's leading cause of disability and death. CVD in the family, low HDL (good) cholesterol, high LDL (bad) cholesterol, hypertension, a high-fat diet, insufficient exercise or inactivity, obesity, smoking (or alcohol), diabetes, age, ethnic background, and gender are all risk factors contributing to CVDs [5]. A few CVDs are listed in Table 15.1.

Table 15.1 Various heart diseases

Sr No.	*Heart disease name*	*Brief description*
1	Coronary heart diseases	Difficulty pumping blood due to high-fat deposits in the arteries.
2	Cardiomyopathy	Difficulty pumping blood to the body because of hereditary weakness of heart muscles.
3	Ischemic Heart diseases	Narrowing of the blood vessels making blood flow less and difficult
4	Cardiac Arrest	Stopping/failure of the heart where it can stop pumping blood temporarily or fatally.
5	Valvular heart disease	One of the heart valves has been damaged or is defective.
6	Inflammatory Heart Disease	Inflammation of heart due to infection which may be viral or bacterial in nature
7	Hypertension	Heart anomalies caused by high blood pressure
8	Cerebrovascular disease	A condition affecting the blood vessels that supply blood to the brain [1]
9	Congenital heart disease	Birth defects that disturb the normal development and functioning of the heart [1].This is referred to as acquired heart disease in later stages of life [6].
10	Stroke	A condition resulting in brain injury due to lack of blood flow in brain
11	Peripheral Heart Disease	Occurs when there is a blockage in the arteries, usually the legs

(Continued)

Table 15.1 (*Continued*) Various heart diseases

Sr No.	*Heart disease name*	*Brief description*
12	Atrial Fibrillation	Causing Irregular heartbeat, rarely lethal but paves wave to other diseases like stroke and heart failure [7].
13	Pulmonary embolism	Blocking of blood vessels in lungs results in oxygen deprivation. [7].
14	Heart infections	Infections of the heart caused by various pathogens [7].
15	Atherosclerosis	Atherosclerosis is a hardening of the arteries [7].

15.2.1 Symptoms of cardiac arrest [8]

i. Pain and discomfort in the chest (angina)
ii. Pain may spread to other areas of the body
iii. Rapid or irregular heartbeat
iv. Fatigue
v. Heartburn or abnormal pain
vi. Shortness of breath and difficulty breathing
vii. Cold sweat and drowsiness
viii. Irregular heartbeat

15.2.2 Prevention

Prevention of a large number of CVDs can be done easily by living a healthy lifestyle and taking care of our body. By remaining fit and active with regular cardio exercise our heart can remain in great shape. The main risks for heart diseases are diabetes, excessive smoking and drinking, high blood pressure, and obesity [9]. Obesity can be prevented by adequate exercise. Harmful substances like alcohol, tobacco, and cigarette should be avoided since these not only harm body organs but also pave wave for other diseases along with them. Adequate sleep and nutritious food are also important for a healthy body. Those with a particularly high risk of CVD can opt for certain medications to reduce such risks.

15.3 MACHINE LEARNING

ML is a subgroup of AI aiming to derive predictions from mathematical models [10] i.e., making a program learn from experience by doing various classes of tasks. It is achieved using two methods, by training and testing an algorithm on a dataset [2]. A dataset is a collection of data about a particular sample consisting of features and examples to train the algorithm or rather give information. Each single value in a dataset is known as datum.

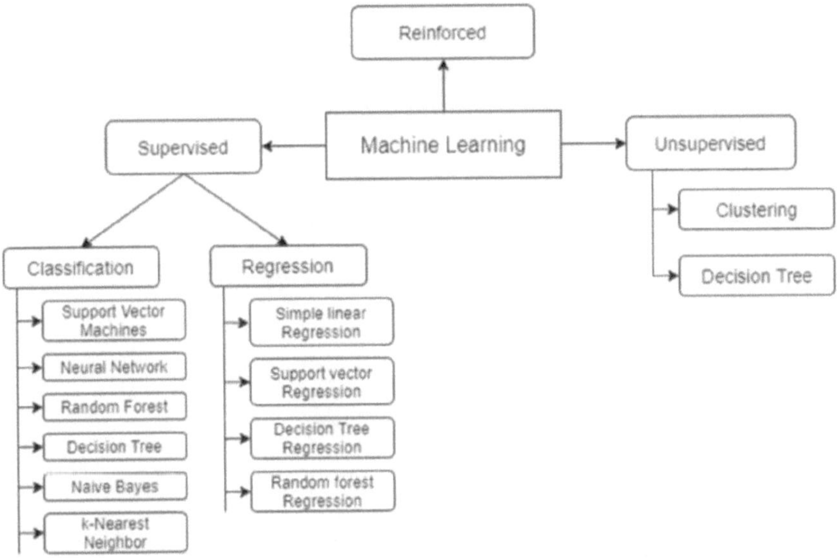

Figure 15.1 Types of machine learning.

Outcome and performance of an algorithm depend heavily on the dataset and its biasness making comparisons difficult for which reason most of the ML algorithms. The Cleveland heart disease dataset is a public dataset that has 303 samples and 76 features which have been used in training ML algorithms [11] (Figure 15.1).

There are three classifications of ML algorithms as represented in Figure 15.1 [2].

15.3.1 Supervised learning

Supervised learning is learning in the presence of a labeled input dataset (data already tagged with correct answer) to obtain experience. The dataset acts as a guide or supervisor and provides important data to the algorithm for training as well as testing. Supervised learning can further be classified into two types of problem algorithms: -

- Classification
- Regression

15.3.1.1 Classification

Classification algorithms are used to predict/classify discrete values apart such as classifying spam from non-spam calls or having heart diseases compared to not having it.

Types of classification algorithm.

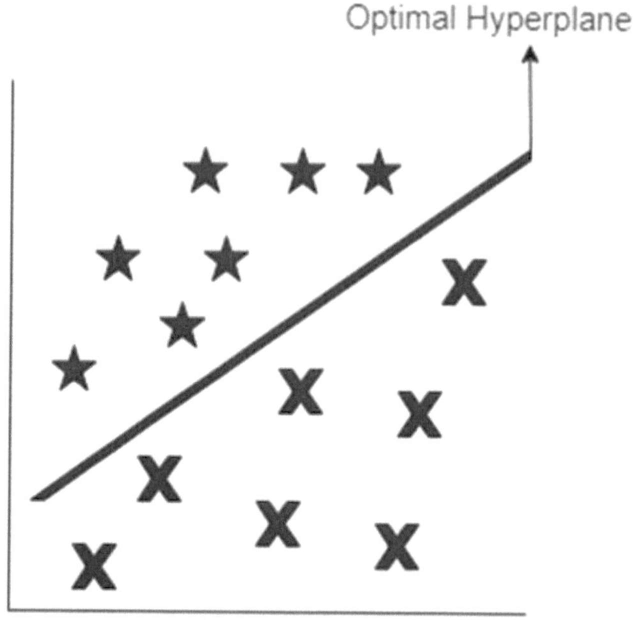

Figure 15.2 Support vector machine visualization.

15.3.1.1.1 Support Vector Machines

Support Vector Machine (SVM) is a simple classification technique. This technique creates a maximal size hyperplane and creates data points for each data item on it for classification separated by margins. The margins are drawn so that the space between the margin and the classes is as little as possible, reducing the classification error. SVM classifiers use various kernel algorithms for data classification, using the kernel trick. SVM can perform both linear and nonlinear classification [12]. SVM is one of the most robust and highly accurate classification algorithms. Figure 15.2 represents working of SVM.

15.3.1.2 Artificial Neural Network

A neural network is a set of algorithms that attempts to recognize underlying relationships in a set of data using a method that mimics how the human brain works. The input's output has already been determined and the anticipated output is compared to the actual output. The parameters are modified based on the error and then feedback into the neural network [12]. The most widely used Artificial Neural Network (ANN) works best with nonlinear datasets. Multilayer Perceptron (MLP) is a feed-forward (direction of information is one-sided i.e. only forward). ANN uses backpropagation (a training

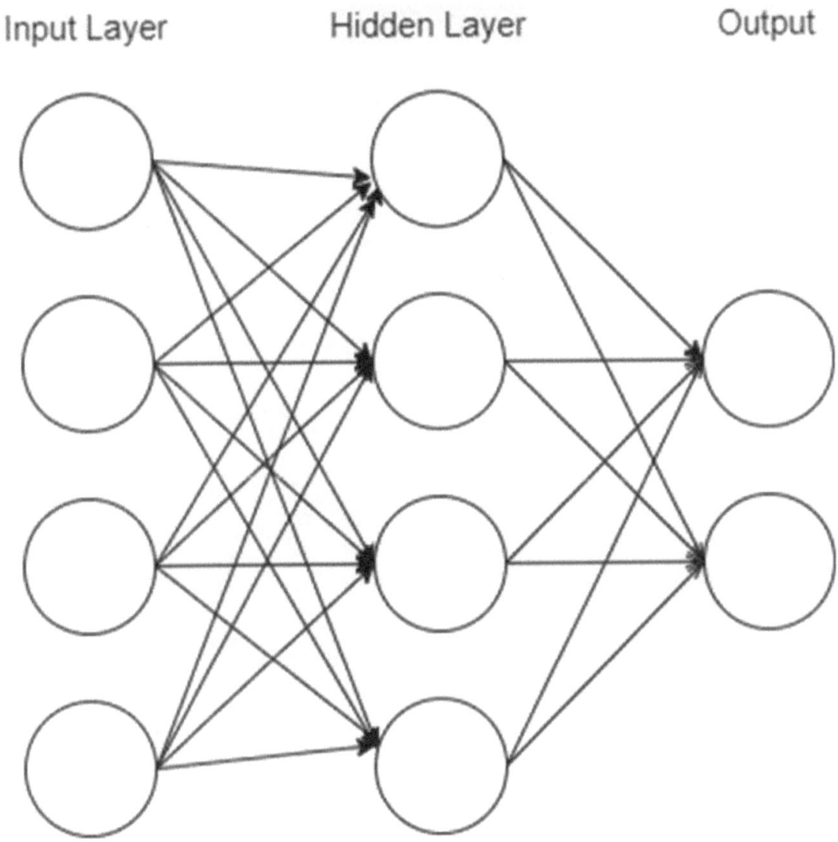

Figure 15.3 Neural network visualization.

algorithm for feed-forward ANN which sends back error) as a supervised ML algorithm. Radial basis function (RBF) is also a type of ANN which is very similar to MLP. Figure 15.3 represents the working of ANN.

15.3.1.3 Decision Tree

The decision tree (DT) classifiers are a group of algorithms using a graph that depicts choices and their outcomes as a tree. Nodes and branches make up each tree. Each branch represents a value that the node can take, and each node represents attributes in a group that need to be classified [7,8]. The attributes are stored in the internal nodes, while the results of each test on each node are represented by the branches. Because it does not require much field knowledge, DT is widely used for classification or parameter setting to function and is very easy to implement. Figure 15.4 represents the working of DT.

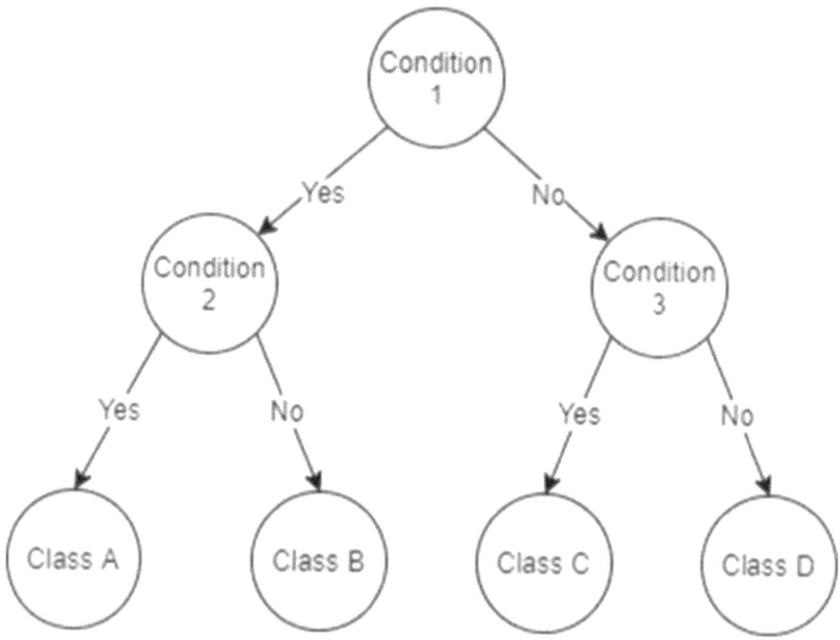

Figure 15.4 Decision tree visualization.

15.3.1.4 Random Forest classification (RF)

A random forest is a type of ensemble classifier that comprises a large number of decision trees. Individual trees represent the output of the classes. This strategy is used in conjunction with a random feature selection to create decision trees with controlled variability [13]. Random Forest uses a bagging ensemble to increase its accuracy. It improves on the limitations of DT to provide more accurate decisions. Figure 15.5 represents working of RF.

15.3.1.5 Naive Bayes

Bayes Theorem forms the basis of Naive Bayes (NB) classification. It is Bayes Theorem is based on conditional probability [14]. The presence of a particular feature in a class is presumed to be unrelated and independent of the presence of any other feature. It can be used for clustering and classification, and it's based on the conditional probability of occurrence [12]. NB is particularly useful for large datasets. Even if there is a dependency, all of these characteristics or attributes contribute to the probability independently, which is why it is called Naive.

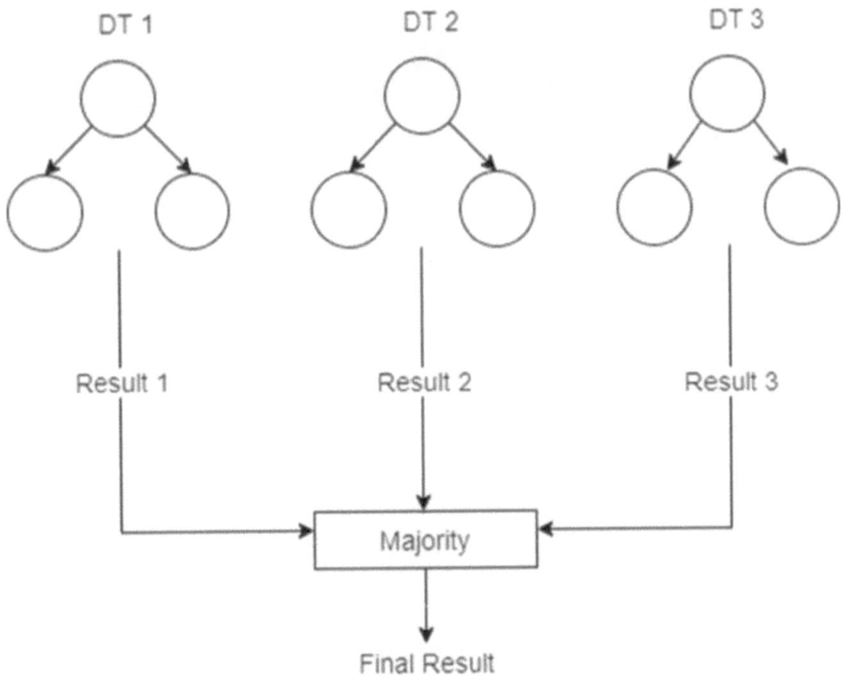

Figure 15.5 Random forest visualization.

15.3.1.6 K-Nearest Neighbor (KNN)

Based on the most votes obtained from its closest neighbors, this algorithm predicts a new instance's class. It calculates an attribute's distance from its neighbors using Euclidean distance. As its name suggests it looks for nearest neighbors in a hyperplane and it assumes they are related to each other. Here k is the chosen number of neighbors. It is called a lazy learning algorithm because it does not learn from data set immediately but instead stores it. Figure 15.6 represents working of KNN.

15.3.1.2 Regression

Regression algorithms are used to predict continuous values such as the stock market. It finds relationships between a target and other variables.

Types of regression algorithm:

- Simple Linear Regression
- Support Vector Regression
- Decision Tree Regression
- Random forest Regression

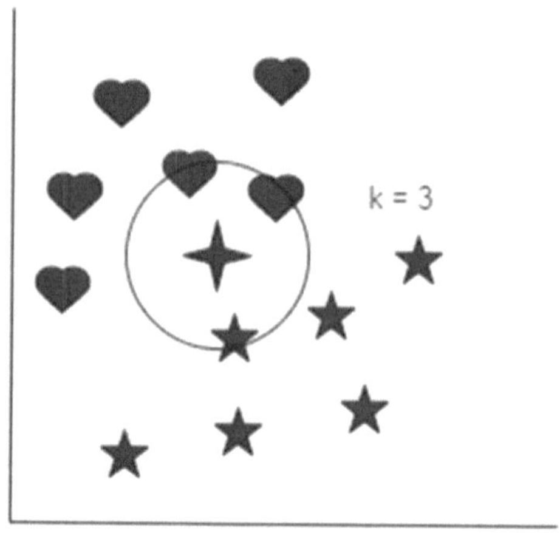

Figure 15.6 K-NN visualization.

15.3.2 Unsupervised ML

Unsupervised ML is learning without guidance. It does not have labeled input data sets. So the algorithm has to classify the dataset and find relation patterns between them similar to the concept of self-study or self-sufficiency [2]. The goal here is to find structure and pattern within data and everything is done by the algorithm. This outcome is used to group or cluster data based on statistical characteristics. It is mainly used for feature detection

Types of unsupervised ML: -

- Clustering
- Association

15.3.2.1 k-means clustering algorithm

The process uses a simple and straightforward approach to classify a given data set into a series of clusters. The main idea is to establish k-centers, one for each cluster classification. These k-centers should be strategically located because different locations lead to different results. Here k is replaced by the number of clusters.

15.3.3 Reinforced ML

Reinforcement ML is a training strategy that incentivizes desirable behaviors while penalizing undesirable ones. The goal here is to get the maximum

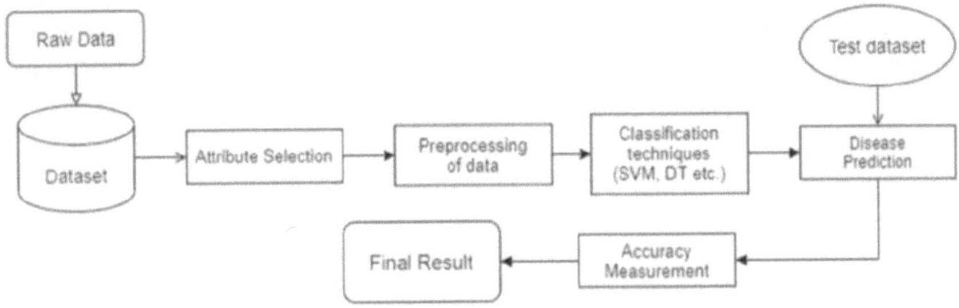

Figure 15.7 Various processes in model classification.

possible reward. Reinforced ML uses data from its environment for processing and learning. It lies in the middle of supervised and unsupervised ML. It is very much dependent on trial and error for training of the prediction algorithm. This technique is mainly used by robots and training of other AI.

15.3.4 Ensemble

Ensemble learning entails merging multiple derived predictions from various learning algorithms to produce a stronger overall prediction and better results. Ensemble itself is supervised ML algorithm that can be trained and then used to make predictions with higher accuracy than the base algorithm provided enough variance in parent algorithms [15]. There are three types of ensemble learning: bagging, boosting, and stacking which have all different effects on the ensembled model. The main goal of ensemble learning is achieving higher accuracy while maintaining the generalization of models. Figure 15.7 represents working of Ensemble learning.

15.4 METHODOLOGY

15.4.1 Data collection

The first step is the collection of data from which the dataset will be derived. The dataset will later be used to train the ML algorithm. In the raw dataset, operations need to be performed and it needs to be converted into a proper dataset before it is suitable for use. At this step, the data is also divided into training and testing set for supervising the algorithm and validating the results respectively.

15.4.2 Attribute selection

Attribute/feature of a dataset is the properties we will use from a dataset for evaluation. This process is also called feature selection and mainly involves reducing the number of input variables since not all the attributes are fit/

relevant for using in the predictive model and may harm the accuracy. Cleaning the real-world data and converting it to a clean dataset which predictive algorithm can benefit from is one of the most important processes since an algorithm relies heavily on dataset.

15.4.3 Pre-processing

Data pre-processing is the procedure for preparing raw information for use in an ML algorithm. Data is modified as per the needs of the predictive algorithm. Tasks such as handling of missing data, noise removal, and conversion of data to machine-understandable format among many others come into play.

15.4.4 Classification techniques

This process involves using ML algorithms on the processed and ready-to-use dataset to get results. Various ML and prediction algorithms such as SVM, NB, NN, etc. can be used. These algorithms are run on the dataset to build a predictive model and tested further.

15.4.5 Result analysis

The result of the predictive algorithm is the accuracy achieved for the prediction of the right outcome. Sensitivity is the percentage of outcomes identified as positive among all actual positive outcomes and specificity is the total percentage of negative outcomes out of all actual negative outcomes. Another factor Precision is total number of correctly specified positives out of all positive predictions. These factors of an algorithm depend on four factors: True positive, false positive, true negative, and false negative.

Here True positive (TP)=correct prediction of having heart disease
True Negative (TP)=correct prediction of not having heart disease
False positive (TP)=wrong prediction of having heart disease
False negative (TP)=wrong prediction of not having heart disease

- Accuracy is (TP+FN)/(TP+TN+FP+FN)
- Specificity is TN/(FP+TN)
- Sensitivity (also known as recall) is TP/(TP+FN)
- F-measure 2 * (precision * recall)/(precision+recall)

15.5 LITERATURE REVIEW

The authors have worked on various ML classification algorithms like SVM, DT, linear regression, k-nearest neighbor (k-NN)on Cleveland UCI dataset having 303 instances using 14 out of 76 features to achieve a highest accuracy of 87% by k-NN while SVM, LR, DT had 83%, 78%, and 79% accuracies respectively. Training dataset 73% and testing dataset was 27% [2].

The authors have worked on Cleveland UCI Heart disease dataset using 303 instances with 50/50 training and testing dataset split. From the 76 total attributes, their experiment uses 19 attributes like chest pain, fasting blood sugar, age, sex, etc. All the feature values are numeric. NB and DT were used as the classification algorithms. The outcome of their experiment indicated that NB performed better than DT. Their work also concludes that NB and DT with information gain calculations perform better than other classifiers but surmises this is due to an increased number of attributes [16].

Using the same Cleveland UCI Heart disease dataset, in another study authors have compared 3DT algorithm J48, RF, and LMT (logistic model Tree) with J48 and LMT using pruning, against each other. The comparisons are done using factors specificity, accuracy, and sensitivity. The original research work also showcases the resultant confusion matrix of the various algorithms with train and test error of 0.142, 0.167 for J48 and 0.166, 0.238 for LMT. Results show that J48 performed better at 56.76% followed with 0.04s to build model and LMT performed worst at 55.77% accuracy with 0.39s build time.

In another study, SVM was applied on dataset containing 500 samples of diabetic patients with 11 features like AIC, LDL and VLDL. Their work uses RBF kernel with SVM classification. The dataset was split using 10-fold cross-validation into training and testing sets (90%/10%). SVM achieved highest of 94.60% accuracy, a high precision of 97.52% recall of 87.10% in case of positive classes [17].

A comparative study using NN, DT, and NB using Cleveland UCI dataset and Stalog heart diseases dataset. Accuracy of 94.44%, 96.66%, 99.25% with 13 features and 90.74%, 99.62%, 100% with 15 features for NB, DT, and NN, respectively [18].

In another study using UCI Cleveland dataset along with 13 features and 297 records after pre-processing was used. Their proposed model uses ANN with backpropagation along with other classifiers such as k-NN, DT, SVM, NN, LR, VOTE, generalized linear model (GLM), deep learning (DL) and Gradient Boosted Trees (GBT) for comparison. The results show that VOTE performs the best accuracy wise at 87.41% followed by DL at 86.4%, DT has the worst performance at 85%. Their proposed model performs relatively well at 88.4% accuracy, 11.6% classification error, 90.1% precision, 92.8% sensitivity, and 82.6% specificity [19].

Another study using UCI Cleveland dataset used SVM, ANN, DT, NB, LR, k-NN, and RF for their work with Relief FS, LASSO, mRMR for feature selection algorithms. SVM with RBF kernel has achieved the highest accuracy with 86% followed by LR at 84% accuracy with ANN and DT having the lowest accuracy of 74%. SVM was the fastest algorithm at 15.23s and NB slowest at 34.10s when using all the features from Cleveland dataset. Their experiment was done using various numbers of features but six features yielded the best result and hence it was the final result. LR with 89% accuracy using Relief FS, NB with 84% using mRMR and LR with 87% using LASSO yielded the best result [20].

Authors have also used SVM, k-NN, NB, RF, and a MLP/ANN optimized by PCO (particle swarm optimization and ACO (ant colony optimization) for redundancy reduction using the same Cleveland UCI dataset with 14 attributes out of 76 total features. Their results concluded that FCBO PCO and ACO optimizations boosted the accuracy of models. K-NN and RF achieved highest of 99.7% and 99.6% accuracy with optimizations respectively [21]. In another study, authors have used a combination of NB, DT, and NN using the same dataset. A total of 909 records were used from Cleveland heart diseases database and were split into equal datasets for training and testing utilizing 13 feature values. Results indicate that NB performs the best with accuracy 86.12% followed by NN at 85.68% and DT at 80.4% [22].

The authors have proposed a cloud computing method for the prediction of heart diseases using HADOOP on Amazon EC2. Their model uses J48, SVM, NB, MLP, RF, AdaBoost, Boosted Tree (BT), and binary discriminant (BD), the best three models from the standalone test have been ensembled to boost the accuracy of the model, the resultant model is tested on cloud and compared with standalone. For cloud analysis Mahout API built on top of HADOOP is used. For performance measures best performing algorithm is NB at 86.42% followed by AdaBoost at 86.21% and BT 85.75%. Another Delhi Health Medicare Health Institute dataset (2009–2015) was used which contained 10,082 samples and 11 attributes. The ensembled model resulted in 92.14% accuracy on standalone and 93.17% on cloud using the Delhi dataset [15].

Authors have performed comparisons between various ML algorithms such as RF, DT, LR and NB using Cleveland UCI dataset comprising 14 attributes. Results showed that RF performed best among the algorithms with 90.16% followed by LR and NB at 85.25% and DT at 80.16%. Random forest also had the highest precision of 0.937 [23]. In one of the studies using the same dataset, accuracy of 91.5% was achieved using MLP [8]. Using DT and NB on the same dataset provided accuracy of 91% for DT and 87% for NB [24].

On the same dataset, DT, NB, MLP, k-NN, SCRL (single conjunctive rule learner), RBF, and SVM are used for comparison along with ensemble such as bagging, boosting, and stacking. Their research concluded that SVM has the best accuracy with 84.15% without any ensembling and SVM remains at the top even with ensembling. Classifiers with low accuracy such as SCRL saw significant improvement after applying ensemble [25].

In another study, the authors used 13 features like blood pressure, thal, blood sugar level, etc. AllPossible-MV algorithm is used to fill the missing values in the dataset. They have used DT algorithms like C4.5 with pruning along with other classifiers such as SVM, 1-NN, PART, MLP, RBF, and TSEAFS along with their proposed model EHDPS (efficient heart disease prediction system) for comparison. Their results indicate that their proposed model performs best among them with an accuracy of 86.7% compared to the closest performing algorithm RBF at 78.85% [26].

The authors have worked on the prediction of ischemic heart disease with their main objective being to improve the specificity and sensitivity of the diagnosis methods (methods for clinical examination like ECG testing) so that the number of patients who need to submit for further testing unnecessarily is less. They have used a dataset consisting of 327 samples (250 males and 77 females) who went through clinical tests (suspected IHD) for feature obtaining. The dataset was from medical examination during 1991–1994 at a Nuclear Medicinal department. The datasets were divided into four parts based on what records were available for each patient from clinical tests and were then compared with NB, semi-NB, backpropagation algorithm based on sensitivity, specificity, accuracy and Inf.score. Their system uses 10-fold stratified cross-validation for split of datasets. All the trained models significantly outperformed clinicians on diagnostic level. These algorithms are optimized for high accuracy but the goal of their work was to improve specificity in which the performance improvement was slight but this can be achieved using variable misclassification cost [9].

The authors discuss various heart diseases and the ways to detect them using algorithms such as LR, k-NN, SVM, NB, DT, RD, MLP, ANN, and DNN (deep neural network). Their work employs two unspecified different datasets from the UCI repository with 14 features, after pre-processing using normalizations and missing values replacement using NAN. Their results showcase that random forest was the best performing algorithm with 95.60% accuracy with logistic regression close behind at 93.40% accuracy, k-NN was the worst performing algorithm with 71.42% accuracy [27].

The authors have used UCI Cleveland dataset which after pre-processing has 300 samples and 14 features like blood pressure, chest pain, etc. They have utilized various classifiers such as NB, k-NN, DT, and RF. Their study has found NB to have the best accuracy at 88.15% in training set followed by RF at 84.2% accuracy, k-NN achieved highest accuracy at $k = 7$ of 90.78% but had very poor performance at $k=2$ with 78.94% [28].

The authors have also done a comparative study between ML classification algorithms such as AdaBoost, NB, SVM, LR, RF for prediction of disease. They have utilized the famous UCI Cleveland dataset with a total of 303 samples and 76 features which after using MICE algorithm for pre-processing and removal of missing values has been reduced to 14 feature set such as ST depression and Serum cholesterol. 8 of these features are categorical and the rest are numerical which were selected using a hybrid of GA and RFE. SVM performed the worst with 79.50% accuracy when run without any pre-processing on feature selection while NB had the best accuracy with 84.79%. After pre-processing with scaling SVM jumped up to 84.79% with NB and after class balancing optimization NB resulted in highest accuracy of 85.07%. After optimization with RandomizedSearchCV () the performance again improved with RF having highest of 86.60% followed closely by AdaBoost at 86.59% [29]. Using the same dataset, a comparative study on various classifiers such as RF, NB, SVM, Hoeffding DT, and logistic model tree (LMT). The dataset they have used is Cleveland UCI dataset comprising 76 features reduced to 13 after pre-processing and 303 samples out of which 80% were used as training dataset and rest as testing dataset. During the initial run, RF and NB performed the best at 88.52% followed by SVM at 86.88%, Hoeffding tree performed the worst at 78.65% accuracy. After further boosting all algorithms observed an increase in their accuracy with RF attaining the maximum accuracy of 95.08% (6.65% increase) followed by NB at 88.52% accuracy [30].

15.6 COMPARATIVE ANALYSIS

A lot of research works has already been done on the topic of Heart Disease Prediction. Table 15.2 summarizes and compares various other studies already conducted and reviewed in this chapter.

Thus from the study, it can be analyzed that very high accuracy has been achieved using ML techniques such as RF and NN. ML has also been applied in many other areas of healthcare, cyber security, IoT, etc [31–33]. The vast majority of studies, according to the extensive literature review, used the Cleveland heart disease dataset, consisting of 303 instances with 14 features. The sample size used is extremely small and geographically limited. A few other studies using different dataset had a single dataset with limited features. As a result, it was impossible to generalize the various classification accuracies obtained in heart disease prediction. Other multiple heart disease datasets from geographically different origins with more features should be explored for developing more efficient ML models in order to obtain a more broad and vague classification and prediction accuracy. This would allow for more accurate classification and early detection of heart diseases, reducing the escalating rates of morbidity and mortality associated with CVDs.

Table 15.2 Comparison of various other studies

S. No.	*Name*	*Year*	*Data set*	*No. of attributes*	*Classification used*	*Accuracy (%)/result*
1	Heart Disease Prediction Using Machine Learning Algorithms [4]	2020	UCI Cleveland dataset	14	SVM: DT: LR: k-NN:	83 79 78 87
2	Prediction of Heart Disease Using Machine Learning Algorithms [16]	2016	Cleveland UCI dataset	19	NB, DT	DT has better accuracy than NB
3	Heart Disease Prediction Using Machine learning and Data Mining Technique [13]	2016	UCI Cleveland dataset	13	J48: LMT:	56.76 55.77
4	Applying Machine Learning Methods in Diagnosing Heart Disease for Diabetic Patients [17]	2012	500 diabetic patients of unknown origin	11	SVM:	94.60
5	Improved Study of Heart Disease Prediction System using Data Mining Classification Techniques [18]	2012	UCI Cleveland and Stalog heart disease dataset	13 and 15	 NB: DT: NN:	13 attr. / 15 attr. 94.44 / 90.74 96.66 / 99.62 99.25 / 100

(*Continued*)

Table 15.2 (Continued) Comparison of various other studies

S. No.	Name	Year	Data set	No. of attributes	Classification used	Accuracy (%)/result			
6	Effective Heart Disease Prediction Using Hybrid Machine Learning Techniques [19]	2019	UCI Cleveland dataset	13	NB:	75.8			
					GLM:	85.1			
					LR:	82.9			
					DL:	87.4			
					DT:	85			
					RF:	86.1			
					GBT:	78.3			
					SVM:	86.1			
					VOTE:	77.41			
					HRFLM:	88.4			
7	A Hybrid Intelligent System Framework for the Prediction of Heart Disease Using Machine Learning Algorithms [20]	2018	UCI Cleveland dataset	13 and 6		13 attr	mRMR (6)	lasso (6)	relief (6)
					LR:	84	78	87	89
					k-NN:	76	62	85	80
					ANN:	74	63	86	77
					SVM (RBF)	86	88	88	87
					SVM (linear):	75	100	84	80
					NB:	83	90	83	85
					DT:	74	60	84	74
					RF:	83	67	83	85

(*Continued*)

Table 15.2 (Continued) Comparison of various other studies

S. No.	*Name*	*Year*	*Data set*	*No. of attributes*	*Classification used*	*Accuracy (%)/result*		
						Not optimized	FCBF	FCBF, PCO, ACO
8	Heart Disease Prediction and Classification Using Machine Learning Algorithms Optimized by Particle Swarm Optimization and Ant Colony Optimization [21]	2018	UCI Cleveland dataset	14	k-NN:	74.8	78.5	99.7
					SVM:	83.6	83.3	83.7
					RF:	81.4	80.3	99.6
					NB:	83.6	84.0	85.9
					MLP:	82.2	84.0	91.1
9	Intelligent Heart Disease Prediction System Using Data Mining Techniques [22]	2008	UCI Cleveland dataset	13	NB:	86.12		
					NN:	85.68		
					DT:	80.4		
10	Intelligent heart disease prediction in cloud environment through ensembling [21]	2017	UCI Cleveland and Delhi heart disease dataset	14 and 11	RF:	81.48		
					J48:	77.78		
					MLP:	83.95		
					NB:	86.42		
					SVM:	85.18		
					AdaBoost:	86.21		
					BT:	85.75		
					BD:	84.26		
					Ensembled (standalone):	92.14		
					Ensembled (cloud):	93.17		

(Continued)

Table 15.2 (Continued) Comparison of various other studies

S. No.	*Name*	*Year*	*Data set*	*No. of attributes*	*Classification used*	*Accuracy (%)/result*
11	IJERT-Heart Disease Prediction using Machine Learning [23]	2020	UCI Cleveland dataset	14	DT: LR: RF: NB:	81.97 85.25 90.16 85.25
12	Prediction of Heart Disease Using Machine Learning [8]	2018	UCI Cleveland dataset	13	MLP:	91.5
13	Prediction of Heart Disease Using Machine Learning Algorithms [24]	2019	UCI Cleveland dataset	14	DT: NB:	91 87
14	A Comprehensive Investigation and Comparison of Machine Learning Techniques in the Domain of Heart Disease [25]	2017	UCI Cleveland heart	14	DT: NB: k-NN (k=1) k-NN (k=3) k-NN (k=9) k-NN (k=15) MLP: RBF: SCRL: SVM:	77.55 83.49 76.23 81.18 83.16 82.83 82.83 83.82 69.96 84.15
15	Efficient Heart Disease Prediction System [26]	2016	UCI Cleveland dataset	13	SVM: C4.5: 1-NN: PART: MLP: RBF: TSEAFS: EHDPS:	70.59 73.53 76.47 73.53 74.85 78.53 77.45 86.7

(Continued)

Table 15.2 (Continued) Comparison of various other studies

S. No.	Name	Year	Data set	No. of attributes	Classification used	Accuracy (%)/result		
16	Analyzing and improving the diagnosis of ischemic heart disease with machine learning [9]	1999	Clinical dataset from 1991 to 1994 Nuclear Medicine department	variable		Accuracy	Specificity	Sensitivity
						(Signs and symptoms)		
					Clinical	0.65	0.76	0.61
					Back-prop.:	0.80	0.76	0.82
					NB:	0.80	0.65	0.92
					Semi-NB:	0.80	0.65	0.92
						(signs, Symptoms, ECG exercise)		
					NB:	0.82	0.62	0.90
					Sem-NB:	0.81	0.59	0.90
						(Myocardial scintigraphy)		
					Clinical:	0.83	0.85	0.83
					Back-prop.:	0.90	0.82	0.93
					Semi-NB:	0.90	0.81	0.94
					Assitant-1:	0.90	0.79	0.94
						(sign symptom ECG myocardial)		
					Back-prop:	0.92	0.84	0.96
					Semi-NB:	0.91	0.81	0.96

(*Continued*)

Table 15.2 (Continued) Comparison of various other studies

S. No.	*Name*	*Year*	*Data set*	*No. of attributes*	*Classification used*	*Accuracy (%)/result*			
17	Machine Learning Techniques for Heart Disease Prediction: A Comparative Study and Analysis [27]	2021	Unspecified dataset from UCI repository	14	LR:	93.40			
					NB:	90.10			
					SVM:	92.30			
					k-NN:	71.42			
					DT:	81.31			
					RF:	95.60			
					ANN:	92.30			
					DNN:	76.92			
					MLP:	75.42			
18	Heart Disease Prediction using Machine Learning Techniques [28]	2020	UCI Cleveland dataset	14	NB (test set):	88.15			
					NB (train set):	81.05			
					k-NN (k=2):	78.94			
					k-NN (k=7):	90.78			
					RF (entropy):	84.21			
					RF (gini):	82.89			
					DT (entropy):	73.68			
					DT (gini):	80.26			
19	A decision support system for heart disease prediction based upon machine learning [29]	2021	UCI Cleveland dataset	14		Normal	Scaling	Class	FS
					NB:	84.79	84.79	85.07	83.55
					SVM:	79.50	84.79	84.16	84.46
					LR:	83.80	82.50	83.24	85.07
					RF:	83.83	83.83	83.85	86.60
					AdaBoost:	82.12	82.12	82.34	86.59
20	Cognitive Approach for Heart Disease Prediction using Machine Learning [30]	2020	UCI Cleveland dataset	13		Initial	Boosting		
					Gaussian NB:	88.52	93.44		
					SVM:	86.88	90.16		
					RF:	88.52	95.08		
					Hoeffding DT:	78.65	81.24		
					LMT:	79.12	80.69		

15.7 CONCLUSION

This study provides a comprehensive and systematic overview of various ML algorithms for predicting cardiac disease which would be very helpful for other researchers working in this area.

REFERENCES

[1] https://www.who.int/health-topics/cardiovascular-diseases#tab=tab_1 (Last accessed 7th April 2024)

[2] Singh, A., & Kumar, R. (2020). Heart disease prediction using machine learning algorithms. *International Conference on Electrical and Electronics Engineering*, ICE3, 452–457. https://doi.org/10.1109/ICE348803.2020.9122958

[3] Singh, A., Chauhan, S., Devasia, T. et al. (2021) Financial burden of heart failure in a developing country: Cost analysis from Manipal Heart Failure Registry, India. *Journal of Public Health (Berl.)*, 29, 585–594. https://doi.org/10.1007/s10389-019-01141-w

[4] Mendis, S., Puska, P., Norrving, B. E., & World Health Organization. (2011). *Global Atlas on Cardiovascular Disease Prevention and Control.* World Health Organization, Geneva.

[5] Yahaya, L., David Oye, N., & Joshua Garba, E. (2020). A comprehensive review on heart disease prediction using data mining and machine learning techniques. *American Journal of Artificial Intelligence*, 4(1), 20. https://doi.org/10.11648/j.ajai.20200401.12

[6] Singh, D., & Samagh, J. S. (2020). A comprehensive review of heart disease prediction using machine learning. *Journal of Critical Reviews*, 7(12), 281–285. https://doi.org/10.31838/jcr.07.12.54

[7] https://www.healthline.com/health/heart-disease#types-of-heart-disease (Last accessed 7th April 2024)

[8] Gavhane, A., Kokkula, G., Pandya, I., & Devadkar, K. (2018). Prediction of heart disease using machine learning. In: *2018 second international conference on electronics, communication and aerospace technology (ICECA)*, pp. 1275–1278. IEEE, New York. https://doi.org/10.1109/ICECA.2018.8474922

[9] Kukar, M., Kononenko, I., Grošelj, C., Kralj, K., & Fettich, J. (1999). Analysing and improving the diagnosis of ischaemic heart disease with machine learning. *Artificial Intelligence in Medicine*, 16(1), 25–50. https://doi.org/10.1016/S0933-3657(98)00063-3

[10] Zhang, X.-D. (2017). Chapter 6 machine learning. *A Matrix Algebra Approach to Artificial Intelligence*, 45, 13. https://doi.org/10.1007/978-981-15-2770-8

[11] https://archive.ics.uci.edu/ml/datasets/heart+disease (Last accessed 7th April 2024)

[12] Mahesh, B. (2019). Machine learning algorithms- A review. *International Journal of Science and Research (IJSR). [Internet]*, 9, 381–386. https://doi.org/10.21275/ART20203995

[13] Patel, J., Tejalupadhyay, S. & Patel, S. (2016). Heart disease prediction using machine learning and data mining technique. *IJCSC*, 7, 129–137. https://doi.org/10.090592/IJCSC.2016.018

[14] Solanki, A., & Barot, M. P. (2019). Study of heart disease diagnosis by comparing various classification algorithms. *International Journal of Engineering and Advanced Technology (IJEAT*), 8(2S2), 40–42.

[15] Gupta, N., Ahuj, N., Malhotra, S., Bala, A., & Kaur, G. (2017). Intelligent heart disease prediction in cloud environment through ensembling. *Expert Systems*, 34, e12207. https://doi.org/10.1111/exsy.12207

[16] Nikhar, S., & Karandikar, A. M. (2016). Prediction of heart disease using machine learning algorithms. *International Journal of Advanced Engineering, Management and Science (IJAEMS)*, 2, 6.

[17] Parthiban, G., & Srivatsa, K. S. (2012). Applying machine learning methods in diagnosing heart disease for diabetic patients. *International Journal of Applied Information Systems*, 3(7), 25–30. https://doi.org/10.5120/ijais12-450593

[18] Dangare, S. C., & Apte, S. S. (2012). Improved study of heart disease prediction system using data mining classification techniques. *International Journal of Computer Applications*, 47(10), 44–48. https://doi.org/10.5120/7228-0076

[19] Mohan, S., Thirumalai, C., & Srivastava, G. (2019). Effective heart disease prediction using hybrid machine learning techniques. *IEEE Access*, 7, 81542–81554. https://doi.org/10.1109/ACCESS.2019.2923707

[20] Haq, A. U., Li, J. P., Memon, M. H., Nazir, S., Sun, R., & Garciá-Magarinõ, I. (2018). A hybrid intelligent system framework for the prediction of heart disease using machine learning algorithms. *Mobile Information Systems*, 2018, 3860146. https://doi.org/10.1155/2018/3860146

[21] Khourdifi, Y., & Bahaj, M. (2019). Heart disease prediction and classification using machine learning algorithms optimized by particle swarm optimization and ant colony optimization. *International Journal of Intelligent Engineering and Systems*, 12(1), 242–252. https://doi.org/10.22266/ijies2019.0228.24

[22] Palaniappan, S., & Awang, R. (2008). Intelligent heart disease prediction system using data mining techniques. In: *AICCSA 08-6th IEEE/ACS International Conference on Computer Systems and Applications,* pp. 108–115. IEEE, New York. https://doi.org/10.1109/AICCSA.2008.4493524

[23] Rajdhan, A., Agarwal, A., Sai, M., Ravi, D., & Ghuli, P. (2020). Heart disease prediction using machine learning, *International Journal of Engineering Research & Technology (IJERT)*, 09, 04. https://doi.org/10.17577/IJERTV9IS040614

[24] Santhana, K. J., & Geetha, S., (2019). Prediction of heart disease using machine learning algorithms. In: *2019 1st International Conference on Innovations in Information and Communication Technology (ICIICT),* pp. 1–5. IEEE, New York. https://doi.org/10.1109/ICIICT1.2019.8741465

[25] Pouriyeh, S., Vahid, S., Sannino, G., De Pietro, G., Arabnia, H., & Gutierrez, J. (2017). A comprehensive investigation and comparison of machine learning techniques in the domain of heart disease. In: *Proceedings - IEEE Symposium on Computers and Communications, ISCC,* pp. 204–207. IEEE, New York. https://doi.org/10.1109/ISCC.2017.8024530

[26] Purushottam Saxena, K., & Sharma, R. (2016). Efficient heart disease prediction system. *Procedia Computer Science*, 85, 962–969. https://doi.org/10.1016/j.procs.2016.05.288
[27] Katarya, R., & Meena, S. K. (2021). Machine learning techniques for heart disease prediction: A comparative study and analysis. *Health and Technology*, 11(1), 87–97. https://doi.org/10.1007/s12553-020-00505-7
[28] Shah, D., Patel, S., & Bharti, S. K. (2020). Heart disease prediction using machine learning techniques. *SN Computer Science*, 1, 345. https://doi.org/10.1007/s42979-020-00365-y
[29] Rani, P., Kumar, R., Ahmed, N. M. O. S. et al. (2021) A decision support system for heart disease prediction based upon machine learning. *Journal on Reliable Intelligent Environments*, 7, 263–275. https://doi.org/10.1007/s40860-021-00133-6
[30] Motarwar, P., Duraphe, A., Suganya, G., & Premalatha, M. (2020). Cognitive approach for heart disease prediction using machine learning. In: *2020 International Conference on Emerging Trends in Information Technology and Engineering (IC-ETITE)*, pp. 1–5. IEEE, New York. https://doi.org/10.1109/ic-ETITE47903.2020.242
[31] Khadidos, A. O., Alyoubi, K. H., Mahato, S., Khadidos, A. O., & Mohanty, S. N. (2023). Computer aided detection of major depressive disorder (MDD) using electroencephalogram signals. *IEEE Access*, 11, 41133–41141 https://doi.org/10.1109/ACCESS.2023.3262930.
[32] Kakandwar, S., Bhushan, B., & Kumar, A. (2023). Integrated machine learning techniques for preserving privacy in internet of things (IOT) systems. In: *Blockchain Technology Solutions for the Security of IoT-Based Healthcare Systems*, pp. 45–75. IEEE, New York. https://doi.org/10.1016/b978-0-323-99199-5.00012-4
[33] Tyagi, N., Bhushan, B. (2023). Natural language processing (NLP) based innovations for smart healthcare applications in healthcare 4.0. In: Ahad, M. A., Casalino, G., Bhushan, B. (eds) *Enabling Technologies for Effective Planning and Management in Sustainable Smart Cities*. Springer, Cham. https://doi.org/10.1007/978-3-031-22922-0_5

Index